Canonical Perturbation Theories
Degenerate Systems and Resonance

ASTROPHYSICS AND SPACE SCIENCE LIBRARY

VOLUME 345

EDITORIAL BOARD
Chairman

W.B. BURTON, National Radio Astronomy Observatory, Charlottesville, Virginia, U.S.A. (bburton@nrao.edu); University of Leiden, The Netherlands (burton@strw.leidenuniv.nl)

MEMBERS

J.M.E. KUIJPERS, *Faculty of Science, Nijmegen, The Netherlands*
F. BERTOLA, *Universitá di Padova, Italy*
J.P. CASSINELLI, *University of Wisconsin, Madison, U.S.A.*
C.J. CESARSKY, *European Southern University, Garching bei München, Germany*
P. EHRENFREUND, *Leiden University, The Netherlands*
O. ENGVOLD, *Institute of Theoretical Astrophysics, University of Oslo, Norway*
A. HECK, *Strassbourg Astronomical Observatory, France*
V.M. KASPI, *McGill University, Montreal, Canada*
P.G. MURDIN, *Institute of Astronomy, Cambridge, U.K.*
F. PACINI, *Istituto Astronomia Arcetri, Firenze, Italy*
V. RADHAKRISHNAN, *Raman Research Institute, Bangalore, India*
F.H. SHU, *University of California, Berkeley, U.S.A.*
B.V. SOMOV, *Astronomical Institute, Moscow State University, Russia*
R.A. SUNYAEV, *Space Research Institute, Moscow, Russia*
E.P.J. VAN DEN HEUVEL, *Astronomical Institute, University of Amsterdam, The Netherlands*
H. VAN DER LAAN, *Astronomical Institute, University of Utrecht, The Netherlands*

Canonical Perturbation Theories
Degenerate Systems and Resonance

Sylvio Ferraz–Mello

Sylvio Ferraz-Mello
Instituto de Astronomia, Geofísicae e Ciências Atmosféricas
Universidade de São Paulo
Rua do Matão, 1226
CEP 05508-900
São Paulo, Brasil
sylvio@usp.br
http://www.astro.iag.usp.br/~sylvio/

Library of Congress Control Number: 2006931783

ISBN-10: 0-387-38900-8   e-ISBN-13: 978-0-387-38905-9
ISBN-13: 978-0-387-38900-4   e-ISBN-10: 0-387-38905-9

Printed on acid-free paper.

© 2007 Springer Science+Business Media, LLC
All rights reserved. This work may not be translated or copied in whole or in part without the written permission of the publisher (Springer Science+Business Media, LLC, 233 Spring Street, New York, NY 10013, USA), except for brief excerpts in connection with reviews or scholarly analysis. Use in connection with any form of information storage and retrieval, electronic adaptation, computer software, or by similar or dissimilar methodology now known or hereafter developed is forbidden.
The use in this publication of trade names, trademarks, service marks and similar terms, even if they are not identified as such, is not to be taken as an expression of opinion as to whether or not they are subject to proprietary rights.

9 8 7 6 5 4 3 2 1

springeronline.com

Preface

> A única maneira de cumprir o trabalho era tê-lo
> como coisa lerda e contínua, mansa, sem começo
> nem fim, as mãos sempre sujas da massa.
>
> João Guimarães Rosa, *Buriti*

The story of this book began in the late 1960s, when Prof. Buarque Borges invited me to give a graduate course at the Aeronautics Institute of Technology. The course was to deal with the perturbation theories used in Celestial Mechanics, but they should be presented in a universal way, so as to be understandable by investigators and students from related fields of science. This hint marked the rest of the story. The course evolved and for the past 30 years was taught almost yearly at the University of São Paulo and, occasionally, in visited institutions abroad. A long visit of Prof. Gen-Ichiro Hori to the University of São Paulo was the occasion for many illuminating discussions on the subject.

Soon, in this story, came the project of a book. But two major obstacles did not allow it to progress at that time. One of them was the concurrence of many other time-consuming duties. The drafts of many chapters could only be written during visits abroad: to Austin, Grasse, La Plata, Oporto, Nice, Paris, Vienna, and the book could only be completed now, after my formal retirement. The other obstacle, more determinant, was the fact that theories able to treat Bohlin's problem, a resonant Hamiltonian system with two degrees of freedom, where the second degree of freedom is degenerate, were not available. So, the book project had to wait for new investigations!

In accordance with the initial proposal, the aim of the book is to present the main canonical perturbation theories used in Celestial Mechanics without any involvement with the particularities of the astronomical problems to which they are applied; one does not need to know Astronomy to read it. The other objective is to provide, in one book, all the information necessary for the application of the theories. For instance, not only is it told how to actually obtain action–angle variables, but they are explicitly given for important dynamical systems such as the simple pendulum, the Ideal Resonance Problem and the first Andoyer Hamiltonian. In addition, every theory presented in the book is followed by case studies and examples able to illustrate the directions for their use in applications. For the sake of making the book useful as a handbook in investigations using perturbation theories, special care was taken to avoid errors in the given equations. All my students have, in the past, communicated to me the errors found in the drafts. I have myself checked every equation, but I am not foolish to say that no flaws remain. Transcriptions, transpositions, and the work on LaTeX source files are non-robust operations that may have added new errors. A Web page will be created to inform readers of any flaws finally remaining in the text.

The book is composed of 10 chapters and four appendices. The two first chapters are devoted to some results of Hamilton–Jacobi theory. This short presentation, where only points of practical interest are given a longer development, is not a substitute for a full text on Analytical Dynamics. Many sections were directly inspired by the seminal classes of the late Prof. Abrahão de Moraes, which I had the privilege of attending in my undergraduate years and by books with which I became acquainted in frequent visits to his personal library. One of them was Charlier's *Die Mechanik des Himmels*, the book referenced in many papers on fundamental Physics in the first decades of past century.

Chapters 3 and 4 are devoted to perturbation theories where canonical transformations are obtained by means of Jacobi's generating function. These chapters include the Poincaré theory for perturbed non-degenerate Hamiltonians, the von Zeipel–Brouwer theory for perturbed degenerate Hamiltonians, the procedures of frequency relocation and quadratic convergence used by Kolmogorov in the proof of his theorem, the theory used in Delaunay's lunar theory and the solution of Garfinkel's Ideal Resonance Problem. It is worth emphasizing that the definition of degeneracy used throughout this book, due to Schwarzschild, is less strict than the definition of degeneracy used in Kolmogorov's theorem.

Chapter 5 introduces Lie mappings and Chap. 6 reconsiders the study of perturbed non-degenerate Hamiltonian systems with canonical transformations written as Lie series. Lie series theories in action–angle variables are completely equivalent to those founded on Jacobian transformations and the choice of one or another is a matter of work economy only. Their comparison is done in two typical examples.

Chapter 6 introduces Hori's theory with unspecified canonical variables and this is the point where the equivalence to the old theories disappears. Hori's theory shows that every perturbation theory has a dynamical kernel, the Hori kernel. From the algorithmic point of view, the Hori kernel is a Hamiltonian system that repeats itself at every order of approximation, and whose Hamilton–Jacobi equation needs to be completely solved. From the dynamical point of view, it forces the solutions given by perturbation theories to have the same topology as the Hori kernel. However, generally, the Hori kernel and the given Hamiltonian have different topologies and this difference gives rise to the well-known small divisors.

In Chap. 7, it is shown how Hori's theory with unspecified canonical variables allows the construction of formal solutions using non-singular Poincaré variables, thus allowing the study of perturbed systems near the singularities of the actions. In Chaps. 8 and 9, the understanding of the role played by the Hori kernel is the key to dealing with resonant systems with two or more degrees of freedom presenting simultaneously resonant and degenerate angles. The Hori kernels in these chapters are systems whose restrictions to one degree of freedom are the simple pendulum and the first Andoyer Hamiltonian, respectively. The techniques discussed in Chap. 2 are used to extend the action–angle of these models to the two-degrees-of-freedom Hori kernel. Finally, in Chap. 10, the theories presented in the previous chapters are applied to some quasiharmonic Hamiltonian systems.

Appendix A is devoted to presenting Bohlin's theory and an extension of Delaunay's theory and to discuss the difficulties presented by these theories when applied to systems with more than one degree of freedom involving simultaneously resonant and degenerate arguments.

Appendices B and C present the complete solutions of two integrable Hamiltonians fundamental in resonance studies: the simple pendulum and the first Andoyer Hamiltonian. The action–angle variables of these two Hamiltonians are constructed with the help of elliptic functions. Expansions in terms of trigonometric functions valid in a neighborhood of the libration center are also given. Appendix C also includes the construction of solutions in the neighborhood of the pendulum separatrix and the associated whisker and standard mappings. Appendix D presents the main features of some higher-order Andoyer Hamiltonians.

One last comment on the contents of this book is that it is not aimed at being an encyclopedia on the subject and does not cover every approach of the problem. On the contrary, several sections and even one chapter not belonging to the backbone of the subject were dropped during the revision. Canonical perturbation theories are an old subject, and many approaches exist that were not even mentioned in the book.

The list of references, at the end of the book, also deserves some comments. One characteristic feature of this list concerns the old references where important concepts in present-day theories were introduced. It is human nature to highlight the more recent contributions showing the importance of some old concepts and to forget the founding fathers that introduced them much earlier. Special attention was paid to give to them the acknowledgement that they deserve and to inform new generations of their achievements. In what concerns the recent references, we included only some items that have a very direct relationship to what is written in this book. We considered it important not to let these few items disappear amid an exhaustive bibliography. This choice was made having in mind that search engines on the internet may give, nowadays, more and better bibliographical information than a long list at the end of a book.

Acknowledgements. I thank my family for continuous support. Many friends and colleagues have given me suggestions that helped to improve the book. I thank all of them and, particularly, Prof. Jean Kovalevsky, who, long ago, introduced me to canonical perturbation theories and Profs. André Brahic, Rudolf Dvorak, Claude Froeschlé, Juan Carlos Muzzio and Bruno Sicardy, who have often invited me to their institutions, allowing me to have time to write. I thank all my students. They have read almost all the drafts of this book and collaborated with valuable comments that resulted in many improvements in the written text. I thank the copy editor Mike Nugent for his invaluable contribution for the editorial quality of this book. During the work on this project, I had the support of USP – University of São Paulo, Observatório Nacional, Bureau des Longitudes (now IMCCE), Observatoire de Paris–Meudon, Wien Universität, FAPESP – Research Foundation of the State of São Paulo and CNPq – National Council for Scientific and Technological Development.

São Paulo, June 2006

Sylvio Ferraz-Mello

Contents

Preface v

1 **The Hamilton–Jacobi Theory** 1
   1.1 Canonical Pertubation Equations 1
   1.2 Hamilton's Principle 2
       1.2.1 Maupertuis' Least Action Principle 4
       1.2.2 Helmholtz Invariant 5
   1.3 Canonical Transformations 6
   1.4 Lagrange Brackets 9
   1.5 Poisson Brackets 11
       1.5.1 Reciprocity Relations 12
   1.6 The Extended Phase Space 13
   1.7 Gyroscopic Systems 15
       1.7.1 Gyroscopic Forces 15
       1.7.2 Example 16
       1.7.3 Rotating Frames 17
       1.7.4 Apparent Forces 17
   1.8 The Partial Differential Equation of Hamilton and Jacobi 18
   1.9 One-Dimensional Motion with a Generic Potential 20
       1.9.1 The Case $m < 0$ 23
       1.9.2 The Harmonic Oscillator 23
   1.10 Involution. Mayer's Lemma. Liouville's Theorem 24

2 **Angle–Action Variables. Separable Systems** 29
   2.1 Periodic Motions 29
       2.1.1 Angle–Action Variables 30
       2.1.2 The Sign of the Action 32
   2.2 Direct Construction of Angle–Action Variables 33
   2.3 Actions in Multiperiodic Systems. Einstein's Theory 35
   2.4 Separable Multiperiodic Systems 37
       2.4.1 Uniformized Angles. Charlier's Theory 37
       2.4.2 The Actions 38
       2.4.3 Algorithms for Construction of the Angles 39

	2.4.4	Angle–Action Variables of $H(q_1, p_1, p_2, \cdots, p_N)$ 40
	2.4.5	Historical Postscript 42
2.5	Simple Separable Systems 42
	2.5.1	Example: Central Motions 43
	2.5.2	Angle–Action Variables of Central Motions 44
2.6	Kepler Motion 47
2.7	Degeneracy 50
	2.7.1	Schwarzschild Transformation 51
	2.7.2	Delaunay Variables 52
2.8	The Separable Cases of Liouville and Stäckel 53
	2.8.1	Example: Liouville Systems 55
	2.8.2	Example: Stäckel Systems 56
	2.8.3	Example: Central Motions 56
2.9	Angle–Action Variables of a Quadratic Hamiltonian 57
	2.9.1	Gyroscopic Systems 60

3 Classical Perturbation Theories 61
3.1 The Problem of Delaunay 61
3.2 The Poincaré Theory 63
   3.2.1 Expansion of H_0 65
   3.2.2 Expansion of H_k 66
   3.2.3 Perturbation Equations 67
3.3 Averaging Rule 68
   3.3.1 Small Divisors. Non-Resonance Condition 69
3.4 Degenerate Systems. The von Zeipel–Brouwer Theory 70
   3.4.1 Expansion of H^* 72
   3.4.2 von Zeipel–Brouwer Perturbation Equations 72
   3.4.3 The von Zeipel Averaging Rule 73
3.5 Small Divisors and Resonance 74
   3.5.1 Elimination of the Non-Critical Short-Period Angles ... 74
3.6 An Example – Part I 77
3.7 Linear Secular Theory 81
3.8 An Example – Part II 83
3.9 Iterative Use of von Zeipel–Brouwer Operations 86
3.10 Divergence of the Series. Poincaré's Theorem 88
3.11 Kolmogorov's Theorem 88
   3.11.1 Frequency Relocation 89
   3.11.2 Convergence 91
   3.11.3 Degenerate Systems 93
   3.11.4 Degeneracy in the Extended Phase Space 94
3.12 Inversion of a Jacobian Transformation 94
   3.12.1 Lagrange Implicit Function Theorem 96
   3.12.2 Practical Considerations 96
3.13 Lindstedt's Direct Calculation of the Series 97

4 Resonance ... 99
- 4.1 The Method of Delaunay's Lunar Theory ... 99
- 4.2 Introduction of the Square Root of the Small Parameter ... 101
  - 4.2.1 Garfinkel's Abnormal Resonance ... 103
- 4.3 Delaunay Theory According to Poincaré ... 103
  - 4.3.1 First-Approximation Solution ... 106
- 4.4 Garfinkel's Ideal Resonance Problem ... 107
  - 4.4.1 Garfinkel–Jupp–Williams Integrals ... 109
  - 4.4.2 Circulation $(E\nu_{11}^* > A^*\nu_{11}^* > 0)$ ... 110
  - 4.4.3 Libration ($|E| < |A^*|$) ... 112
  - 4.4.4 Asymptotic Motions $(E = A^*)$ ... 114
- 4.5 Angle–Action Variables of the Ideal Resonance Problem ... 115
  - 4.5.1 Circulation ... 115
  - 4.5.2 Libration ... 116
  - 4.5.3 Small-Amplitude Librations ... 117
- 4.6 Morbidelli's Successive Elimination of Harmonics ... 118
  - 4.6.1 An Example ... 120

5 Lie Mappings ... 127
- 5.1 Lie Transformations ... 127
  - 5.1.1 Infinitesimal Canonical Transformations ... 127
- 5.2 Lie Derivatives ... 130
- 5.3 Lie Series ... 131
- 5.4 Inversion of a Lie Mapping ... 134
- 5.5 Lie Series Expansions ... 135
  - 5.5.1 Lie Series Expansion of f ... 136
  - 5.5.2 Deprit's Recursion Formula ... 137

6 Lie Series Perturbation Theory ... 139
- 6.1 Introduction ... 139
- 6.2 Lie Series Theory with Angle–Action Variables ... 140
  - 6.2.1 Averaging ... 142
  - 6.2.2 High-Order Theories ... 143
- 6.3 Comparison to Poincaré Theory. Example I ... 144
- 6.4 Comparison to Poincaré Theory. Example II ... 147
- 6.5 Hori's General Theory. Hori Kernel and Averaging ... 151
  - 6.5.1 Cauchy–Darboux Theory of Characteristics ... 154
- 6.6 Topology and Small Divisors ... 155
  - 6.6.1 Topological Constraint. The Rise of Small Divisors ... 156
- 6.7 Hori's Formal First Integral ... 157
- 6.8 "Average" Hamiltonians ... 158
  - 6.8.1 On Secular Theories and Proper Elements ... 159

7 Non-Singular Canonical Variables ... 161
- 7.1 Singularities of the Actions ... 161
- 7.2 Poincaré Non-Singular Variables ... 162
- 7.3 The d'Alembert Property ... 164
- 7.4 Regular Integrable Hamiltonians ... 165
- 7.5 Lie Series Expansions About the Origin ... 167
- 7.6 Lie Series Perturbation Theory in Non-Singular Variables ... 169
  - 7.6.1 Solutions Close to the Origin (Case $J_1 < 0$) ... 172
  - 7.6.2 Angle–Action Variables of H_2^* (Case $J_1 < 0$) ... 173
- 7.7 The Non-Resonance Condition ... 173
- 7.8 Example ... 175

8 Lie Series Theory for Resonant Systems ... 181
- 8.1 Bohlin's Problem (The Single-Resonance Problem) ... 181
- 8.2 Outline of the Solution ... 182
- 8.3 Functions Expansions ... 185
- 8.4 Perturbation Equations ... 188
- 8.5 Averaging ... 190
- 8.6 An Example ... 190
- 8.7 Example with a Separated Hori Kernel ... 198
- 8.8 One Degree of Freedom ... 204
  - 8.8.1 Garfinkel's Ideal Resonance Problem ... 204

9 Single Resonance near a Singularity ... 209
- 9.1 Resonances Near the Origin: Real and Virtual ... 209
- 9.2 One Degree of Freedom ... 210
- 9.3 Many Degrees of Freedom. One Single Resonance ... 213
- 9.4 A First-Order Resonance Case Study ... 216
  - 9.4.1 The Hori Kernel ... 218
  - 9.4.2 First Perturbation Equation ... 219
  - 9.4.3 Averaging ... 220
  - 9.4.4 The Post-Harmonic Solution ... 221
  - 9.4.5 Secular Resonance ... 223
  - 9.4.6 Secondary Resonances ... 224
  - 9.4.7 Initial Conditions Diagram ... 225
- 9.5 Sessin Transformation and Integral ... 227
  - 9.5.1 The Restricted (Asteroidal) Case ... 229

10 Nonlinear Oscillators ... 231
- 10.1 Quasiharmonic Hamiltonian Systems ... 231
- 10.2 Formal Solutions. General Case ... 232
- 10.3 Exact Commensurability of Frequencies (Resonance) ... 234
- 10.4 Birkhoff Normalization ... 236
  - 10.4.1 A Formal Extension Including One Single Resonance ... 240
  - 10.4.2 The Comensurabilities of Lower Order ... 242

10.5 The Restricted Three-Body Problem 242
    10.5.1 Equations of the Motion Around the Lagrangian Point $\mathcal{L}_4$ 244
    10.5.2 Internal 2:1 Resonance 246
    10.5.3 Internal 3:1 Resonance 247
    10.5.4 Other Internal Resonances..........................249
10.6 The Hénon–Heiles Hamiltonian...............................250
    10.6.1 The Toda Lattice Hamiltonian 252
10.7 Systems with Multiple Commensurabilities 253
    10.7.1 The Ford–Lunsford Hamiltonian. 1:2:3 Resonance 255
10.8 Parametrically Excited Systems 255
    10.8.1 A Nonlinear Extension 260

A **Bohlin Theory** 263
  A.1 Bohlin's Resonance Problem 263
  A.2 Bohlin's Perturbation Equations 265
  A.3 Poincaré Singularity 268
  A.4 An Extension of Delaunay Theory 269

B **The Simple Pendulum** 271
  B.1 Equations of Motion 271
      B.1.1 Circulation 273
      B.1.2 Libration 274
      B.1.3 The Separatrix....................................276
  B.2 Angle–Action Variables of the Pendulum 277
      B.2.1 Circulation 277
      B.2.2 Libration 278
  B.3 Small Oscillations of the Pendulum 279
      B.3.1 Angle–Action Variables 280
  B.4 Direct Construction of Angle–Action Variables 281
  B.5 The Neighborhood of the Pendulum Separatrix..............283
      B.5.1 Motion near the Separatrix 285
  B.6 The Separatrix or Whisker Map 286
  B.7 The Standard Map 288

C **Andoyer Hamiltonian with $k = 1$** 289
  C.1 Andoyer Hamiltonians 289
  C.2 Centers and Saddle Points 290
      C.2.1 The Case $k = 1$ 292
  C.3 Morphogenesis 293
  C.4 Width of the Libration Zone 296
  C.5 Integration 298
      C.5.1 The Case $\Delta > 0$..............................301
      C.5.2 The Case $\Delta < 0$..............................302
      C.5.3 The Separatrices 303

		C.5.4 The Angle σ 304
	C.6	Equilibrium Points 305
		C.6.1 The Inner Circulations Center 306
		C.6.2 The Libration Center 306
	C.7	Proper Periods 306
		C.7.1 Inner Circulations 307
		C.7.2 Librations 307
	C.8	The Angle Variable w 308
	C.9	Small-Amplitude Librations.............................. 308
		C.9.1 The Action Λ 312
		C.9.2 The New Hamiltonian 312
D	**Andoyer Hamiltonians with $k \geq 2$** 315	
	D.1	Introduction 315
	D.2	The Case $k = 2$ 315
		D.2.1 Morphogenesis 316
		D.2.2 Width of the Libration Zone 318
	D.3	The Case $k = 3$ 320
		D.3.1 Morphogenesis 321
		D.3.2 Width of the Libration Zone 323
	D.4	The Case $k = 4$ 325
		D.4.1 Morphogenesis 327
		D.4.2 Width of the Libration Zone 327
	D.5	Comparative Analysis 328
		D.5.1 Virtual Resonances 329

References 331

Index 337

1
The Hamilton–Jacobi Theory

1.1 Canonical Pertubation Equations

Astronomers in the nineteenth century found that the form of the Lagrange–Laplace equations for the perturbed Keplerian motion becomes very simple when the set of variables known as *Delaunay variables*,

$$\begin{aligned}
\ell &= \text{mean anomaly,} & L &= \sqrt{\mu a}, \\
g &= \text{argument of the periapsis,} & G &= L\sqrt{1-e^2}, \\
h &= \text{longitude of the node,} & H &= G\cos i,
\end{aligned} \qquad (1.1)$$

is used (see [15]). Here, μ is the product of the gravitational constant and the mass of the central body, a the semi-major axis, e the orbital eccentricity and i the inclination of the orbit over the reference plane.

With these variables, the equations of variation of the orbital elements are the *Delaunay equations*

$$\begin{aligned}
\frac{d\ell}{dt} &= \frac{\partial \mathcal{F}}{\partial L} & \frac{dL}{dt} &= -\frac{\partial \mathcal{F}}{\partial \ell} \\
\frac{dg}{dt} &= \frac{\partial \mathcal{F}}{\partial G} & \frac{dG}{dt} &= -\frac{\partial \mathcal{F}}{\partial g} \\
\frac{dh}{dt} &= \frac{\partial \mathcal{F}}{\partial H} & \frac{dH}{dt} &= -\frac{\partial \mathcal{F}}{\partial h},
\end{aligned} \qquad (1.2)$$

where

$$\mathcal{F} = -\frac{\mu^2}{2L^2} + R(L, G, H, \ell, g, h). \qquad (1.3)$$

In (1.3), R is the potential of the disturbing forces expressed with the Delaunay variables. The variational equations are in *canonical form*[1].

---

[1] In accordance with the conventions adopted in this book, the minus sign always appears in the differential equations for the second set of variables (momenta).

Delaunay soon discovered that the canonical form of the perturbation equations make easier the research of their solution. He also introduced important new ideas in his lunar theories [22], which became the first instance in which a canonical perturbation theory was used to obtain the "averaged" solution of a cumbersome dynamical system. Later, many other problems in Physics were brought to this form and the whole discipline of Analytical Dynamics was established. A detailed account of this is not included here[2]. However, some basic results and some details important in perturbation and averaging theories are gathered in this chapter and in the next one. It is worth emphasizing that this short introduction, say, for pedestrians, where only points of practical interest are given longer development, may not replace a full text on Analytical Dynamics. In addition, many results are deeply rooted in the theory of first-order partial differential equations. The learning of a few rules may not replace a correct understanding of the theories of Lagrange, Hamilton, Jacobi, Lie and others.

It must be emphasized that the signs in (1.2) and (1.3) are not the same as often seen in Celestial Mechanics books and papers. In fact, it is traditional in Celestial Mechanics, as well as in Mathematics, to use, instead of the energy of the system, its opposite. Also, instead of the potential, usually the so-called force function, which is its opposite, is used. In the study of actual problems, this ambiguity in convention is a frequent source of errors. Formally, both practices are equivalent; however, energy and potential are not just two arbitrary functions but quantities with well-established physical meanings. Since we have to make one choice, we choose that which is correct for everybody. Thus, the signs in this book are those adopted in Physics and in Mechanics.

1.2 Hamilton's Principle

Let us first introduce the usual concepts. We consider only unconstrained dynamical systems whose configuration is completely defined by N generalized coordinates q_i ($i = 1, 2, \cdots, N$). This system is said to have N *degrees of freedom*. The state of motion of the system is given by the generalized velocities $\dot{q}_i$ ($i = 1, 2, \cdots, N$).

Let T be the kinetic energy defined by a function of the generalized velocities $\dot{q}_i$ ($i = 1, 2, \cdots, N$) whose actual expression depends on the particular geometry of the configuration space. For example, if q_i are Cartesian coordinates, then the kinetic energy is given by the quadratic form $T = \frac{1}{2} \sum_{i=1}^{N} m_i \dot{q}_i^2$, where m_i are the masses of the particles (which, of course, must be the same for groups of subscripts indicating coordinates of the same particle).

Let $V(q_i, t)$ be the potential energy of the system. We assume that the acting forces derive from a velocity-independent potential and this function

---

[2] For an outstanding conceptual presentation of Dynamics, see [59].

is defined in such a way that the generalized forces are obtained by means of $\boldsymbol{f} = -\mathrm{grad}\, V$. We introduce also the generalized momenta,

$$p_i = \frac{\partial T}{\partial \dot{q}_i}, \tag{1.4}$$

and the Hamiltonian function

$$H = \sum_{i=1}^{N} p_i \dot{q}_i - T(q_i, \dot{q}_i) + V(q_i, t). \tag{1.5}$$

Principle (Hamilton). *The action of the system between t_1 and t_2, defined by the definite integral*

$$A = \int_{t_1}^{t_2} \left[\sum_{i=1}^{N} p_i \dot{q}_i - H(q_i, p_i, t) \right] dt, \tag{1.6}$$

is stationary for arbitrary variations of the solutions between the initial and final states.

□

In the usual notation, we may write

$$\delta A = \delta \int_{t_1}^{t_2} L\, dt = 0, \tag{1.7}$$

where

$$L = T(q_i, \dot{q}_i) - V(q_i, t) = \sum_{i=1}^{N} p_i \dot{q}_i - H(q_i, p_i, t) \tag{1.8}$$

is the Lagrangian function associated to H.

This is a simple variational problem in $2N$-dimensional phase space[3]. Using classical Euler–Lagrange equations for the solution of a variational problem, we obtain a system of $2N$ differential equations:

$$\begin{cases} \dfrac{d}{dt}\dfrac{\partial L}{\partial \dot{q}_i} - \dfrac{\partial L}{\partial q_i} = \dfrac{dp_i}{dt} + \dfrac{\partial H}{\partial q_i} = 0 \\[2mm] \dfrac{d}{dt}\dfrac{\partial L}{\partial \dot{p}_i} - \dfrac{\partial L}{\partial p_i} = 0 - \dot{q}_i + \dfrac{\partial H}{\partial p_i} = 0, \end{cases} \tag{1.9}$$

that is,

$$\dot{q}_i = \frac{\partial H}{\partial p_i}, \qquad \dot{p}_i = -\frac{\partial H}{\partial q_i}. \tag{1.10}$$

---

[3] We are considering Hamilton's principle in $2N$ dimensions instead of the more usual formulation in N dimensions with $L(q, \dot{q}, t)$ as integrand. Both formulations are equivalent (see [59], Chap. 6).

These equations are the *Hamilton equations* of the given dynamical system.

This system of equations has the same simple formulation as (1.2) and its main formal property is the complete definition of the right-hand sides of the equations by the function H alone

However, notwithstanding this simple structure, there is no general technique for the integration of these equations. As for Lagrangian equations, an important role is played by the operation known as Routhian reduction (see [97]). In the case of Hamiltonian equations, the Routhian reduction is immediate. Indeed, if the coordinate q_ϱ is *cyclic* (also called *ignorable*), that is, if the Hamiltonian does not depend on q_ϱ, then $\partial H/\partial q_\varrho = 0$ and $\dot{p}_\varrho = 0$, that is, the momentum p_ϱ is constant. Since $p_\varrho = c$, it may be replaced by c in the Hamiltonian function and the reduced Hamiltonian function is a function only of the q_i, p_i ($i \neq \varrho$). Thus, the given dynamical system is reduced to $N - 1$ degrees of freedom. When this reduced system is solved, the complete integration is achieved afterwards by means of the integral

$$q_\varrho = \int \frac{\partial H}{\partial c}\, dt. \tag{1.11}$$

This procedure is easily extended to other cyclic or ignorable variables.

Exercise 1.2.1. Show that, if T is a purely quadratic kinetic energy (that is, a homogeneous function of degree 2 in the $\dot{q}_i$), then the Hamiltonian is the total mechanical energy $H = T + V$. *Hint.* Use the definition of generalized momenta and Euler's homogeneous functions theorem.

Exercise 1.2.2 (Conservative systems). Show that $H = E$ (constant) when H is time-independent.

Exercise 1.2.3. Show that one rigid system formed by M particles ($M \geq 3$) has six degrees of freedom. *Hint:* Each particle has three coordinates but, since the system is rigid, they are not independent. The distances between the particles are constants unaltered by the motion of the system.

1.2.1 Maupertuis' Least Action Principle

Principle (Maupertuis). *The action*

$$S = \int_{t_1}^{t_2} \sum_{i=1}^{N} p_i \dot{q}_i\, dt \tag{1.12}$$

of a conservative dynamical system is stationary for arbitrary variations of the solutions between the initial and final states.

□

Indeed, if the system is conservative, $\delta \int_{t_1}^{t_2} H\, dt = 0$ because H is constant and, then, $\delta \int_{t_1}^{t_2} L\, dt = \delta \int_{t_1}^{t_2} \sum_{i=1}^{N} p_i \dot{q}_i\, dt$.

This principle, stated one century before Hamilton's principle, is usually written, in the case of one particle, as

$$\delta \int_{t_1}^{t_2} 2T \, dt = 0$$

which is equivalent to $\delta S = 0$ when T is assumed to be a quadratic function of the velocity.

1.2.2 Helmholtz Invariant

Lemma 1.2.1. *If we consider two isochronous solutions of a conservative Hamiltonian system whose initial states are infinitesimally close, the difference in the actions A of the system between initial and final instants $t = 0$ and $t = \tau$, along the two solutions, is*

$$\delta A = \left[\sum_{i=1}^{N} p_i \, dq_i \right]_0^\tau. \tag{1.13}$$

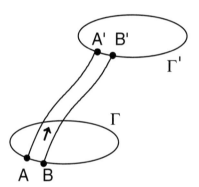

Fig. 1.1. Isochronous solutions starting at two neighboring points of Γ

The proof of this lemma is trivial. From Hamilton's principle, we know that the action of the system between 0 and τ is stationary for arbitrary variations of one solution between the initial and final states, say, between A and A' (see Fig. 1.1). Therefore, if we change the integration path from AA' to ABB'A', the result of the integral is the same. The decomposition of the integral along the path ABB'A' in its three parts and the substitution of the integrals over the arcs AB and A'B', respectively, by $\left[\sum_{i=1}^{N} p_i \, dq_i\right]_{t=0}$ and $\left[\sum_{i=1}^{N} p_i \, dq_i\right]_{t=\tau}$ gives

$$\int_{AA'} \sum_{i=1}^{N} p_i \, dq_i - \int_{BB'} \sum_{i=1}^{N} p_i \, dq_i = \left[\sum_{i=1}^{N} p_i \, dq_i \right]_0^\tau$$

as stated. □

Theorem 1.2.1 (Helmholtz). *If Γ is an arbitrary closed curve in phase space, the quantity*

$$J = \oint_\Gamma \sum_{i=1}^N p_i \, dq_i, \qquad (1.14)$$

is an invariant of the motion.

Proof. [59] In order to prove this theorem, let us consider a curve Γ and all solutions whose initial state lies on Γ. Let us consider two solutions whose initial state lies on neighboring points of Γ, say A and B (see Fig. 1.1). The difference of the actions on isochronous motions starting from the end points of the element is given by (1.13). If we divide Γ into a succession of infinitesimal arcs and add the contribution coming from each arc, the net result is null, that is,

$$0 = \oint \left[\sum_{i=1}^N p_i \, dq_i \right]_0^\tau. \qquad (1.15)$$

Therefore, the integral of the function enclosed with brackets is the same at $t = 0$ and at $t = \tau$, that is,

$$\oint_\Gamma \sum_{i=1}^N p_i \, dq_i = \oint_{\Gamma'} \sum_{i=1}^N p_i \, dq_i,$$

where Γ' is the closed curve into which Γ is transported by the solutions in the time τ. □

1.3 Canonical Transformations

The Routhian reduction is one of the basic steps in the Hamilton–Jacobi theory and in the perturbation theories discussed in this book: It is the search for one transformation leading to a new set of variables such that the canonical form of the equations is preserved, but some of the coordinates become ignorable. Transformations preserving the canonical form of the equations are called *canonical transformations*.

If we consider a change of the given variables q_i, p_i into a new set q_i^*, p_i^* defined by a system of $2N$ equations:

$$q_j^* = q_j^*(q_i, p_i, t), \qquad p_j^* = p_j^*(q_i, p_i, t), \qquad (1.16)$$

this transformation is said to be canonical if it preserves the canonical form of any given canonical system, that is, if, under the transformation, (1.10) becomes

$$\dot{q}_i^* = \frac{\partial H^*}{\partial p_i^*}, \qquad \dot{p}_i^* = -\frac{\partial H^*}{\partial q_i^*}, \qquad (1.17)$$

where H^* is a new function of q_i^*, p_i^*. An equivalent definition is obtained by saying that canonical transformations preserve Hamilton's principle (see 1.7). Thus, in the new variables,

$$\delta \int_{t_1}^{t_2} \left[\sum_{i=1}^{N} p_i^* \dot{q}_i^* - H^*(q_i^*, p_i^*, t) \right] dt = 0. \tag{1.18}$$

Comparing to (1.7), we obtain

$$\delta \int_{t_1}^{t_2} \left[\sum_{i=1}^{N} p_i \dot{q}_i - \sum_{i=1}^{N} p_i^* \dot{q}_i^* - (H - H^*) \right] dt = 0. \tag{1.19}$$

One solution of the given problem is obtained by making the function under the integral sign equal to zero. Moreover, the introduction under the integral sign of an arbitrary exact differential does not alter the result, because

$$\delta \int_{t_1}^{t_2} \dot{S} \, dt = 0 \tag{1.20}$$

for any function S of the considered variables. As the $4N$ variables q_i, p_i, q_i^*, p_i^* are not independent, we select $2N$ of them, for instance, q_i, q_i^* and write the solution of (1.19) as

$$\sum_{i=1}^{N} p_i dq_i - \sum_{i=1}^{N} p_i^* dq_i^* = dS(q_i, q_i^*, t) + (H - H^*) dt. \tag{1.21}$$

It is worth recalling that our dynamical system is, by assumption, unconstrained; otherwise, we should take into account the non-independence of the variables q_i, p_i. We recall that energies in this book are defined as in Physics. When H is the opposite of the energy, the sign in front of the last parenthesis in the above equation should be changed.

The function S is known as the generating function of the canonical transformation. Later on, in this book, we will refer to it as the Jacobian generating function, to distinguish it from the generating function due to Sophus Lie. It completely characterizes the transformation. From (1.21), we obtain the equations of the canonical transformation of variables:

$$p_i = \frac{\partial S}{\partial q_i}, \qquad p_i^* = -\frac{\partial S}{\partial q_i^*}. \tag{1.22}$$

The relation between the Hamiltonian functions before and after the transformation is

$$H^* = H + \frac{\partial S}{\partial t}. \tag{1.23}$$

When S is time-independent, it follows that $H^* = H$ and the transformation is said to be conservative.

Equations (1.22) are not unique in many respects. For example, if we introduce the exact differential

$$\mathrm{d}\sum_{i=1}^{N} q_i^* p_i^* = \sum_{i=1}^{N} p_i^* \mathrm{d}q_i^* + \sum_{i=1}^{N} q_i^* \mathrm{d}p_i^* \qquad (1.24)$$

into (1.21), it becomes

$$\sum_{i=1}^{N} p_i \mathrm{d}q_i + \sum_{i=1}^{N} q_i^* \mathrm{d}p_i^* = \mathrm{d}S' + (H - H^*)\,\mathrm{d}t, \qquad (1.25)$$

where the new generating function, $S' = S + \sum_{i=1}^{N} q_i^* p_i^*$, may be considered as a function of the variables (q_i, p_i^*) and, instead of (1.22), we obtain

$$p_i = \frac{\partial S'}{\partial q_i}, \qquad q_i^* = \frac{\partial S'}{\partial p_i^*}. \qquad (1.26)$$

The relation between the Hamiltonians H and H' is the same as before (with S' instead of S).

Many different combinations are possible. However, we cannot escape the fate of always having half of the equations defining the transformation solved with respect to the old variables and half of them solved with respect to the new ones. An algebraic inversion is always needed to obtain the transformation in explicit form.

In (1.21) and (1.25), the time variation is important as it shows how the Hamiltonian will be changed in a time-dependent transformation. However, in what concerns the canonical condition itself, instead of the actual displacements of the system, only the *virtual* displacements, or *variations* δq_i and δq_i^* at a fixed instant are considered. The only requirement is that these displacements are possible; for instance, if the dynamical system were assumed to be constrained, the virtual displacements should obey the system constraints. Using variations instead of differentials, (1.21) becomes

$$\sum_{i=1}^{N} p_i \delta q_i - \sum_{i=1}^{N} p_i^* \delta q_i^* = \delta S(q_i, q_i^*, t). \qquad (1.27)$$

The relationship between variations and differentials is obvious. For instance,

$$\mathrm{d}S = \delta S + \frac{\partial S}{\partial t}\,\mathrm{d}t.$$

Exercise 1.3.1. Show that the composition of two canonical transformations is canonical.

Exercise 1.3.2. We could have introduced a *valence* (or *multiplier*) λ and written

$$\delta \int_{t_1}^{t_2} \left[\sum_{i=1}^{N} p_i \dot{q}_i - \lambda \sum_{i=1}^{N} p_i^* \dot{q}_i^* - (H - \lambda H^*) \right] dt = 0 \qquad (1.28)$$

instead of (1.19) (since that equation results from the comparison of (1.7) and (1.18), both equal to zero). Construct the equations giving the canonical transformation and the new *conservation* equation

$$H(q, p) = \lambda H^*(q^*, p^*). \qquad (1.29)$$

1.4 Lagrange Brackets

When, in (1.21), S is forced to be a function of q_i, p_i, the calculations are less immediate. Let us consider the transformation in its explicit form

$$q_k = q_k(q_i^*, p_i^*, t), \qquad p_k = p_k(q_i^*, p_i^*, t) \qquad (1.30)$$

and let us calculate the variation δq_k corresponding to an arbitrary change $\delta q_i^*, \delta p_i^*$. δq_k is a linear differential form in $\delta q_i^*, \delta p_i^*$, and the time-independent part of (1.21) becomes

$$\sum_{k=1}^{N} p_k \sum_{i=1}^{N} \left(\frac{\partial q_k}{\partial q_i^*} \delta q_i^* + \frac{\partial q_k}{\partial p_i^*} \delta p_i^* \right) - \sum_{i=1}^{N} p_i^* \delta q_i^* = \delta S. \qquad (1.31)$$

If the transformation is canonical, (1.31) must be an exact form. Thus, it may satisfy the conditions for exact differential forms, which, after some calculations, give

$$\sum_{k=1}^{N} \left(\frac{\partial q_k}{\partial q_i^*} \frac{\partial p_k}{\partial p_j^*} - \frac{\partial q_k}{\partial p_j^*} \frac{\partial p_k}{\partial q_i^*} \right) = \delta_{ij}$$

$$\sum_{k=1}^{N} \left(\frac{\partial q_k}{\partial q_i^*} \frac{\partial p_k}{\partial q_j^*} - \frac{\partial q_k}{\partial q_j^*} \frac{\partial p_k}{\partial q_i^*} \right) = 0 \qquad (1.32)$$

$$\sum_{k=1}^{N} \left(\frac{\partial q_k}{\partial p_i^*} \frac{\partial p_k}{\partial p_j^*} - \frac{\partial q_k}{\partial p_j^*} \frac{\partial p_k}{\partial p_i^*} \right) = 0,$$

where δ_{ij} is the Kronecker symbol: $\delta_{ij} = 0$ for $i \neq j$ and $\delta_{ij} = 1$ for $i = j$ $(i, j = 1, \cdots, N)$. Using the Lagrange brackets

$$[f, g] \stackrel{\text{def}}{=} \sum_{k=1}^{N} \left(\frac{\partial q_k}{\partial f} \frac{\partial p_k}{\partial g} - \frac{\partial q_k}{\partial g} \frac{\partial p_k}{\partial f} \right), \qquad (1.33)$$

equations (1.32) are written

$$[q_i^*, p_j^*] = \delta_{ij}, \qquad [q_i^*, q_j^*] = 0, \qquad [p_i^*, p_j^*] = 0. \qquad (1.34)$$

Since

$$\sum_{k=1}^{N} \delta q_k \wedge \delta p_k = \sum_{i<j} [q_i^*, q_j^*] \delta q_i^* \wedge \delta q_j^* + \sum_{i<j} [p_i^*, p_j^*] \delta p_i^* \wedge \delta p_j^*$$
$$+ \sum_{i=1}^{N} \sum_{j=1}^{N} [q_i^*, p_j^*] \delta q_i^* \wedge \delta p_j^*$$
$$= \sum_{i=1}^{N} \delta q_i^* \wedge \delta p_i^*,$$

the form of the canonical condition given by (1.34) is often expressed by saying that the differential form $\sum_{i=1}^{N} \delta q_i \wedge \delta p_i$ is invariant under the transformation.

Equations (1.32) are relations among the elements of the Jacobian matrix of the transformation $(q^*, p^*) \Rightarrow (q, p)$:

$$\mathsf{M} = \begin{pmatrix} \left(\dfrac{\partial q_i}{\partial q_j^*}\right) & \left(\dfrac{\partial q_i}{\partial p_j^*}\right) \\ \left(\dfrac{\partial p_i}{\partial q_j^*}\right) & \left(\dfrac{\partial p_i}{\partial p_j^*}\right) \end{pmatrix} \tag{1.35}$$

and are equivalent to the matrix equation

$$\mathsf{M}' \mathsf{J} \mathsf{M} = \mathsf{J}, \tag{1.36}$$

where M' is the transpose of M. J is the symplectic unit matrix[4] of rank $2N$

$$\mathsf{J} = \begin{pmatrix} 0 & -\mathsf{E} \\ \mathsf{E} & 0 \end{pmatrix}, \tag{1.37}$$

where E is the unit matrix of rank N. Since the determinant of a product of matrices is equal to the product of the determinants of the matrices being multiplied and $\det \mathsf{J} = 1$, we find from (1.36) that $(\det \mathsf{M})^2 = 1$. The proof that $\det \mathsf{M} = +1$ requires further considerations (see [99]). One must first show that canonical transformations must be decomposed into a canonical transformation whose Jacobian matrix is positive definite and one orthogonal canonical transformation (that is, one canonical transformation whose Jacobian matrix O is such that $\mathsf{O}' = \mathsf{O}^{-1}$). In the case of the orthogonal canonical

---

[4] With the above definition of J, the energy sign and the $q - p$ order adopted in this book, the Hamilton equations are:

$$\dot{z} = -\mathsf{J} \frac{\partial H}{\partial z} \qquad z \equiv (q, p).$$

In some other books, other conventions are adopted, changing the minus sign in the above equation into plus.

transformation, the canonical condition may be written as $\mathsf{JO} = \mathsf{OJ}$, and this implies that O is of the form

$$\begin{pmatrix} \mathsf{A} & \mathsf{B} \\ -\mathsf{B} & \mathsf{A} \end{pmatrix}.$$

Then, one introduces the unitary complex matrix

$$\mathsf{F} = \frac{1}{2N} \begin{pmatrix} i\mathsf{E} & \mathsf{E} \\ \mathsf{E} & i\mathsf{E} \end{pmatrix}$$

and shows that $\det(\mathsf{FOF}^{-1})$ is the product of the two complex numbers $\det(\mathsf{A} \pm i\mathsf{B})$ and so cannot be negative. Since $\det(\mathsf{FOF}^{-1}) = \det \mathsf{O}$, the proof is complete.

1.5 Poisson Brackets

The Poisson bracket of two differentiable functions of the canonical variables:

$$f = f(q, p)$$
$$g = g(q, p)$$

is the bilinear operation

$$\{f, g\} \stackrel{\text{def}}{=} \sum_{i=1}^{N} \left(\frac{\partial f}{\partial q_i} \frac{\partial g}{\partial p_i} - \frac{\partial g}{\partial q_i} \frac{\partial f}{\partial p_i} \right). \tag{1.38}$$

The usefulness of these brackets comes from the fact that the canonical equations (1.10) are simply the Poisson brackets of the Hamiltonian function $H(q, p)$ and the variables:

$$\begin{aligned} \dot{q}_i &= \{q_i, H\} \\ \dot{p}_i &= \{p_i, H\}. \end{aligned} \tag{1.39}$$

It is worth mentioning that Poisson brackets may be written as

$$\{f, g\} = \operatorname{grad} f \cdot (\mathsf{J} \cdot \operatorname{grad} g);$$

that is, the scalar product of the gradient of f and the symplectic rotation of the gradient of g.

Exercise 1.5.1 (Invariance to Canonical Transformations). Consider two differentiable functions $f_k = f_k(q_i, p_i)$ $(k = 1, 2; i = 1, 2, \cdots, N)$ and the canonical transformation

$$\begin{aligned} q_i &= q_i(q_j^*, p_j^*) \\ p_i &= p_i(q_j^*, p_j^*) \end{aligned} \qquad (i, j = 1, 2, \cdots, N).$$

Show that the Poisson bracket of the functions f_1 and f_2 is the same, no matter whether it is calculated with the variables q_i, p_i or the variables q_i^*, p_i^*; that is

$$\{f_1(q_i(q_j^*,p_j^*),p_i(q_j^*,p_j^*)), f_2(q_i(q_j^*,p_j^*),p_i(q_j^*,p_j^*))\} = \{f_1(q_i,p_i), f_2(q_i,p_i)\}.$$

Exercise 1.5.2. Consider $2N$ differentiable functions $f_i = f_i(q_j, p_j)$ $(i, j = 1, 2, \cdots, 2N)$. Show that

$$\sum_{k=1}^{2N} [f_k, f_i]\{f_k, f_\ell\} = \delta_{i\ell}. \tag{1.40}$$

Although cumbersome, the proof of this result is straightforward.

Exercise 1.5.3 (Canonical condition). Consider the canonical transformation

$$\begin{aligned} q_i^* &= q_i^*(q_j, p_j) \\ p_i^* &= p_i^*(q_j, p_j) \end{aligned} \qquad (i, j = 1, 2, \cdots, N) \tag{1.41}$$

and show that, in this case,

$$\{q_i^*, p_j^*\} = \delta_{ij}, \qquad \{q_i^*, q_j^*\} = 0, \qquad \{p_i^*, p_j^*\} = 0. \tag{1.42}$$

This new form of the canonical condition is an immediate consequence of (1.40) and (1.34).

1.5.1 Reciprocity Relations

In the case of the above canonical transformation, in addition to (1.40), we may establish some useful one-to-one relations between the mutual derivatives of the two sets of canonical variables.

Proposition 1.5.1. *Given a conservative canonical transformation*

$$\begin{aligned} q_i^* &= q_i^*(q_j, p_j) \\ p_i^* &= p_i^*(q_j, p_j) \end{aligned} \qquad (i, j = 1, 2, \cdots, N) \tag{1.43}$$

and its inverse

$$\begin{aligned} q_j &= q_j(q_i^*, p_i^*) \\ p_j &= p_j(q_i^*, p_i^*) \end{aligned} \qquad (i, j = 1, 2, \cdots, N) \tag{1.44}$$

then, for any i, j:

$$\begin{aligned} \frac{\partial q_i}{\partial q_j^*} &= \frac{\partial p_j^*}{\partial p_i} & \frac{\partial p_i}{\partial q_j^*} &= -\frac{\partial p_j^*}{\partial q_i} \\ \frac{\partial q_i}{\partial p_j^*} &= -\frac{\partial q_j^*}{\partial p_i} & \frac{\partial p_i}{\partial p_j^*} &= \frac{\partial q_j^*}{\partial q_i}. \end{aligned} \tag{1.45}$$

The proof of this statement is very simple. Let us first consider the canonical transformation given by (1.22), derived from the Jacobian generating function $S(q, q^*)$. From (1.22) we have

$$\frac{\partial p_i}{\partial q_j^*} = \frac{\partial^2 S}{\partial q_i \partial q_j^*} = -\frac{\partial p_j^*}{\partial q_i}$$

and the second of equations (1.45) is proved. We may, then, repeat the same calculations with the canonical transformation given by (1.26), derived from $S'(q, p^*)$ and prove the fourth of equations (1.45) (as well as the first one in which just the direction of the transformation is changed). The third of equations (1.45) is similarly proved using transformations derived from the generating functions $S''(p, p^*)$.

Exercise 1.5.4. Show that $\mathsf{M}^{-1} = -(\mathsf{JMJ})'$ and that the reciprocity relations may be obtained by comparing the elements of both sides of this equation.

1.6 The Extended Phase Space

Time-dependent Hamiltonians and time-dependent canonical transformations are not separately considered in this book. Time-dependent Hamiltonian dynamics is a *particular case* of time-independent Hamiltonian dynamics. To see this, let the canonical equations of a time-dependent system be formulated in parametric form. Let us introduce a parameter τ and let us consider the time t not as the independent variable, but as one of the $N+1$ generalized coordinates $q_1, q_2, \cdots, q_N, t$ given as functions of the parameter τ. The system now has $N+1$ degrees of freedom and the $2N$ former equations become

$$\frac{\mathrm{d}q_i}{\mathrm{d}\tau} = t'\frac{\partial H}{\partial p_i} \qquad \frac{\mathrm{d}p_i}{\mathrm{d}\tau} = -t'\frac{\partial H}{\partial q_i} \qquad (i = 1, 2, \cdots, N), \qquad (1.46)$$

where t' denotes the derivative of t with respect to τ, which is considered as a known function of τ. As t' is independent of the variables q_i, p_i, we may write

$$\frac{\mathrm{d}q_i}{\mathrm{d}\tau} = \frac{\partial(Ht')}{\partial p_i}, \qquad \frac{\mathrm{d}p_i}{\mathrm{d}\tau} = -\frac{\partial(Ht')}{\partial q_i}. \qquad (1.47)$$

If p_t is the momentum conjugate to t, we may introduce one complementary differential equation:

$$t' \stackrel{\text{def}}{=} \frac{\mathrm{d}t}{\mathrm{d}\tau} = \frac{\partial(t'p_t)}{\partial p_t}. \qquad (1.48)$$

These equations may be written in the unified form

14   1 The Hamilton–Jacobi Theory

$$\frac{dq_i}{d\tau} = \frac{\partial K}{\partial p_i} \qquad \frac{dp_i}{d\tau} = -\frac{\partial K}{\partial q_i}$$
$$\frac{dt}{d\tau} = \frac{\partial K}{\partial p_t} \qquad \frac{dp_t}{d\tau} = -\frac{\partial K}{\partial t}, \qquad (1.49)$$

where

$$K(q_i, t, p_i, p_t) = t'H + t'p_t \qquad (1.50)$$

is the Hamiltonian of the given system in the extended phase space. The system was completed by the addition of one equation for the momentum p_t.

Generally, knowledge of the meaning of p_t is not needed. It may be kept in the equations as an extra unknown function that, automatically, disappears when we go back to the $2N$-dimensional phase space.

To understand the meaning of the momentum p_t we have to construct the Lagrangian function associated with K and to calculate its derivative with respect to t'. This Lagrangian function is obtained from the Hamiltonian by means of Legendre's dual transformation:

$$\mathcal{L} = \sum_{i=1}^{N} p_i q'_i + p_t t' - K, \qquad (1.51)$$

where primes denote derivatives with respect to τ. If we introduce in this definition the above expression for K and note that t' appears in $\mathcal{L}$ only as a factor of some terms (two of which are opposite), we obtain

$$p_t = \frac{\partial \mathcal{L}}{\partial t'} = -H, \qquad (1.52)$$

that is, the momentum conjugate to time is the opposite of the energy.

An immediate consequence is that the numerical value of K is zero. Since K is independent of τ, the extended system is conservative and has the integral $K = \text{const}$. Thus, the condition $K = 0$ is permanently satisfied. The introduction of t as an $(N+1)^{\text{th}}$ generalized coordinate leads to a new mechanical system with $N+1$ degrees of freedom, always conservative. The only difference from the usual conservative systems lies on the fact that the extended energy K cannot take arbitrary values. It is necessarily equal to zero (or to another fixed constant, since the addition of a constant to K does not alter the equations).

In practical applications, the relationship between the time t and the parameter τ is a mere identity. So, usually $t' = 1$, and t is written as an independent variable, instead of τ, in the equations. In this case, the extended Hamiltonian is, simply,

$$K(q_i, t, p_i, p_t) = H(q_i, p_i, t) + p_t. \qquad (1.53)$$

Another frequent choice is to introduce as a new generalized coordinate a linear function of the time (a *mean longitude*) $\lambda = \nu t + \text{const}$, instead of the time itself. The transformation from the above case to this one is trivial and the extended energy, now, is:

$$K(q_i, \lambda, p_i, p_\lambda) = H(q_i, p_i, \lambda) + \nu p_\lambda. \tag{1.54}$$

We still have $K = 0$ and the new generalized momentum p_λ is related to the energy through $p_\lambda = -E/\nu$.

1.7 Gyroscopic Systems

The word *gyroscopic* is often used to designate terms in the kinetic energy that are linear in the velocity components. In this book, we use it to designate systems whose Lagrangian has linear terms in the velocity, of the form $\boldsymbol{r} \times \boldsymbol{v}$. These terms may be introduced through a velocity-dependent potential energy, as in the case of charged particles under the action of magnetic forces, or through the kinetic energy, as in the case of a motion relative to a rotating frame.

1.7.1 Gyroscopic Forces

In the topics studied in previous sections, the potential energy was velocity-independent. Let us consider, now, a system of N points $P_i \in \mathbf{R}^3$ with masses m_i, let $\boldsymbol{r}_i$ and $\boldsymbol{v}_i$ be the position vector and the velocity of P_i with respect to an inertial frame and let us assume that the system is submitted to gyroscopic forces arising from a generalized potential energy $W(\boldsymbol{r}_i, \boldsymbol{v}_i)$. How does the generalized potential energy relate to the forces applied on the particles? The corresponding Lagrangian equations may be written as

$$\frac{d}{dt}\left(\frac{\partial T}{\partial \boldsymbol{v}_i}\right) - \frac{\partial T}{\partial \boldsymbol{r}_i} = \frac{d}{dt}\left(\frac{\partial W}{\partial \boldsymbol{v}_i}\right) - \frac{\partial W}{\partial \boldsymbol{r}_i}. \tag{1.55}$$

Since $T = \frac{1}{2}\sum m_i v_i^2$, the above equations are equivalent to

$$m_i \dot{\boldsymbol{v}}_i = \frac{d}{dt}\left(\frac{\partial W}{\partial \boldsymbol{v}_i}\right) - \frac{\partial W}{\partial \boldsymbol{r}_i} \stackrel{\text{def}}{=} \boldsymbol{F}_i, \tag{1.56}$$

showing that the right-hand side of (1.55) expresses the forces applied on the particles. One may note that this equation generalizes the usual $\boldsymbol{F}_i = -\partial V/\partial \boldsymbol{r} = -\operatorname{grad}_{P_i} V$ of the velocity-independent case. The momenta of P_i are, now, given by

$$\boldsymbol{p}_i = \frac{\partial (T - W)}{\partial \boldsymbol{v}_i} = m_i \boldsymbol{v}_i - \frac{\partial W}{\partial \boldsymbol{v}_i}. \tag{1.57}$$

and the corresponding Hamiltonian is given by

$$H = \sum_{i=1}^{N} \boldsymbol{p}_i \cdot \boldsymbol{v}_i - T + W, \tag{1.58}$$

that is,
$$H = \sum_{i=1}^{N} \frac{p_i^2}{2m_i} + Y(r_i, p_i), \tag{1.59}$$

where
$$Y(r_i, p_i) = W - \sum_{i=1}^{N} \frac{1}{2m_i} \left(\frac{\partial W}{\partial v_i}\right)^2. \tag{1.60}$$

It is easy to see, by using the Hamiltonian equations instead of the Lagrangian ones, that the expression of the force applied on the particles in terms of $Y(r_i, p_i)$ is

$$F_i = m_i \frac{\mathrm{d}}{\mathrm{d}t}\left(\frac{\partial H}{\partial p_i}\right) = m_i \frac{\mathrm{d}}{\mathrm{d}t}\left(\frac{\partial Y}{\partial p_i}\right) - \frac{\partial Y}{\partial r_i}. \tag{1.61}$$

1.7.2 Example

Let us consider the simple generalized potential energy[5]

$$W = \sum_{i=1}^{N} m_i [A, r_i, v_i].$$

To get the right physical dimension of W, A needs to have the dimension of an angular velocity. W is a scalar quantity; it is equal to $|A|$ times the projection of the angular momentum of the system on the direction of the vector A. From (1.56) and (1.60), we obtain

$$F_i = 2m_i A \times v_i$$

and

$$Y = W - \sum_{i=1}^{N} \frac{A^2}{2} m_i [r_i^2 - (r_i \cdot u)^2],$$

where u is the unit vector in the direction of A. It is worth noting that $[r_i^2 - (r_i \cdot u)^2]$ is the square of the distance of the particle P_i to the axis defined by u. Therefore, we may also write $Y = W - A^2 \mathcal{I}/2$, where $\mathcal{I}$ is the moment of inertia of the system with respect to the axis defined by u.

An example of a force of this kind, in nature, is the force acting on an electric charge moving in a magnetic field: $F = -\frac{e}{c} B \times v$. ($e$ is the electric charge, c the velocity of light and B is the magnetic induction.) The corresponding generalized potential energy is $W = -\frac{e}{2c}[B, r, v]$.

---

[5] $[a, b, c]$ denotes the triple scalar product $a \cdot (b \times c)$. Two elementary rules used in this section are the invariance of the triple scalar product to a circular permutation of the operands and Lagrange's identity for the triple vector product: $a \times (b \times c) = (a \cdot c)b - (a \cdot b)c$. These two rules combine to give $(a \times b)^2 = a^2 b^2 - (a \cdot b)^2$.

1.7.3 Rotating Frames

Let us consider a system under the action of applied forces depending on a velocity-independent potential $U(m_i, r_i)$, but in a frame rotating with angular velocity Ω around an axis directed along a given unit vector $\boldsymbol{u}$. Let r_i and v_i be the coordinates and velocity components of P_i with respect to the rotating frame. With respect to an inertial (non-rotating) frame, the velocity is given by $v_i + \boldsymbol{\Omega} \times r_i$ (where $\boldsymbol{\Omega} = \Omega \boldsymbol{u}$) and the kinetic energy of the system is

$$T = \frac{1}{2} \sum_{i=1}^{N} m_i (v_i + \boldsymbol{\Omega} \times r_i)^2 \qquad (1.62)$$

or

$$T = \frac{1}{2} \sum_{i=1}^{N} m_i \left[v_i^2 + 2[v_i, \boldsymbol{\Omega}, r_i] + \Omega^2 r_i^2 - (\boldsymbol{\Omega} \cdot r_i)^2 \right]. \qquad (1.63)$$

The momenta conjugated to the relative vector radii r_i are

$$\boldsymbol{p}_i = \frac{\partial T}{\partial v_i} = m_i v_i + m_i \boldsymbol{\Omega} \times r_i \qquad (1.64)$$

and the Hamiltonian function is given by

$$H = \sum_{i=1}^{N} \boldsymbol{p}_i \cdot v_i - T + mU, \qquad (1.65)$$

that is, in terms of the canonical variables r_i, p_i,

$$H = \sum_{i=1}^{N} \left(\frac{p_i^2}{2m_i} - [\boldsymbol{\Omega}, r_i, p_i] \right) + mU(r). \qquad (1.66)$$

1.7.4 Apparent Forces

An observer fixed in the rotating frame will perceive modifications in the motion of a point or system of points as if the system were under the action of apparent forces corresponding to the "potential energy" $-\sum [\boldsymbol{\Omega}, r_i, p_i]$ (in addition to $mU(r)$). To determine which forces these are, we substitute $Y = -\sum [\boldsymbol{\Omega}, r_i, p_i] + mU(r)$ into (1.61) and obtain

$$F_i = 2m_i v_i \times \boldsymbol{\Omega} + m_i \Omega^2 \left[r_i - (r_i.\boldsymbol{u})\boldsymbol{u} \right] - m_i \mathrm{grad}_{P_i} U.$$

The term $m_i \Omega^2 \left[r_i - (r_i.\boldsymbol{u})\boldsymbol{u} \right]$ is the *centrifugal force*. Note that $r_i - (r_i.\boldsymbol{u})\boldsymbol{u} = \rho_i$ is a vector perpendicular to the rotation axis going from the axis to the point P_i.

The term $2m_i v_i \times \boldsymbol{\Omega}$ is the so-called *Coriolis force*. The Coriolis force tends to make the free motion of one particle on a rotating frame deviate from a straight line. On the surface of the Earth, the deviation of the moving particle due to the Coriolis force is to the right in the Northern hemisphere and to the left in the Southern hemisphere (see [94]).

1.8 The Partial Differential Equation of Hamilton and Jacobi

The Hamilton–Jacobi theory is the cornerstone of Analytical Mechanics. The entirely new understanding of the problems of mechanics that it allowed was impressive. In the last quarter of the nineteenth century, it experienced an enormous development in close relationship with the theory of first-order partial differential equations and the introduction, by Sophus Lie, of contact transformations. In modern Physics, it was introduced around 1916 through the work of Schwarzschild, Epstein and Sommerfeld into the mechanics of the atom and provided the basis of the old Quantum Theory and, a few years later, of the new Quantum Mechanics of Schrödinger, Heisenberg and Born. Much of that work was directly inspired from Celestial Mechanics and a key reference in all papers of this period is Charlier's *Die Mechanik des Himmels*. Charlier's book [20] included one chapter on the Staude–Stäckel theory of conditionally periodic systems and the construction of the uniformized angle variables, which would be later called, with their conjugates, angle–action variables (see Chap. 2). More or less at the same time, Hamilton–Jacobi theory played an essential role in the geometrization of dynamics and its variational principles are the basis of Einstein's theory of general relativity.

The equation nowadays named after Hamilton and Jacobi is due to Jacobi and is a modification of the equation published by Hamilton in 1834. It was an extension to dynamics of the partial differential equation discovered by Hamilton 10 years before, in optics. Using canonical transformations, this equation is easily obtained.

Let us consider a conservative mechanical system with N degrees of freedom, a time-independent Hamiltonian function $H(q,p)$, ($q \equiv q_1, q_2, \cdots, q_N$; $p \equiv p_1, p_2, \cdots, p_N$) and one canonical transformation $\phi : (q,p) \Rightarrow (q^*, p^*)$.

The main idea of Hamiltonian theories is, in general, to seek a canonical transformation such that the Hamiltonian of the resulting system is as simple as possible. In the earlier Hamilton's theory, the sought canonical transformation was expected to lead to a Hamiltonian independent of all variables. In the current Hamilton–Jacobi theory, it is required that the Hamiltonian obtained by applying this transformation to $H(q,p)$ shall be equal to one of the new variables, say:

$$H^*(q^*, p^*) = p_1^*. \tag{1.67}$$

This new Hamiltonian system is trivial. The energy integral is $E = p_1^*$ and the system has the general solution

$$\begin{aligned} q_1^* &= t + \alpha_1 & p_1^* &= \beta_1 = E \\ q_\varrho^* &= \alpha_\varrho & p_\varrho^* &= \beta_\varrho & (\varrho = 2, \cdots, N), \end{aligned} \tag{1.68}$$

where α_i, β_i ($i = 1, 2, \cdots, N$) are integration constants.

The equation found by Jacobi is the partial differential equation giving the function that generates the canonical transformation. To obtain it, we use the

1.8 The Partial Differential Equation of Hamilton and Jacobi

fact that the canonical transformation ϕ is conservative. As a consequence, the Hamiltonians, before and after the transformation, must be equal, that is,

$$H(q,p) = H^*(q^*,p^*). \tag{1.69}$$

Let $S(q,p^*)$ be the generating function of the canonical transformation ϕ. Introducing the transformation equations

$$q_i^* = \frac{\partial S}{\partial p_i^*}, \qquad p_i = \frac{\partial S}{\partial q_i}, \qquad (i = 1, 2, \cdots, N) \tag{1.70}$$

into the conservation equation, we obtain the *Hamilton–Jacobi equation*

$$H\left(q_i, \frac{\partial S}{\partial q_i}\right) = p_1^* = E. \tag{1.71}$$

This equation is a first-order partial differential equation for the function $S(q_1, q_2, \cdots, q_N)$. E is just a parameter. To use the solution of the Hamilton–Jacobi equation as a generating function of a canonical transformation, it may be a function of $2N$ variables, $S = S(q, p^*)$. The role of p_i^* may be played by N integration constants β_i.

Any solution of a first-order partial differential equation, containing as many integration constants as there are variables, is called a *complete solution*. The complete solution of the Hamilton–Jacobi equation generates a canonical transformation flattening out the surfaces $H(q,p) = E$(constant) of the given phase space into the parallel planes $p_1^* = E$ (Fig. 1.2).

When a complete solution $S(q, \beta)$ is known, the solution of the dynamical system is given by

$$\begin{aligned} t + \alpha_1 &= q_1^* = \frac{\partial S(q,\beta)}{\partial \beta_1} \\ \alpha_\varrho &= q_\varrho^* = \frac{\partial S(q,\beta)}{\partial \beta_\varrho} \qquad (\varrho = 2, \cdots, N). \end{aligned} \tag{1.72}$$

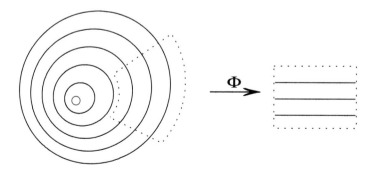

Fig. 1.2. A simple Hamilton–Jacobi mapping

Hamilton–Jacobi mappings, when they exist, are finite transformations generally including singularities due to the topological differences between the surfaces $H(q,p) = E$ and the planes $p_1^* = E$. As a consequence, they are more general than the Lie mappings that will be introduced in Chap. 5. All Lie mappings may be written as Hamilton–Jacobi mappings, but the converse is not true. Lie mappings are infinitesimal (near-identity) homeomorphisms of the phase space into itself.

1.9 One-Dimensional Motion with a Generic Potential

Let us consider the problem of the motion of one particle of mass m on a straight line under the action of a generic velocity-independent potential $U(q_1)$. Newton's equations of motion are

$$\ddot{q}_1 = -\ \mathrm{grad}\ U(q_1) = -\frac{\mathrm{d}U}{\mathrm{d}q_1} \tag{1.73}$$

and the corresponding Hamiltonian is

$$H = \frac{p_1^2}{2m} + mU(q_1), \tag{1.74}$$

where $p_1 = m\dot{q}_1$ is the momentum of the particle. The Hamilton–Jacobi equation is

$$\frac{1}{2m}\left(\frac{\partial S}{\partial q_1}\right)^2 + mU(q_1) = E. \tag{1.75}$$

The (complete) solution of (1.75) is

$$S(q_1, E) = \sqrt{2}\,m \int \sqrt{\frac{E}{m} - U(q_1)}\,\mathrm{d}q_1 \tag{1.76}$$

and the equations of the motion are given by

$$t + \alpha_1 = q_1^* = \frac{\partial S}{\partial E} = \sqrt{\frac{1}{2}} \int \frac{\mathrm{d}q_1}{\sqrt{\frac{E}{m} - U(q_1)}}. \tag{1.77}$$

Because of the square root in (1.77), real solutions may exist only for the values of q_1 such that $\frac{E}{m} - U(q_1) > 0$. Moreover, for each such value, we have two solutions: one, prograde, corresponding to the choice of the positive branch of the square root, and the other, retrograde, corresponding to the choice of the negative branch. When $\frac{E}{m} - U(q_1) > 0$ for all q_1 (Fig. 1.3a), the resulting function $q_1(t)$ is monotonic. The variable q_1 is continuously increasing or decreasing. The motion is unbounded. In the case of q_1 defined on a circle (for instance, from 0 to 2π), the solution $q_1(t)$ is a periodic function of t whose period is given by

1.9 One-Dimensional Motion with a Generic Potential

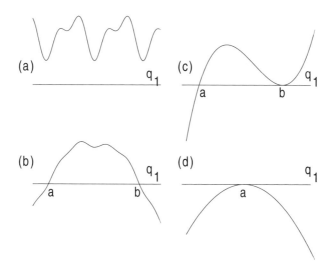

Fig. 1.3. Examples of functions $\frac{E}{m} - U(q_1)$

$$T = \sqrt{\frac{1}{2}} \int_0^{2\pi} \frac{dq_1}{\sqrt{\frac{E}{m} - U(q_1)}}. \tag{1.78}$$

In this case, the periodicity of the motion is only due to the angular nature of the variable q_1 and the motion is called a *circulation*. One simple example is the motion of a pendulum whose energy is large enough to allow the weight to reach the highest point of its trajectory with a non-zero velocity.

When the constant E is such that the function $\frac{E}{m} - U(q_1)$ is positive for some values of q_1 and negative for others, the motion is possible only for those q_1 for which $\frac{E}{m} - U(q_1) \geq 0$. The boundaries of the regions of possible motion are the roots of the equation $E - mU(q_1) = 0$. To understand this motion, let us start with a simple case. Let a and b ($a < b$) be two simple roots of $E - mU(q_1) = 0$ and let $\frac{E}{m} - U(q_1)$ be positive in the whole interval between these roots (Fig. 1.3b). It is easy to prove that, in this case, the motion is a periodic oscillation between the boundaries a and b. In what follows, we assume that the motion starts at the time $t = t_0$ at a point q_{10} between a and b with a positive speed and, then, we choose the positive branch of the square root in (1.77). Then, q_1 will increase continuously ($\dot{q}_1 > 0$) to reach b.

Proposition 1.9.1. *b is reached in a finite time.*

The time in which b is reached is given by

$$t_b = t_0 + \sqrt{\frac{1}{2}} \int_{q_{10}}^{b} \frac{dq_1}{\sqrt{\frac{E}{m} - U(q_1)}}. \tag{1.79}$$

This integral is improper, since the function under the integral sign goes to infinity when $q_1 \to b$. To circumvent this difficulty, we may separately consider a small neighborhood of the root b; then, we replace the function $\frac{E}{m} - U(q_1)$ by its linear approximation $-U'(b)(q_1 - b)$. The improper integral, in this neighborhood, will be approximated by

$$\frac{1}{\sqrt{U'(b)}} \int^b \frac{dq_1}{\sqrt{b - q_1}}, \tag{1.80}$$

which is a classical example of a convergent integral. Thus, the result of (1.79) is finite and the point b is reached in a finite time, as proposed.

Proposition 1.9.2. *The motion is symmetrically reflected in b.*

Indeed, in b we have $\dot{q}_1 = 0$ and $\ddot{q}_1 \neq 0$. Then $\dot{q}_1$ changes sign at $t = t_b$ and we have to change the branch of the square root (for $t > t_b$), that is, to put a minus sign in (1.77). As the only difference in the equations before and after the instant t_b is the sign of the square root, the motion after t_b is an exact reflection of the motion before t_b and the function $q_1(t)$ is even with respect to $t = t_b$.

Proposition 1.9.3. *The motion is periodic with period*

$$T = \sqrt{2} \int_a^b \frac{dq_1}{\sqrt{\frac{E}{m} - U(q_1)}}. \tag{1.81}$$

The proof of this proposition is immediate from the two preceding ones and their extension to the reflection of the motion at the point $q_1 = a$. This motion is, therefore, an oscillation between a and b and is usually called a *libration*. The motion of a pendulum, when the energy is not sufficient to allow the weight to reach the highest point of the circle, is an example of an oscillation of this kind.

To complete this analysis, we may consider the limiting case in which one of the roots of $E - mU(q_1) = 0$ is double (Fig. 1.3c). Let us assume that the function $\frac{E}{m} - U(q_1)$ has a minimum equal to zero in b. That is, $\frac{E}{m} - U(b) = 0$ and $-U'(b) = 0$.

The analysis follows the same steps in proposition 1.9.1 up to (1.79). As before, we separate a small neighborhood at the left of the root b and, in this neighborhood, replace the function $\frac{E}{m} - U(q_1)$ by an approximation. However, the approximation used before does not work since, now, $U'(b) = 0$. We then use the second-order Taylor approximation $-\frac{1}{2}U''(b)(q_1-b)^2$ and the improper integral, in this neighborhood, becomes

$$\sqrt{\frac{2}{-U''(b)}} \int^b \frac{dq_1}{b - q_1}, \tag{1.82}$$

which is divergent. Thus, the result of (1.79), in this case, is infinite and the point b is never reached. The motion tends asymptotically to b but never reaches it. If the motion were retrograde at the initial instant, b would be its limit for $t \to -\infty$. The point b is, itself, an unstable equilibrium point since a small displacement from this position will make the point drift away from it.

The particular case where the function $\frac{E}{m} - U(q_1)$ has a maximum equal to zero at a point is trivial. The solution may only exist at this point and this point is a *stable* equilibrium point (Fig. 1.3d).

1.9.1 The Case $m < 0$

The equations of this section were written in such a way that they still remain valid when $m < 0$. This was done because, in perturbation theories, we often have approximations corresponding to dynamical systems like that given by (1.74) whose parameters, including m, may be either positive or negative. The change of sign of m does not affect the dynamics and is equivalent to a time reversal. The trajectory in the phase space is not changed when m changes sign, but the direction of the motion is reversed.

It is worth recalling that we are not just changing the mass sign in an equation like Newton's $\boldsymbol{f} = m\,\boldsymbol{a}$, which would change an attractive action into a repulsive one, or vice versa. In our case, $\boldsymbol{f} = -m\,\mathrm{grad}\,U$ and, thus, the actual equation is $m\,\boldsymbol{a} = -m\,\mathrm{grad}\,U$, which does not depend on m. The only real change will be in the sign of the momentum since $p_1 = m\dot{q}_1$, which (because of the symmetric reflection) is equivalent to the above mentioned reversal in the direction of the motion.

1.9.2 The Harmonic Oscillator

Let us consider the simple dynamical system defined by the potential $U = \frac{k}{2}q_1^2$ ($k > 0$). The Hamiltonian is

$$H = \frac{p_1^2}{2m} + \frac{mk}{2}q_1^2 \qquad (m \in \mathbf{R})$$

and the trajectories, in the phase plane, are the ellipses $H = \mathrm{const}$. We note that the assumption $m > 0$ is not done.

The new Hamiltonian is $H^*(q_1^*, p_1^*) = p_1^*$ whose trajectories are straight lines. The Hamilton–Jacobi mapping of one domain of the plane (q_1, p_1), not containing the origin, into the plane (q_1^*, p_1^*) is a homeomorphism (as in the case shown in Fig. 1.2). However, they are no longer homeomorphic when the origin is included in the domain. Indeed, the function S, solution of the Hamilton–Jacobi equation, is singular at this point.

Equation (1.77), giving the solution of the system now becomes

$$q_1^* = t + \alpha_1 = \sqrt{\frac{1}{2}} \int \frac{dq_1}{\sqrt{\frac{E}{m} - \frac{k}{2}q_1^2}}. \tag{1.83}$$

For $E/m < 0$, there is no real solution; for $E = 0$, there is just a stable equilibrium point at $q_1 = 0$, and, for $E/m > 0$, the solutions are oscillations in the interval $[-\sqrt{2E/mk}, \sqrt{2E/mk}]$. In the later case, the integral is easily solved giving

$$t + \alpha_1 = \sqrt{\frac{1}{k}} \arcsin \sqrt{\frac{mk}{2E}} q_1 \tag{1.84}$$

or

$$q_1 = \sqrt{\frac{2E}{mk}} \sin \sqrt{k}(t + \alpha_1). \tag{1.85}$$

Hence,

$$p_1 = m\sqrt{\frac{2E}{m}} \cos \sqrt{k}(t + \alpha_1). \tag{1.86}$$

The coefficient in the last equation was not simplified because the simplification depends on the sign of m. In fact, this is the only point where a change is verified when the sign of m changes.

The motion is an oscillation with period $2\pi/\sqrt{k}$. It is worth emphasizing that the frequency (period) of the oscillation does not depend on the amplitude of the oscillations, a situation physically acceptable only for small-amplitude oscillations.

Exercise 1.9.1. Consider the harmonic oscillator with an additional repulsive cubic force, whose potential is $U = \frac{k}{2}q_1^2 - k'q_1^4$ ($k, k' > 0$), and study all possible solutions, periodic and non-periodic.

1.10 Involution. Mayer's Lemma. Liouville's Theorem

Definition 1.10.1 (Involution). *Let $f_1, f_2, \cdots, f_M$ ($M \leq N$) denote M functions of the $2N$ variables q_i, p_i; if the Poisson brackets $\{f_i, f_j\}$ are all zero, the functions f_i, $i = 1, \cdots, M$ are said to be pairwise in involution.*

Lemma 1.10.1 (Mayer). *Given N functions*

$$p_i^* = f_i(q, p) \qquad (i = 1, \cdots, N) \tag{1.87}$$

that are pairwise in involution and such that the functional determinant $\det(\partial f_i/\partial p_j)$ is not zero, there exist N functions $q_i^*(q, p)$ ($i = 1, \cdots, N$) such that the transformation $(q, p) \to (q^*, p^*)$ is canonical.

Proof. [18] From (1.87), we obtain

$$p_i = \phi_i(q, p^*). \tag{1.88}$$

When (q, p^*) are considered as independent variables and (1.87) is differentiated with respect to q_k, we obtain[6]

---

[6] All sums are taken over subscripts from 1 to N.

1.10 Involution. Mayer's Lemma. Liouville's Theorem

$$\frac{\partial f_i}{\partial q_k} + \sum_j \frac{\partial f_i}{\partial p_j}\frac{\partial \phi_j}{\partial q_k} = 0$$

By convolution of this equation with $\partial f_m/\partial p_k$, we obtain

$$\sum_k \frac{\partial f_m}{\partial p_k}\frac{\partial f_i}{\partial q_k} = -\sum_j \sum_k \frac{\partial f_m}{\partial p_k}\frac{\partial f_i}{\partial p_j}\frac{\partial \phi_j}{\partial q_k}.$$

In an analogous way,

$$\sum_k \frac{\partial f_i}{\partial p_k}\frac{\partial f_m}{\partial q_k} = -\sum_j \sum_k \frac{\partial f_i}{\partial p_j}\frac{\partial f_m}{\partial p_k}\frac{\partial \phi_k}{\partial q_j},$$

where the role of the subscripts i and m was interchanged, as well as, in the last summations, the repeated subscripts k and j. Introducing these equations in the definition of the Poisson bracket $\{f_i, f_m\}$, we obtain

$$\{f_i, f_m\} = \sum_j \sum_k \frac{\partial f_i}{\partial p_j}\frac{\partial f_m}{\partial p_k}\left(\frac{\partial \phi_k}{\partial q_j} - \frac{\partial \phi_j}{\partial q_k}\right) = 0$$

which is equal to zero because, by hypothesis, the functions f_i are pairwise in involution. Introducing the auxiliary functions

$$\sum_k \frac{\partial f_m}{\partial p_k}\left(\frac{\partial \phi_k}{\partial q_j} - \frac{\partial \phi_j}{\partial q_k}\right) = \psi_{mj}$$

the previous equation becomes

$$\sum_j \frac{\partial f_i}{\partial p_j}\psi_{mj} = 0.$$

This is a system of linear equations in the unknowns ψ_{mj} ($j = 1, \cdots, N$). Since $\det(\partial f_i/\partial p_j)$ is the non-zero Jacobian of the functions f_i with respect to the p_j, it follows that $\psi_{mj} = 0$, that is

$$\sum_k \frac{\partial f_m}{\partial p_k}\left(\frac{\partial \phi_k}{\partial q_j} - \frac{\partial \phi_j}{\partial q_k}\right) = 0.$$

This is again a system of linear equations and as $\det(\partial f_m/\partial p_k)$ is a non-zero Jacobian, the solution is

$$\left(\frac{\partial \phi_k}{\partial q_j} - \frac{\partial \phi_j}{\partial q_k}\right) = 0. \tag{1.89}$$

This result means that the differential form

$$\sum_i \phi_i(q, f)\, dq_i$$

is exact and one function $S(q, f)$ exists such that

$$p_i = \phi_i(q, f) = \frac{\partial S}{\partial q_i}. \qquad (1.90)$$

If we introduce N new variables

$$q_i^* = +\frac{\partial S}{\partial f_i}, \qquad (1.91)$$

(1.90) and (1.91) define a canonical transformation $(q, p) \to (q^*, p^*)$. □

The lemma is also valid in the case where the given functions are M coordinates and $N - M$ momenta provided that the involution property is preserved (i.e. that there are no conjugate pairs among the given coordinates and momenta).

The difference in this case is that when the given function f_i is a coordinate, the $+$ sign in (1.91) may be changed into $-$ for the subscripts $i \leq M$.

Theorem 1.10.1 (Liouville). *If a canonical system of N degrees of freedom admits N integrals*

$$f_i(q, p) = c_i = \text{const} \qquad (i = 1, \cdots, N) \qquad (1.92)$$

which are independent, pairwise in involution and can be solved for the momenta p_i, then the system is completely integrable and the general solution can be constructed by means of quadratures.

Proof. [51][7] Since the conditions of Mayer's lemma are satisfied, there exist N functions q_i^* ($i = 1, \cdots, N$) such that the transformation $(q, p) \to (q^*, f)$ is canonical.

With the variables (q^*, f), the equations of motion are

$$\frac{df_i}{dt} = 0 \qquad \frac{dq_i^*}{dt} = \frac{\partial H^*}{\partial f_i}, \qquad (1.93)$$

where H^* is the Hamiltonian as a function of (q^*, f). The quadratures

$$q_i^* = \int \frac{\partial H^*}{\partial f_i} \, dt \qquad (1.94)$$

complete the integration of the system

□

If one of the integrals, say, f_1, is the energy integral, then $H^* = f_1$ and the derivatives $\partial H^*/\partial f_i$ are zero for all $i \geq 2$. The solution of the system is then reduced just to the trivial quadrature:

$$q_1^* = \int dt = t + \text{const}. \qquad (1.95)$$

---

[7] For the classical proof using Jacobi's lemma, see [45].

The variables (q^*, f) are akin to the variables (α, β) introduced in the solution of the Hamilton–Jacobi equation.

When the number of known integrals in involution is not enough to guarantee the complete integrability of the system, they may be used to reduce the number of degrees of freedom according to the following similar theorem.

Theorem 1.10.2 (Lie). *If a canonical system of N degrees of freedom admits M ($M < N$) integrals $f_j(q,p) =$ const that are independent, in involution and can be solved for M momenta, then we can reduce the number of degrees of freedom of the system to $N - M$.*

To prove this theorem we need to complete the involution system with $N-M$ functions $\widehat{f}_j$ and proceed as before. The analysis of the resulting Hamiltonian is trivial, the main difference being that, now, the aditional functions $\widehat{f}_j$ ($j = M+1, \cdots, N$) are not integrals of the motion and will not allow the corresponding $N - M$ degrees of freedom to be trivially solved.

2
Angle–Action Variables. Separable Systems

2.1 Periodic Motions

The trajectories of systems with one degree of freedom are the curves $H(q_1, p_1) = E$. As shown in Sect. 1.8, the equations of the motion are given by

$$t + \alpha_1 = q_1^* = \frac{\partial S}{\partial E} \stackrel{\text{def}}{=} F_{11}(q_1), \qquad (2.1)$$

where α_1 is a constant and $S = S(q_1, E)$ is the solution of the Hamilton–Jacobi equation.

In the particular examples given in Sect. 1.9, we have found two kinds of periodic solutions:

- *Circulatory Motions.* Motions occurring when the variable q_1 is defined on a circle (for instance, q_1 is an angle defined from 0 to 2π) and is always increasing or decreasing (see Fig. 2.1a). The periodicity of the motion is due to the angular nature of q_1. The phase space of this system is a cylinder and the circulations are solutions closing on themselves after a complete tour encircling the cylinder.
- *Oscillatory Motions* or *Librations.* Motions occurring when the variable q_1 oscillates periodically between two boundaries a and b (see Fig. 2.1b). The variable q_1 may be either an angle (as in the pendulum) or a length (as in the harmonic oscillator). Accordingly, the phase space is either a cylinder (if $q_1 \in \mathbf{S}$) or a plane (if $q_1 \in \mathbf{R}$). Librations are closed curves with the particular property, in the case $q_1 \in \mathbf{S}$, that they close on themselves without encircling the cylinder.

It is not difficult to see that all bounded solutions of a Hamiltonian system with one degree of freedom are either periodic or asymptotic to an unstable equilibrium point. It is enough to remember that, since the phase flow preserves volumes in phase space (see [5], Chap. 1, Sect. 3.6), the only ordinary singular points allowed in the two-dimensional phase space of a Hamiltonian

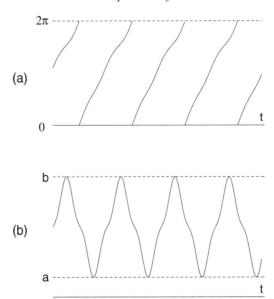

Fig. 2.1. Functions $q_1(t)$: (a) circulation; (b) libration

system are centers and saddle points. All bounded curves in this space that do not start or end in a saddle point correspond to a periodic motion.

2.1.1 Angle–Action Variables[1]

The equations resulting from the transformation $(q_1, p_1) \Rightarrow (q_1^*, p_1^*)$ are

$$\begin{aligned} q_1^* &= t + \alpha_1 \\ p_1^* &= \beta_1 = E, \end{aligned} \quad (2.2)$$

where α_1 and β_1 are constants. The phase space (q_1^*, p_1^*) is either a plane or a cylinder as discussed above. The phase trajectories are the lines $p_1^* = \beta_1$ and the phase velocity is $\dot{q}_1^* = 1$ on all trajectories. There are no explicit constraints imposed on α_1, β_1, which, however, exist and may be found by the analysis of $S(q_1, E)$. For instance, in the harmonic oscillator (Sect. 1.9.2), the solutions exist only in the domain formed by the upper half-plane $E/m \geq 0$. Another property not appearing in the functional expression of the Hamiltonian $H^* = p_1^*$ is the possible periodicity of the solutions (or of one set of solutions). For instance, in the harmonic oscillator, all solutions for $E/m > 0$ are periodic with period $T = 2\pi/\sqrt{k}$, that is, $q_1^* \in \mathbf{R}/T\mathbf{Z}$.

---

[1] Throughout this book, the order coordinate–momentum is adopted. Thus, we shall refer to these variables as "angle–action" variables, instead of "action–angle" as usually done everywhere.

2.1 Periodic Motions

To correct this lack of topological information on the motion in the phase space q_1^*, p_1^*, we introduce, in the case of periodic motions, a new angular variable $w_1 \in \mathbf{S}^1$. By definition, it increases 2π when q_1 performs a complete circulation or libration. From (2.1) we have

$$\begin{aligned} q_1^* &= t + \alpha_1 &&= F_{11}(q_1) \\ q_1^* + T &= t + \alpha_1 + T &&= F_{11}(q_1 + \oint dq_1), \end{aligned} \quad (2.3)$$

where $\oint dq_1$ means a complete circulation or libration of q_1. Then, in order to have, instead of q_1^*, a uniformized variable, it is enough to define[2]

$$w_1 = 2\pi \frac{t + \alpha_1}{T} = 2\pi \frac{q_1^*}{T}. \quad (2.4)$$

Obviously, the period T is the same for all initial conditions on a periodic orbit, but it is worth keeping in mind that it is not the same for all periodic solutions of a given system.

The momentum conjugate to w_1 may be easily obtained in terms of q_1, p_1. Let $\tilde{S}(q_1, J_1)$ be the Jacobian generating function of the canonical transformation $\tilde{\phi} : (q_1, p_1) \Rightarrow (w_1, J_1)$. Hence,

$$w_1 = \frac{\partial \tilde{S}}{\partial J_1} \qquad p_1 = \frac{\partial \tilde{S}}{\partial q_1}. \quad (2.5)$$

The following chain of calculations is simple and just uses elementary calculus:

$$\frac{dw_1}{dt} = \frac{d}{dt}\left(\frac{\partial \tilde{S}}{\partial J_1}\right) = \frac{\partial}{\partial q_1}\left(\frac{\partial \tilde{S}}{\partial J_1}\right) \dot{q}_1 = \frac{\partial^2 \tilde{S}}{\partial J_1 \partial q_1} \dot{q}_1$$

and

$$2\pi = \int_t^{t+T} \frac{dw_1}{dt} dt = \oint \frac{\partial^2 \tilde{S}}{\partial J_1 \partial q_1} dq_1 = \frac{\partial}{\partial J_1} \oint \frac{\partial \tilde{S}}{\partial q_1} dq_1 = \frac{\partial}{\partial J_1} \oint p_1 \, dq_1.$$

Hence

$$J_1 = \frac{1}{2\pi} \oint p_1 \, dq_1, \quad (2.6)$$

except for an arbitrary integration constant (of the integration in J_1). The quantity J_1 has the dimension of angular momentum or action and is an invariant of the motion (see Sect. 1.2.2). It is equal to the variation of the action when the solution performs a complete circulation or libration. Because of this property, it was called *modulus of periodicity of the action* [93] or *modulus of variation of the action* [11]. Since it gives the area delimited by the trajectory in the phase plane, it was also called *phase integral*. The adoption of these variables in the old Quantum Theory was first proposed by Sommerfeld.

---

[2] We adopted $\oint dw_1 = 2\pi$. In many classical texts, $\oint dw_1 = 1$.

The conjugate variables w_1, J_1 were called *angle* and *action* variables, a denomination that became standard after its adoption in Born's *Atommechanik* [12]. This denomination is the one currently used. The corresponding canonical equations are

$$\dot{w}_1 = \frac{\partial E}{\partial J_1} = \frac{2\pi}{T}, \qquad \dot{J}_1 = -\frac{\partial E}{\partial w_1} = 0. \tag{2.7}$$

Finally, let us note that, when the periodic motion is a libration, the quantity defined by (2.6) is singular when $J_1 = 0$. Indeed, the integral gives the area enclosed by the libration orbit and the singularity $J_1 = 0$ is a consequence of the fact that the direction of the motion in the phase space (q_1, p_1) cannot be reversed. Examples and consequences of this singularity in Celestial Mechanics will be extensively considered in Chap. 7

2.1.2 The Sign of the Action

We shall emphasize that the result of the operation defining the actions may be either positive or negative. To avoid any ambiguity, it is enough to write the definition of the action as

$$J_1 = \frac{1}{2\pi} \int_t^{t+T} p_1 \dot{q}_1 \, dt. \tag{2.8}$$

For instance, in the simple pendulum solutions, J_1 is positive if $m > 0$ or negative if $m < 0$ (see Fig. B.1). We recall that w_1 is, by definition, always such that $\dot{w}_1 > 0$.

Exercise 2.1.1 (Angle–Action Variables of the Harmonic Oscillator).

1. Show that the angle–action variables of the harmonic oscillator defined by $U = \frac{k}{2} q_1^2$ $(k > 0)$ are

$$w_1 = \arcsin \sqrt{\frac{mk}{2E}} q_1 = \sqrt{k}(t + \alpha_1), \tag{2.9}$$

$$J_1 = \frac{E}{\sqrt{k}}. \tag{2.10}$$

α_1 is a constant. *Hint:* $H = \dfrac{p_1^2}{2m} + \dfrac{km}{2} q_1^2$.

2. Show that

$$p_1 = \sqrt{2mE} \cos w_1. \tag{2.11}$$

2.2 Direct Construction of Angle–Action Variables

It is possible to rearrange the theory to directly obtain the angle–action variables. We may start from

$$J_1 = \frac{1}{2\pi} \oint p_1(q_1, E)\,dq_1, \qquad (2.12)$$

where $p_1(q_1, E)$ is obtained from the inversion of the energy integral $E = E(q_1, p_1)$. If the given Hamiltonian is quadratic in p_1, like in Sect. 1.9, this is an Abelian integral whose solution may benefit from some usual transformations and, when necessary, the use of the theory of residues[3]. The other basic equations are

$$\tilde{S} = \int p_1(q_1, E)\,dq_1 \qquad (2.13)$$

and

$$w_1 = \frac{\partial \tilde{S}}{\partial J_1} = \int \frac{\partial}{\partial J_1} p_1(q_1, E)\,dq_1. \qquad (2.14)$$

This step depends on the algebraic inversion of the solution of (2.12) to obtain $E = E(J_1)$. Another possibility is to take, instead of (2.14),

$$w_1 = \frac{\partial \tilde{S}}{\partial E}\left(\frac{dJ_1}{dE}\right)^{-1} = \left(\frac{dJ_1}{dE}\right)^{-1} \int \frac{\partial}{\partial E} p_1(q_1, E)\,dq_1. \qquad (2.15)$$

The replacement of dE/dJ_1 by $(dJ_1/dE)^{-1}$, which may be directly obtained from (2.12) without the need of any algebraic inversion, is always possible as long as $E(J_1)$ is a monotonic function.

However, these tasks are often made very difficult or even impossible to accomplish analytically because of the reasonably complex forms of $p_1(E, q_1)$.

There are ways of overcoming this situation. One of them, used in this book to obtain angle–action variables for the small oscillations of the pendulum (Sect. B.4) and of the Andoyer Hamiltonian (Sect. C.9), is founded on the fact that we are dealing with periodic solutions of the given Hamiltonian system, which may be represented by Fourier series. There are many different ways of calculating these series. In this book, we limit ourselves to the neighborhood of the equilibrium solutions. The solutions of the given system are represented by the series

$$q_1 = a_0 + \sum_{i=1}^{n} a_i \gamma^i,$$

where a_i are undetermined periodic functions of the angle w_1 and γ is a free parameter of the order of the amplitude of the oscillations. ($\gamma = 0$ corresponds to the stable equilibrium solution $q_1 = a_0$.) It is important to keep in mind that we need to construct the whole family of periodic solutions and that $\dot{w}_1$

---

[3] For some specific examples, see [93], Note 6.

is not the same for all solutions but is itself also a function of the parameter γ. It is assumed to be a power series in γ with undetermined coefficients:

$$\dot w_1 = \omega_0 + \sum_{i=1}^{n} o_i \gamma^i.$$

$p_1(w_1)$ is constructed using the equations of the motion or the energy integral. The angle–action variables are w_1 and

$$J_1 \stackrel{\text{def}}{=} \frac{1}{2\pi} \int_0^{2\pi} p_1 \frac{dq_1}{dw_1} dw_1.$$

The order n of the solution may be chosen according to the practical needs of the problem being solved and the means available for the calculation. Existing algebraic manipulators allow high orders to be considered. The practical steps of this construction may be seen in the cases presented in Sects. B.4 and C.9.

A different method is the numerical construction of the angle–action variables [50]. Let $H(q_1, p_1)$ be the Hamiltonian of an autonomous system and

$$\begin{aligned} q_1 &= q_1(q_0, p_0, t) \\ p_1 &= p_1(q_0, p_0, t) \end{aligned} \tag{2.16}$$

its solution for a given initial condition (q_0, p_0) and let $T(q_0, p_0)$ be the period of this solution.

The corresponding angle–action variables are

$$w_1 \stackrel{\text{def}}{=} \frac{2\pi}{T} t \tag{2.17}$$

$$J_1 \stackrel{\text{def}}{=} \frac{1}{2\pi} \int_0^T p_1 \frac{dq_1}{dt} dt = -\frac{1}{2\pi} \int_0^T q_1 \frac{dp_1}{dt} dt$$

and the inverses of these definitions are

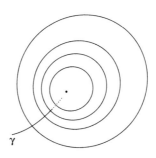

Fig. 2.2. Orbits transverse to a curve γ

$$q_0 = q_0(w_1, J_1)$$
$$p_0 = p_0(w_1, J_1). \qquad (2.18)$$

In this technique, all functions are constructed numerically. It is noteworthy that the last inversion may be more economically done when, beforehand, one has constructed the derivatives $\partial J_1/\partial q_0$ and $\partial J_1/\partial p_0$.

A last practical point to be noted is that we need just to numerically integrate from initial conditions lying on a given curve (γ) transverse to the orbits (and passing through the center of the orbits if we intend to include in the study also its immediate neighborhood) (Fig. 2.2). The extension of the solutions of (2.17) to the other points on each orbit is immediate.

The algorithms provided by Mayer's lemma (Sect. 1.10) allow the above construction to be extended to obtain a canonical transformation including other degrees of freedom. (see Sect. 2.4.4)

2.3 Actions in Multiperiodic Systems. Einstein's Theory

Let us consider a conservative Hamiltonian system with N degrees of freedom. It was shown in Sect. 1.2.2 that the action

$$J = \oint \sum_{i=1}^{N} p_i \, dq_i \qquad (2.19)$$

is an invariant of the motion (Helmholtz invariant).

If $S(q, \beta)$ is a solution of the Hamilton–Jacobi equation, then $p_i = \partial S/\partial q_i$,

$$\sum_{i=1}^{N} p_i \, dq_i = dS(q, \beta)$$

is an exact differential and the integral (2.19) has the same value for all closed curves that may be continuously deformed into one another. In particular, for all curves that may be reduced to one point by means of a continuous deformation, we have $J = 0$. When the solutions lie on a multiply connected manifold, there are closed curves that cannot be reduced to one point by continuous deformation (see [42]). This property was used by Einstein [26] to prove that, when the Hamiltonian is integrable, it is possible to construct N independent actions.

The multiperiodic solutions of a conservative integrable Hamiltonian corresponding to N constants β_i form a surface homeomorphic to $\mathbf{T}^N$. An N-dimensional torus is an N times connected surface and then we can find N different closed curves Γ_k that cannot be pairwise deformed into one another or reduced to one point and that may serve to uniquely define N independent actions

$$J_k = \oint_{\Gamma_k} \sum_{i=1}^{N} p_i \, dq_i. \qquad (2.20)$$

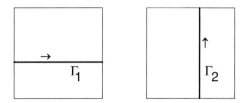

Fig. 2.3. Curves on tori $\mathbf{T}^2$ (the tori are obtained by joining the opposite sides of each square)

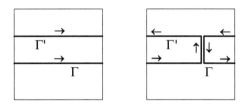

Fig. 2.4. Left: Integration paths Γ and Γ'. Right: Integration path obtained by introducing a cut between them and inverting the direction of Γ'

Let us consider the particular case $N = 2$. In $\mathbf{T}^2$, there are two types of closed curves that cannot be reduced to one point or transformed into one another by continuous deformation. They are shown in Fig. 2.3. All other closed curves on the surface of the torus can, by means of continuous deformations, be reduced to one point or transformed into one or more loops of the curves Γ_1 and Γ_2. To the closed curves Γ_1 and Γ_2, there correspond two independent actions J_1 and J_2.

In order that the definitions of J_k ($k = 1, 2$) have a meaning, the values of J obtained from all closed curves that can be continuously transformed into one another may be the same. Let Γ and Γ' be two oriented closed curves that may be transformed into one another (Fig. 2.4, left). We may prove that the resulting actions J and J' are such that $J = J'$. To show this, we calculate $J - J'$. First, a cut joining Γ and Γ' is introduced. (The cut is shown in Fig. 2.4, right, as a pair of infinitesimally separated segments.) The resulting path is a curve drawn <u>on</u> the torus without encircling it and which may be reduced to one point; the integral over this path is then equal to zero. If we note that the integrals over the cut are opposite and cancel each other and that the integral over Γ' is done in a direction contrary to that used to define J', it follows that $J - J' = 0$. □

The actions constructed with Einstein's theory may be completed by angles defined by $w_k = \partial \tilde{S}/\partial J_k$ where $\tilde{S}(q, J) = S(q, \beta(J))$ [4].

---

[4] For a modern and rigorous definition of the angle–action variables of an integrable system, see [4], Sect. 50.

2.4 Separable Multiperiodic Systems

There are no general methods for the solution of the Hamilton–Jacobi equation in the case of more than one degree of freedom. The use of general theories, such as Cauchy characteristics, just recovers the given Hamiltonian system. However, under certain special conditions, for some important problems such as Keplerian motion (or the Rutherford–Bohr atom), a complete solution of the Hamilton–Jacobi equation may be obtained. In these very particular cases, one partial differential equation in N variables can be replaced by N separate ordinary differential equations, one for each variable, and the complete integration of the equation is achieved.

Generally speaking, a problem is said to be *separable* when the corresponding Hamilton–Jacobi equation has a <u>complete</u> integral $S(q, \beta)$ which may be separated as

$$S(q, \beta) = S_1(q_1, \beta) + S_2(q_2, \beta) + \cdots + S_N(q_N, \beta), \tag{2.21}$$

where each term $S_k = S_k(q_k, \beta)$ is independent of the q_j ($j \neq k$).

In this case, the equations of the motion are given by

$$\begin{aligned} t + \alpha_1 = q_1^* &= \frac{\partial S}{\partial \beta_1} = \sum_{k=1}^{N} \frac{\partial S_k(q_k, \beta)}{\partial \beta_1} = \sum_{k=1}^{N} F_{1k}(q_k) \\ \alpha_\varrho = q_\varrho^* &= \frac{\partial S}{\partial \beta_\varrho} = \sum_{k=1}^{N} \frac{\partial S_k(q_k, \beta)}{\partial \beta_\varrho} = \sum_{k=1}^{N} F_{\varrho k}(q_k), \end{aligned} \tag{2.22}$$

($\varrho = 2, 3, \cdots, N$), where we have introduced the functions

$$F_{jk}(q_k) \stackrel{\text{def}}{=} \frac{\partial S_k(q_k, \beta)}{\partial \beta_j}. \tag{2.23}$$

The other equations, completing the transformation, are

$$p_k = \frac{\partial S}{\partial q_k} = \frac{\partial S_k(q_k, \beta)}{\partial q_k} \qquad (k = 1, \cdots, N). \tag{2.24}$$

Equations (2.24) show that the trajectories projected in the phase subspaces q_k, p_k are mutually independent. The law of motion along the projected trajectories may be obtained by solving the equations of the motion, (2.22), with respect to the q_k. As in Sect. 2.1, these projected periodic motions may be either circulations or librations.

2.4.1 Uniformized Angles. Charlier's Theory

The generalization of the angle variables of Sect. 2.1.1 to N degrees of freedom may be done following the same principle as there. We define a *partial cyclic variation* in which the corresponding variable q_i performs a complete circulation or libration while the other variables $q_k (k \neq i)$ are kept unaltered. We then define a set of N angle variables $w_i \in \mathbf{S}$ such that in a partial cyclic variation of q_i, the corresponding w_i increases 2π while the other angles $w_k (k \neq i)$ are not affected. Such angle variables are said to be uniformized.

In a partial cyclic variation of q_i, the functions F_{ji} change while all functions $F_{jk} (k \neq i)$ remain unchanged. Let γ_{ji} be the increment of the functions $F_{ji}(q_i, \beta)$ in a partial cyclic variation of q_i:

$$\gamma_{ji} = F_{ji}(q_i + \oint \mathrm{d}q_i) - F_{ji}(q_i), \tag{2.25}$$

where $\oint \mathrm{d}q_i$ denotes the partial cyclic variation of q_i. It is important to keep in mind that the resulting *repetition numbers* γ_{ji} are not independent of the initial values of the q_i (as T is not independent of the initial q_1 in the case of one degree of freedom).

Proposition 2.4.1 (Charlier [20]). *If $\det(\gamma_{ji}) \neq 0$, the variables w_i defined by the equations*

$$q_j^* = \frac{1}{2\pi} \sum_{\ell=1}^{N} \gamma_{j\ell} w_\ell \tag{2.26}$$

are uniformized angle variables.

Proof. Let us introduce the inverse matrix of (γ_{ji}) and denote its elements by γ_{ji}^{-1}. If $\det(\gamma_{ji}) \neq 0$, (2.26) may be inverted, giving

$$w_k = 2\pi \sum_{j=1}^{N} \gamma_{kj}^{-1} q_j^*. \tag{2.27}$$

In a partial cyclic variation of q_i, the variation of w_k is

$$\delta_i w_k = 2\pi \sum_{j=1}^{N} \gamma_{kj}^{-1} \delta_i q_j^* = 2\pi \delta_{ki} \tag{2.28}$$

(by construction, $\delta_i q_j^* = \gamma_{ji}$). Therefore, in a partial cyclic variation of q_i, w_i increases of 2π while the others w_k ($k \neq i$) remain unchanged. □

2.4.2 The Actions

The next step is to find the action variables J_k canonically conjugate to the angle variables w_k. To do this, we introduce the Jacobian generating function of the canonical transformation $\tilde{\phi} : (q, p) \Rightarrow (w, J)$, namely $\tilde{S}(q, J)$. We then have

$$w_k = \frac{\partial \tilde{S}}{\partial J_k} \tag{2.29}$$

and
$$dw_k = \sum_{j=1}^{N} \frac{\partial^2 \tilde{S}}{\partial q_j \partial J_k} dq_j + \sum_{j=1}^{N} \frac{\partial^2 \tilde{S}}{\partial J_j \partial J_k} dJ_j. \tag{2.30}$$

In a partial cyclic variation of q_k, w_k increases 2π. Besides, along the given path, $dq_j = 0$ for $j \neq k$ and $dJ_i = 0$. The above equation then reduces to

$$2\pi = \oint \frac{\partial^2 \tilde{S}}{\partial q_k \partial J_k} dq_k.$$

A trivial calculation, similar to that of Sect. 2.1.1, gives

$$J_k = \frac{1}{2\pi} \oint p_k \, dq_k \tag{2.31}$$

for every $k \in \{1, \cdots, N\}$.

2.4.3 Algorithms for Construction of the Angles

In practice, we use some straightforward approaches to obtain the angles. The separation of the Hamilton–Jacobi equation leads us to obtain $p_j = p_j(q_j, \beta)$ and the solution

$$S(q, \beta) = \sum_{j=1}^{N} S_j(q_j, \beta) = \sum_{j=1}^{N} \int p_j(q_j, \beta) \, dq_j. \tag{2.32}$$

We may also solve (2.31) with $p_k = p_k(q_k, \beta)$ to obtain the actions J_k as functions of the constants β_i.

The Jacobian generating function $\tilde{S}(q, J)$ may be obtained, now, from $\tilde{S}(q, J) = S(q, \beta(J))$ and (2.29) gives the angles:

$$w_k = \frac{\partial \tilde{S}}{\partial J_k} = \sum_{i=1}^{N} \frac{\partial S}{\partial \beta_i} \frac{\partial \beta_i}{\partial J_k} = \sum_{i=1}^{N} \frac{\partial \beta_i}{\partial J_k} \sum_{j=1}^{N} \int \frac{\partial p_j}{\partial \beta_i} dq_j. \tag{2.33}$$

These equations are akin to the equations

$$w_k = \frac{\partial}{\partial J_k} \int \sum_{j=1}^{N} \widehat{p}_j(q, J) \, dq_j, \tag{2.34}$$

which would result if the canonical transformation of Mayer's lemma (Sect. 1.10) were used in this case. The conditions under which that transformation was established (involution of the functions $J_i(q, p)$ and possibility of inversion to obtain the functions $\widehat{p}_i(q, J)$) are satisfied and it can be used. Equation (2.34) transforms itself into (2.33) in the separable case in which every term p_j depends only on the corresponding q_j.

2.4.4 Angle–Action Variables of $H(q_1, p_1, p_2, \cdots, p_N)$

Let us consider the case of a Hamiltonian having the form $H(q_1, p_1, p_2, \cdots, p_N)$, where the coordinates $q_2, \cdots, q_N$ are ignorable and the momenta $p_2, \cdots, p_N$ are constants. Because of frequent applications, it is worth having the algorithm of the previous section explicitly given in this case.

The Hamiltonian H is reducible to one degree of freedom and the angle–action variables of the reduced Hamiltonian may be obtained with one of the algorithms discussed in Sect. 2.2. The results of the previous section allow the one-degree-of-freedom transformation $(q_1, p_1) \to (w_1, J_1)$, thus obtained, to be embedded into a more general transformation $(q, p) \to (w, J)$ that considers also the remaining degrees of freedom of the given Hamiltonian. To do this, we consider as given the N functions

$$J_1 = f_1(q_1, p) \qquad (2.35)$$
$$J_\varrho = p_\varrho \equiv f_\varrho(q, p) \qquad (\varrho = 2, \cdots, N).$$

These functions are pairwise in involution and may be solved for the momenta. Because of the particular form of the functions f_ϱ, the inversion is trivial, giving $p = \widehat{p}(q_1, J)$. The resulting generating function of the transformation (2.35) is simply

$$\sum_{j=1}^{N} \int \widehat{p}_j(q_1, J) \, dq_j = \int \widehat{p}_1(q_1, J) \, dq_1 + \sum_{\varrho=2}^{N} J_\varrho q_\varrho. \qquad (2.36)$$

We then have

$$w_1 = \Xi_1$$
$$w_\varrho = q_\varrho + \Xi_\varrho \qquad (\varrho \geq 2), \qquad (2.37)$$

where

$$\Xi_k = \frac{\partial}{\partial J_k} \int \widehat{p}_1(q_1, J) \, dq_1 \qquad (k \geq 1). \qquad (2.38)$$

We note that (2.36) comes from the integration of an exact differential form in dq_j and that we may add to the generating function any arbitrary function of J.

The one-degree-of-freedom canonical transformation $(q_1, p_1) \to (w_1, J_1)$ is often given in the inverted form

$$q_1 = Q_1(w_1, J)$$
$$p_1 = P_1(w_1, J). \qquad (2.39)$$

In this case, (2.38) may be written

$$\Xi_k = \int \left[\frac{\partial \widehat{p}_1(q_1, J)}{\partial J_k} \right]_{q_1 = Q_1} \frac{\partial Q_1}{\partial w_1} \, dw_1. \qquad (2.40)$$

It is worth emphasizing that the substitution $q_1 = Q_1(w_1, J)$ may be done after the differentiation and that it is no longer possible to permute the differentiation with respect to J_k and the integration.

If the differentials of $\widehat{p}_1(q_1, J)$ and $P_1(w_1, J)$ are compared, we obtain

$$\frac{\partial P_1}{\partial J_k} = \frac{\partial \widehat{p}_1}{\partial J_k} + \frac{\partial \widehat{p}_1}{\partial q_1}\frac{\partial Q_1}{\partial J_k}$$

$$\frac{\partial P_1}{\partial w_1} = \frac{\partial \widehat{p}_1}{\partial q_1}\frac{\partial Q_1}{\partial w_1},$$

which, substituted in (2.40), give the equivalent result

$$\Xi_\varrho = \int_0^{w_1} \left(\frac{\partial Q_1}{\partial w_1}\frac{\partial P_1}{\partial J_\varrho} - \frac{\partial Q_1}{\partial J_\varrho}\frac{\partial P_1}{\partial w_1}\right) dw_1, \qquad (2.41)$$

obtained by Henrard and Lemaitre [50]. We also have the trivial relation $\Xi_1 = w_1$, since the integrand in this case is the one-dimensional Lagrange bracket $[w_1, J_1]$ which is equal to 1 because the given transformation $(q_1, p_1) \to (w_1, J_1)$ is canonical.

In this section, we have considered a Hamiltonian independent of the coordinates q_ϱ ($\varrho = 2, \cdots, N$). The algorithms derived from Mayer's lemma are valid in more general circumstances, but the results are not angle–action variables of the given Hamiltonian when H depends on the q_ϱ. If, for instance, a general H may be decomposed into two parts: $H = H_a(q_1, p_1) + H_b(q_\varrho, p_\varrho)$, and the formula is used to extend the angle–action variables of H_a, the resulting variables w, J are not angle–action variables of H. It is easy to see that the calculations to obtain the w_ϱ, J_ϱ are the same for any H_b and, thus, we cannot expect that it eliminates the angles from H_b. This comment is somewhat obvious, but is useful to avoid pitfalls.

Exercise 2.4.1. By construction, the functions $q_1(w_1, J)$ and $p_1(w_1, J)$ are 2π-periodic in the angle variable w_1. Under which conditions may we guarantee that the functions Ξ_ϱ are also 2π-periodic in w_1?

Exercise 2.4.2. Find the angle variable conjugate to $J_1 = \frac{1}{2}(q_1^2 + p_1^2)$. Check the result with $\{w_1, J_1\} = 1$.

Exercise 2.4.3. Find a set of angle–action variables for the Hamiltonian

$$H = \frac{1}{2}(p_1^2 + p_2^2) + \frac{1}{2}\lambda^2 q_1^2,$$

where $\lambda = \lambda(p_2)$. Check the results with $\{w_1, J_j\} = \delta_{1j}$. *Hint:* See Exercise 2.1.1.

2.4.5 Historical Postscript

The given definitions of angle–action variables follow those found in several classical texts on Celestial Mechanics and on the old Quantum Theory[5]. Angles and actions appeared separately. The *angles* were first introduced by Charlier [20] as a complement to what was then called *Staude–Stäckel theory*. They came as a result of an application of Weierstrass' theory of multiperiodic functions to the solutions of the Hamilton–Jacobi equation of a separable system. The *actions* evolved from the quantity defined in Malpertuis' least action principle (see Sect. 1.2.1), quantized in the theories of Planck and Bohr, to their definition for separable multiperiodic Hamiltonians given by Sommerfeld [92] and Epstein [28]. The introduction of the angles as variables canonically conjugate to the actions through a Jacobian generating function $\tilde{S}(q, J)$ is due to Kramers (cf. [10], Note 24). The definition of the actions of an integrable Hamiltonian system without recourse to the separability hypothesis is due to Einstein [26]. (An alternative construction was presented, at the same time, by Burgers; cf. [12].) The introduction of invariant tori in modern theory is due to Arnold [3]. It is worth mentioning that Einstein's construction of invariant tori is very different from that adopted in the modern theory of Hamiltonian systems. Einstein considered one example (central motions in a plane) and used the fact that, for given β_i, the phase space may be seen as a vector field on a Riemann surface formed by two annular sheets joined by their edges (which is homeomorphic to $\mathbf{T}^2$).

The angle–action variables ℓ, g, h, L, G, H obtained in Sect. 2.7.2 as an application of the Schwarzschild transformation to the angle–action variables of Keplerian motion, were actually discovered by Delaunay a long time before and fully employed in his (canonical) *Théorie de la Lune* [22].

It is worth emphasizing that the introduction of angle–action variables in Delaunay's work, as well as in the work of Sommerfeld and his contemporaries, resulted from specific needs for the actual solution of problems in Astronomy and Physics. The hiatus between the results of old Quantum Theory (before 1920) and modern theories (ca. 1960) has an explanation. The construction of action variables was the central point of the Bohr–Sommerfeld quantum condition. With the foundation of Quantum Mechanics, in the early 1920s, the actions lost their position in center stage. KAM theory has again made angle and action variables central concepts in Physics and Dynamics.

2.5 Simple Separable Systems

We only know some sets of sufficient conditions for separability. Some simple cases are the dynamical systems whose Hamiltonians have special structures, such as

---

[5] Specifically, we mention (in chronological order) Charlier [20], Schwarzschild [84], Einstein [26], Sommerfeld [93], Born [12], and Boll and Salomon [11].

$$H = G[f_1(q_1,p_1),\cdots,f_N(q_N,p_N)] \tag{2.42}$$

and

$$H = f_1\{q_1,p_1,f_2[q_2,p_2,f_3(\cdots,f_N(q_N,p_N))]\}. \tag{2.43}$$

In the first case, the variables in the expression for the function H are separated, i.e., only one pair of conjugate variables q_i, p_i enters into each function f_i. The Hamilton–Jacobi equation corresponding to this case is

$$G\left[f_1\left(q_1,\frac{\partial S}{\partial q_1}\right),\cdots,f_N\left(q_N,\frac{\partial S}{\partial q_N}\right)\right] = E. \tag{2.44}$$

After the introduction of $S = \sum_{i=1}^{N} S_i(q_i)$, this equation is separated into N equations

$$f_i\left(q_i,\frac{dS_i}{dq_i}\right) = \beta_i, \tag{2.45}$$

the integration constants β_i being such that

$$E = G(\beta_1,\cdots,\beta_N). \tag{2.46}$$

In the second case, the variables appear in a hierarchical disposition and the corresponding Hamilton–Jacobi equation,

$$f_1\left\{q_1,\frac{\partial S}{\partial q_1},f_2\left[q_2,\frac{\partial S}{\partial q_2},f_3\left(\cdots,f_N\left(q_N,\frac{\partial S}{\partial q_N}\right)\right)\right]\right\} = E, \tag{2.47}$$

after the introduction of $S = \sum_{i=1}^{N} S_i(q_i)$, is separated into N equations

$$\begin{aligned} f_N\left(q_N,\frac{dS_N}{dq_N}\right) &= \beta_N \\ f_i\left(q_i,\frac{dS_i}{dq_i},\beta_{i+1}\right) &= \beta_i \quad (i=1,\cdots,N-1), \end{aligned} \tag{2.48}$$

with $\beta_1 = E$. If we assume that $\partial f_i/\partial p_i \neq 0$ for all $i = 1,\cdots,N$, these equations can be solved to give

$$\begin{aligned} \frac{dS_i}{dq_i} &= G_i(q_i,\beta_{i+1},\beta_i) \quad (i=1,\cdots,N-1) \\ \frac{dS_N}{dq_N} &= G_N(q_N,\beta_N). \end{aligned} \tag{2.49}$$

2.5.1 Example: Central Motions

The classical example of a separable system of this kind is the motion of a particle in a central force field. In spherical coordinates, the total energy of the particle is

44   2 Angle–Action Variables. Separable Systems

$$H = T + mU(r) = \frac{m}{2}(\dot{r}^2 + r^2\dot{\theta}^2 + r^2\sin^2\theta\,\dot{\phi}^2) + mU(r) \tag{2.50}$$

or, introducing the generalized momenta

$$\begin{aligned}
p_r &= \frac{\partial T}{\partial \dot{r}} = m\dot{r} \\
p_\theta &= \frac{\partial T}{\partial \dot{\theta}} = mr^2\dot{\theta} \\
p_\phi &= \frac{\partial T}{\partial \dot{\phi}} = mr^2\sin^2\theta\,\dot{\phi},
\end{aligned} \tag{2.51}$$

we obtain

$$H = \frac{1}{2m}\left[p_r^2 + \frac{1}{r^2}\left(p_\theta^2 + \frac{p_\phi^2}{\sin^2\theta}\right)\right] + mU(r). \tag{2.52}$$

The above Hamiltonian has the special structure of (2.43) with

$$\begin{aligned}
f_1 &= \frac{1}{2m}\left(p_r^2 + \frac{f_2}{r^2}\right) + mU(r) = E \\
f_2 &= p_\theta^2 + \frac{f_3}{\sin^2\theta} = \beta_2 \\
f_3 &= p_\phi^2 = \beta_3,
\end{aligned} \tag{2.53}$$

and an application of (2.49) gives

$$\begin{aligned}
p_r &= \frac{dS_1}{dr} = \sqrt{2mE - \frac{\beta_2}{r^2} - 2m^2 U(r)} \\
p_\theta &= \frac{dS_2}{d\theta} = \sqrt{\beta_2 - \frac{\beta_3}{\sin^2\theta}} \\
p_\phi &= \frac{dS_3}{d\phi} = \sqrt{\beta_3}.
\end{aligned} \tag{2.54}$$

2.5.2 Angle–Action Variables of Central Motions

Let us calculate the angle–action variables of the central motions, starting with J_ϕ. A short chain of elementary calculations gives

$$J_\phi = \frac{1}{2\pi}\oint \sqrt{\beta_3}\,d\phi = \frac{1}{2\pi}\int_0^{2\pi}\sqrt{\beta_3}\,d\phi = \sqrt{\beta_3} = p_\phi. \tag{2.55}$$

The integration of the next one is also elementary, but not as immediate:

$$J_\theta = \frac{1}{2\pi}\oint p_\theta\,d\theta = \frac{1}{2\pi}\oint \sqrt{\beta_2 - \frac{\beta_3}{\sin^2\theta}}\,d\theta.$$

It may, however, be avoided by noting that when the plane of the motion is chosen as the fundamental reference plane, we have, in analogy with the previous result,

$$J_\psi = p_\psi = mr^2\dot\psi, \tag{2.56}$$

where ψ denotes the longitude reckoned on the plane of motion. Taking into account (2.53) and $\dot\psi^2 = \dot\theta^2 + \sin^2\theta\, \dot\phi^2$, we obtain

$$J_\psi = \sqrt{\beta_2}. \tag{2.57}$$

Comparing, now, the kinetic energies in the two reference systems:

$$p_r \dot r + p_\theta \dot\theta + p_\phi \dot\phi = p_r \dot r + p_\psi \dot\psi, \tag{2.58}$$

it follows that

$$J_\theta = \frac{1}{2\pi} \oint p_\theta\, d\theta = \frac{1}{2\pi} \oint (p_\psi\, d\psi - p_\phi\, d\phi) = J_\psi - J_\phi \tag{2.59}$$

and then[6]

$$J_\theta = \sqrt{\beta_2} - \sqrt{\beta_3}. \tag{2.60}$$

The radial action

$$J_r = \frac{1}{2\pi} \oint \sqrt{2mE - 2m^2 U(r) - \frac{\beta_2}{r^2}}\, dr \tag{2.61}$$

cannot be calculated now since the potential $U(r)$ has not yet been given (see next section).

To obtain the angle variables w_k, we follow the procedure given in Sect. 2.4.1. We first write

$$S(q,\beta) = \int p_r\, dr + \int p_\theta\, d\theta + \int p_\phi\, d\phi \tag{2.62}$$

and then introduce

$$\begin{aligned}\beta_1 &= E = E(J_r, J_\theta, J_\phi) \\ \beta_2 &= (J_\theta + J_\phi)^2 \\ \beta_3 &= J_\phi^2,\end{aligned} \tag{2.63}$$

where we have to keep in mind that the function $E = E(J_r, J_\theta, J_\phi)$ may be known only when the potential $U(r)$, of the central force, is given. We then have

---

[6] $J_\psi \geq J_\phi > 0$, $\beta_2 \geq \beta_3 > 0$ and $J_\theta \geq 0$.

$$w_r = \frac{\partial \tilde{S}}{\partial J_r} = \frac{\partial S}{\partial E}\frac{\partial E}{\partial J_r}$$
$$w_\theta = \frac{\partial \tilde{S}}{\partial J_\theta} = \frac{\partial S}{\partial E}\frac{\partial E}{\partial J_\theta} + \frac{\partial S}{\partial \beta_2} 2(J_\theta + J_\phi) \qquad (2.64)$$
$$w_\phi = \frac{\partial \tilde{S}}{\partial J_\phi} = \frac{\partial S}{\partial E}\frac{\partial E}{\partial J_\phi} + \frac{\partial S}{\partial \beta_2} 2(J_\theta + J_\phi) + \frac{\partial S}{\partial \beta_3} 2J_\phi,$$

where

$$\frac{\partial S}{\partial E} = \int \frac{\partial p_r}{\partial E} dr$$
$$\frac{\partial S}{\partial \beta_2} = \int \frac{\partial p_r}{\partial \beta_2} dr + \int \frac{\partial p_\theta}{\partial \beta_2} d\theta \qquad (2.65)$$
$$\frac{\partial S}{\partial \beta_3} = \int \frac{\partial p_\theta}{\partial \beta_3} d\theta + \int \frac{\partial p_\phi}{\partial \beta_3} d\phi.$$

As for the actions, some of these integral cannot be calculated, because $U(r)$ has not yet been given. Those that may be calculated are

$$\int \frac{\partial p_\theta}{\partial \beta_2} d\theta = \int \frac{d\theta}{2p_\theta}$$
$$\int \frac{\partial p_\theta}{\partial \beta_3} d\theta = -\int \frac{d\theta}{2p_\theta \sin^2\theta} \qquad (2.66)$$

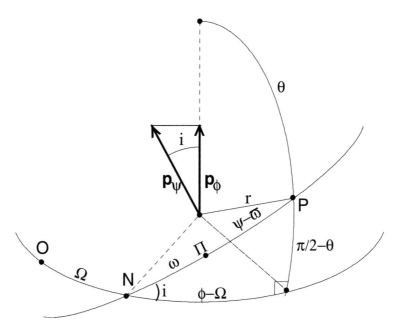

Fig. 2.5. Geometry of central motions ($\varpi = \Omega + \omega$)

$$\int \frac{\partial p_\phi}{\partial \beta_3} \, d\phi = \int \frac{d\phi}{2p_\phi}.$$

The third of these integrals is trivial since $p_\phi = J_\phi = \text{const}$. Omitting the integration constant,

$$\int \frac{\partial p_\phi}{\partial \beta_3} \, d\phi = \frac{\phi}{2J_\phi}.$$

The two other integrals are also easy to calculate using an immediate relation between the angles and momenta ($\dot\theta/p_\theta = \dot\psi/p_\psi = \dot\phi \sin^2\theta/p_\phi$) and recalling that $p_\phi = J_\phi$ and $p_\psi = J_\psi = J_\phi + J_\theta$ are constants. We thus obtain

$$\int \frac{\partial p_\theta}{\partial \beta_2} \, d\theta = \int \frac{d\theta}{2p_\theta} = \int \frac{d\psi}{2p_\psi} = \frac{\psi}{2(J_\phi + J_\theta)}$$

and

$$\int \frac{\partial p_\theta}{\partial \beta_3} \, d\theta = -\int \frac{d\theta}{2p_\theta \sin^2\theta} = -\int \frac{d\phi}{2p_\phi} = -\frac{\phi - \Omega}{2J_\phi}.$$

In general, we have omitted integration constants, since this arbitrariness is intrinsic to the definition of the angles w_i. However, in the last equation, to shift the x-axis to the ascending node (N) of the orbit (see Fig. 2.5), we have introduced the integration constant $\Omega/2J_\phi$.

We may summarize the results by writing

$$w_r = \frac{\partial E}{\partial J_r} \int \frac{\partial p_r}{\partial E} \, dr \tag{2.67}$$

$$w_\theta = \frac{\partial E}{\partial J_\theta} \int \frac{\partial p_r}{\partial E} \, dr + 2(J_\theta + J_\phi) \int \frac{\partial p_r}{\partial \beta_2} \, dr + \psi$$

$$w_\phi = \frac{\partial E}{\partial J_\phi} \int \frac{\partial p_r}{\partial E} \, dr + 2(J_\theta + J_\phi) \int \frac{\partial p_r}{\partial \beta_2} \, dr + \psi + \Omega.$$

2.6 Kepler Motion

In the case of the heliocentric motion of a planet, we have

$$U(r) = -\frac{\mu}{r},$$

where $\mu = G(M + m)$; G is the universal gravitation constant and M is the mass of the Sun. Now, we can consider the several integrals left uncalculated in the last section. The first one is the radial action J_r (see 2.61). We have

$$J_r = \frac{1}{2\pi} \oint \frac{1}{r}\sqrt{2mEr^2 + 2\mu m^2 r - \beta_2} \, dr. \tag{2.68}$$

The radicand has the real roots

$$r_{1,2} = \frac{-\mu m}{2E}\left(1 \pm \sqrt{1 + \frac{2E\beta_2}{\mu^2 m^3}}\right). \tag{2.69}$$

One may note that, if $E > 0$, the two roots are real, but one is negative. In this case, the motion is only possible for r larger than the positive root and has no upper bound. For $-\mu^2 m^3/2\beta_2 < E < 0$, the two roots are real and positive, say $r_1 < r_2$; the motion is periodic and is a libration between the two roots. In this case, we may calculate the action J_r. The integral of (2.68) may be done along a path in a two-sheet Riemann surface enclosing the two branch points r_1, r_2. It has been thoroughly studied by Sommerfeld (see [93], Note 6). The sophisticated procedure idealized by Sommerfeld has, since then, been reproduced in many treatises on Mechanics. However, there is a simpler way of doing it. We introduce the *mean distance* to the force center

$$a \stackrel{\text{def}}{=} \frac{r_1 + r_2}{2} = -\frac{\mu m}{2E}, \tag{2.70}$$

the *eccentricity*

$$e \stackrel{\text{def}}{=} \frac{r_2 - r_1}{2a} = \sqrt{1 + \frac{2E\beta_2}{\mu^2 m^3}} \tag{2.71}$$

and the angle u (*eccentric anomaly*) defined through

$$r = a(1 - e\cos u). \tag{2.72}$$

A lengthy but elementary calculation gives

$$p_r = \sqrt{-2mE}\,\frac{e\sin u}{1 - e\cos u}$$

and the given integral becomes

$$J_r = \frac{1}{2\pi} ae^2 \sqrt{-2mE} \int_0^{2\pi} \frac{\sin^2 u \, du}{1 - e\cos u}. \tag{2.73}$$

The integral to be solved is trivial. We may just introduce $z = e^{iu}$ and perform the integration along the circle $|z| = 1$ in the complex plane, with recourse to the theory of residues. We obtain[7]

$$\int_0^{2\pi} \frac{\sin^2 u \, du}{1 - e\cos u} = \frac{2\pi}{e^2}\left(1 - \sqrt{1 - e^2}\right). \tag{2.74}$$

After some elementary calculations, we obtain

$$J_r = \sqrt{-2mE}\, a\left(1 - \sqrt{1 - e^2}\right) = \mu m\sqrt{\frac{m}{-2E}} - \sqrt{\beta_2} \tag{2.75}$$

---

[7] This integral is also found in tables, e.g. [25], [41].

and the inversion of this equation gives

$$E = -\frac{\mu^2 m^3}{2(J_r + J_\theta + J_\phi)^2} \tag{2.76}$$

(since $\sqrt{\beta_2} = J_\psi = J_\theta + J_\phi$).

We may now proceed with the remaining integrals. They are

$$\int \frac{\partial p_r}{\partial E} dr = \int \frac{m \, dr}{p_r},$$

$$\int \frac{\partial p_r}{\partial \beta_2} dr = -\int \frac{dr}{2p_r r^2}.$$

Introducing the eccentric anomaly u, these integrals are changed into elementary ones. The first one is

$$\int \frac{m \, dr}{p_r} = \frac{a^{3/2}}{\sqrt{\mu}} \int (1 - e \cos u) \, du = \frac{a^{3/2}}{\sqrt{\mu}} (u - e \sin u) \tag{2.77}$$

which introduces the *mean anomaly*

$$\ell \stackrel{\text{def}}{=} u - e \sin u.$$

The second one is

$$-\int \frac{dr}{2p_r r^2} = -\frac{1}{2m\sqrt{\mu a}} \int \frac{du}{1 - e \cos u} \tag{2.78}$$

$$= -\frac{1}{m\sqrt{\mu a(1 - e^2)}} \arctan \sqrt{\frac{1+e}{1-e}} \tan \frac{u}{2}$$

which introduces the *true anomaly*

$$v \stackrel{\text{def}}{=} 2 \arctan \sqrt{\frac{1+e}{1-e}} \tan \frac{u}{2}. \tag{2.79}$$

The two remaining integrals are, then,

$$\int \frac{\partial p_r}{\partial E} dr = \frac{a^{3/2} \ell}{\sqrt{\mu}}$$

$$\int \frac{\partial p_r}{\partial \beta_2} dr = -\frac{v}{2m\sqrt{\mu a(1 - e^2)}}.$$

Substituting these integrals into (2.67), and noting that

$$\frac{\partial E}{\partial J_r} = \frac{\partial E}{\partial J_\theta} = \frac{\partial E}{\partial J_\phi} = \frac{\mu^2 m^3}{(J_r + J_\theta + J_\phi)^3}$$

$$J_r + J_\theta + J_\phi = m\sqrt{\mu a} \tag{2.80}$$

and

$$J_\theta + J_\phi = \sqrt{\beta_2} = m\sqrt{\mu a(1-e^2)}$$

it follows that

$$\begin{aligned} w_r &= \ell \\ w_\theta &= \ell - v + \psi &= \ell + \omega \\ w_\phi &= \ell - v + \psi + \Omega = \ell + \omega + \Omega, \end{aligned} \tag{2.81}$$

where we have introduced the so-called argument of pericenter $\omega = \psi - v$, giving the distance of the pericenter (Π) to the ascending node (N) (see Fig. 2.5).

To complete the definition of the angle–action variables of the Kepler motion, we write[8]

$$J_\phi = m\sqrt{\mu a(1-e^2)}\cos i. \tag{2.82}$$

2.7 Degeneracy

In the example studied in the previous section, the three frequencies

$$\nu_k = \frac{\partial E}{\partial J_k} \tag{2.83}$$

of the system are equal. We follow Schwarzschild and call this case *degenerate*. In general, degeneracy is said to occur when there exists a commensurability relation

$$(h \mid \nu) = \sum_{k=1}^{N} h_k \nu_k = 0 \qquad h \in \mathbf{Z}^N \tag{2.84}$$

amongst the frequencies of the system. Degeneracy may be essential or accidental. A degeneracy is said to be *essential* when it does not depend on the initial conditions. We shall stress that this does not mean that the frequencies themselves are independent of the initial conditions. The Keplerian motion is a good example: the frequencies $\nu_r, \nu_\theta, \nu_\phi$ (defined by the derivatives of E with respect to J_r, J_θ, J_ϕ) depend on the initial conditions but they are always equal, regardless of the initial conditions.

Otherwise, a degeneracy is called *accidental* when it only occurs for some particular values of the initial conditions. One example is the motion of an asteroid in an orbit whose period is commensurable with Jupiter's. In this case, the commensurability relation ceases to exist if the asteroid orbit is moved inward (or outward). The main consequence of an accidental degeneracy is

---

[8] The inclination is introduced by the fact that p_ψ is the angular momentum of the motion and p_ϕ is the angular momentum of the motion projected on the reference plane: $p_\phi = p_\psi \cos i$.

the appearance of small divisors, which impair the performance of perturbation theories. Motions affected by accidental commensurabilities are called *resonant* and are the subject of several of the next chapters.

A separable multiperiodic system may be such that multiple commensurability relations exist. Degeneracy affects the degree of periodicity of the solutions: the solutions of a degenerate separable multiperiodic system with N degrees of freedom and D independent commensurability relations are $(N-D)$-periodic. When $D = N - 1$, the system is said to be *completely degenerate*. For instance, the degeneracy of the Keplerian motion is complete, since we may write two independent commensurability relations, viz. $\nu_\theta - \nu_r = 0$ and $\nu_\phi - \nu_\theta = 0$. As a consequence, the Keplerian motion is periodic. The central motions of Sect. 2.5 are always degenerate, since $\nu_\phi - \nu_\theta = 0$. However, they are not completely degenerate, except in some particular cases such as Keplerian motion and the harmonic oscillator (Bertrand's theorem). In these cases, besides $\nu_\phi - \nu_\theta = 0$, we also have $\nu_\theta - \nu_r = 0$. For other laws of force, a second commensurability relation may only occur for given initial conditions (accidental degeneracy or resonance).

In Kolmogorv's theorem, the non-degeneracy of an integrable Hamiltonian $H(J)$ is defined as

$$\det\left(\frac{\partial^2 H}{\partial J_i \partial J_j}\right) \neq 0, \tag{2.85}$$

which guarantees the reversibility of the transformation from actions to frequencies. This definition is more restrictive than Schwarzschild's. Indeed, all Hamiltonians linear in one of the actions are degenerate in Kolmogorov's sense[9]. For these Hamiltonians, one whole row of the Hessian determinant consists of zeros. It happens that a common operation in the applications of Hamiltonian Mechanics to Astronomy is the extension of the phase space, because of time-dependent applied forces. In such an extension, a new generalized momentum (or action) is added to the given Hamiltonian. The extended Hamiltonian will always be such that the Hessian determinant is zero. If the condition given by (2.85) were a universal restriction, almost all dynamical systems of Astronomy would be excluded from the possibility of application of the theories discussed in this book. However, when frequency relocation is not done, the most general non-degeneracy condition is Schwarzschild's, that is, $(h \mid \nu) \neq 0$ for all $h \in \mathbf{D_k} \subset \mathbf{Z}^N \backslash 0$.

2.7.1 Schwarzschild Transformation

In the study of degenerate systems, it is often convenient to redefine angles and actions to introduce angles whose frequencies are equal to zero. Let a separable multiperiodic system of N degrees of freedom have L essential commensurability relations

---

[9] For a more accurate discussion, see Sect. 3.11.4.

2 Angle–Action Variables. Separable Systems

$$\sum_{k=1}^{N} h_k^{(\varrho)} \nu_k = 0 \qquad \varrho = N - L + 1, \cdots, N \qquad (2.86)$$

and let us introduce the point transformation of the angles,

$$\begin{aligned}
w_1 &= \ell_1 \\
w_2 &= \ell_2 \\
&\cdots \\
w_M &= \ell_M \\
\sum_k h_k^{(M+1)} w_k &= \ell_{M+1} \\
&\cdots \\
\sum_k h_k^{(N)} w_k &= \ell_N,
\end{aligned} \qquad (2.87)$$

where, for simplicity, we have introduced $M = N - L$. Extending this transformation to the momenta, we obtain

$$\begin{aligned}
J_1 &= x_1 + \sum_\varrho h_1^{(\varrho)} x_\varrho \\
J_2 &= x_2 + \sum_\varrho h_2^{(\varrho)} x_\varrho \\
&\cdots \\
J_M &= x_M + \sum_\varrho h_M^{(\varrho)} x_\varrho \\
J_{M+1} &= \sum_\varrho h_{M+1}^{(\varrho)} x_\varrho \\
&\cdots \\
J_N &= \sum_\varrho h_N^{(\varrho)} x_\varrho,
\end{aligned} \qquad (2.88)$$

where the x_k are the momenta conjugate to the new angles ℓ_k.

The angles ℓ_μ ($\mu = 1, \cdots, M$) are called *non-degenerate*[10] while the remaining ones, ℓ_ϱ ($\varrho = M + 1, \cdots, N$), are called *degenerate*.

With the new variables, the Hamiltonian depends only on the actions conjugate to non-degenerate angles. Thus, the frequencies of the degenerate angles are

$$\tilde{\nu}_\varrho = \frac{\mathrm{d}\ell_\varrho}{\mathrm{d}t} = \frac{\partial \tilde{H}(x)}{\partial x_\varrho} = 0 \qquad (\varrho = M + 1, \cdots, N). \qquad (2.89)$$

The equations $\tilde{\nu}_\varrho = 0$ are the new commensurability relations.

2.7.2 Delaunay Variables

The usual angle–action variables of the Keplerian motion, the Delaunay variables, are the result of the application of the Schwarzschild transformation

---

[10] The actions conjugate to non-degenerate angles are sometimes called *proper*. However, the word *proper* is used, in this book, to indicate the almost constant actions resulting from an averaging process. Thus, to avoid ambiguities, the word proper will not be used to mean *non-degenerate*.

to the angle–action variables obtained in Sect. 2.6. Indeed, in this case, the commensurabilities are

$$\nu_\theta - \nu_r = 0$$
$$\nu_\phi - \nu_\theta = 0. \qquad (2.90)$$

Then
$$\ell_1 = w_r \qquad\quad = \ell$$
$$\ell_2 = w_\theta - w_r = \omega \qquad (2.91)$$
$$\ell_3 = w_\phi - w_\theta = \Omega$$

and
$$J_r = x_1 - x_2$$
$$J_\theta = x_2 - x_3 \qquad (2.92)$$
$$J_\phi = x_3$$

or
$$x_1 = J_r + J_\theta + J_\phi = m\sqrt{\mu a}$$
$$x_2 = J_\theta + J_\phi \qquad = m\sqrt{\mu a(1-e^2)} \qquad (2.93)$$
$$x_3 = J_\phi \qquad\qquad = m\sqrt{\mu a(1-e^2)}\cos i.$$

For $m = 1$, these variables are exactly the variables ℓ, g, h, L, G, H of Delaunay. Indeed, point dynamics problems often are such that the mass of the moving particle cancels in the equations and does not affect the results. In this case, energies, momenta and actions are considered per unit mass and we write

$$x_1 = \sqrt{\mu a}$$
$$x_2 = x_1\sqrt{1-e^2} \qquad (2.94)$$
$$x_3 = x_2 \cos i$$

and
$$E = -\frac{\mu^2}{2x_1^2}. \qquad (2.95)$$

2.8 The Separable Cases of Liouville and Stäckel

Autonomous systems whose energy consists of a kinetic energy quadratic in the velocities and a potential energy independent of the velocities have been thoroughly studied in the past. Sufficient conditions for their separability were established by Liouville and Stäckel. These cases are generally presented as sets of conditions for the potential and kinetic energies, separately. In what follows, kinetic and potential energies are considered together to give a set of conditions for the Hamiltonian; this choice is more appropriate for the scope of this book.

Theorem 2.8.1 (Liouville). *The dynamical systems whose Hamiltonian may be written as*

$$H = \frac{f_1(q_1,p_1) + \cdots + f_N(q_N,p_N)}{g_1(q_1,p_1) + \cdots + g_N(q_N,p_N)} \tag{2.96}$$

are separable.

The Hamilton–Jacobi equation in this case is

$$\sum_{i=1}^{N} f_i\left(q_i, \frac{\partial S}{\partial q_i}\right) = E \sum_{i=1}^{N} g_i\left(q_i, \frac{\partial S}{\partial q_i}\right) \tag{2.97}$$

which, after the introduction of $S = \sum_i S_i(q_i)$, may be separated into N equations

$$f_i\left(q_i, \frac{dS_i}{dq_i}\right) - E g_i\left(q_i, \frac{dS_i}{dq_i}\right) = \beta_i, \tag{2.98}$$

the integration constants β_i being such that $\sum_i \beta_i = 0$. These equations may be solved with respect to dS_i/dq_i when

$$\left(\frac{\partial f_i}{\partial p_i}\right) - E\left(\frac{\partial g_i}{\partial p_i}\right) \neq 0 \qquad \text{for all } i.$$

$\square$

Theorem 2.8.2 (Stäckel). *The dynamical systems whose Hamiltonian may be written as*

$$H = \frac{1}{\Delta} \sum_{i=1}^{N} A_i f_i(q_i, p_i), \tag{2.99}$$

where Δ is the determinant of a square matrix of rank N in which each column depends only on the coordinate of the same subscript as the column:

$$\Delta = \det\left(a_{ji}(q_i)\right), \tag{2.100}$$

and the A_i are the cofactors of the elements of any of the rows of the matrix, are separable.

The Hamilton–Jacobi equation in this case is

$$\sum_{i=1}^{N} A_i f_i\left(q_i, \frac{\partial S}{\partial q_1}\right) = E\Delta. \tag{2.101}$$

This partial differential equation has a complete integral of the form $S = \sum_i S_i(q_i)$. If we assume, for instance, that the A_i are the cofactors of the elements of the first row, the theorems of Laplace allow us to write

2.8 The Separable Cases of Liouville and Stäckel

$$\sum_{i=1}^{N} a_{1i} A_i = \Delta \tag{2.102}$$

$$\sum_{i=1}^{N} a_{\varrho i} A_i = 0 \qquad (\varrho = 2, \cdots, N).$$

Because of these relations, the Hamilton–Jacobi equation is not affected when we introduce the sum

$$-\sum_{i=1}^{N} A_i \sum_{\varrho=2}^{N} \beta_\varrho a_{\varrho i}(q_i),$$

where the β_ϱ are $N-1$ arbitrary constants. Using also the Laplacian expression for Δ, the Hamilton–Jacobi equation becomes

$$\sum_{i=1}^{N} A_i \left(f_i \left(q_i, \frac{\partial S}{\partial q_i} \right) - \sum_{\varrho=2}^{N} \beta_\varrho a_{\varrho i}(q_i) \right) = E \sum_{i=1}^{N} a_{1i}(q_i) A_i, \tag{2.103}$$

which may be separated into N equations

$$f_i \left(q_i, \frac{\mathrm{d}S_i}{\mathrm{d}q_i} \right) = \sum_{\varrho=2}^{N} \beta_\varrho a_{\varrho i}(q_i) + E a_{1i}(q_i). \tag{2.104}$$

These equations may be solved with respect to $\mathrm{d}S_i/\mathrm{d}q_i$ if

$$\frac{\partial f_i}{\partial p_i} \neq 0 \qquad \text{for all } i.$$

$\square$

2.8.1 Example: Liouville Systems

The original form of Liouville's separability conditions says that the kinetic and potential energies may be written, respectively, as

$$T = \frac{1}{2}(A_1 + A_2 + \cdots + A_N)(B_1 \dot{q}_1^2 + B_2 \dot{q}_2^2 + \cdots + B_N \dot{q}_N^2) \tag{2.105}$$

and

$$V = \frac{V_1 + V_2 + \cdots + V_N}{A_1 + A_2 + \cdots + A_N}, \tag{2.106}$$

where $A_i = A_i(q_i)$, $B_i = B_i(q_i)$ and $V_i = V_i(q_i)$. (The function with subscript i depends only on the generalized coordinate q_i.) A simple calculation shows that the energy $H = T + V$ has the form given in the above theorem and that the Hamilton–Jacobi equation is separated into the N equations:

$$\frac{1}{2B_k} \left(\frac{\mathrm{d}S_k}{\mathrm{d}q_k} \right)^2 = E A_k + \beta_k - V_k. \tag{2.107}$$

2.8.2 Example: Stäckel Systems

The original form of Stäckel's separability conditions says that the kinetic and potential energies must be, respectively,

$$T = \frac{1}{2}\Delta \left(\frac{\dot{q}_1^2}{A_1} + \frac{\dot{q}_2^2}{A_2} + \cdots + \frac{\dot{q}_N^2}{A_N} \right) \tag{2.108}$$

and

$$V = \frac{1}{\Delta} \sum_{i=i}^{N} g_i(q_i) A_i, \tag{2.109}$$

where Δ and A_i are the same as in the given theorem. The resulting energy $H = T + V$ has the form as given in the theorem and the Hamilton–Jacobi equation is separated into the N equations:

$$\frac{1}{2}\left(\frac{\mathrm{d}S_i}{\mathrm{d}q_i}\right)^2 = \sum_{\varrho=2}^{N} \beta_\varrho a_{\varrho i}(q_i) + E a_{1i}(q_i) - g_i(q_i). \tag{2.110}$$

2.8.3 Example: Central Motions

The example of the motion of a particle in a central force field, considered in the previous section, is also an example of a separable Stäckel system. The Hamiltonian of this system is (see 2.52):

$$H = \frac{1}{2m}\left[p_r^2 + \frac{1}{r^2}\left(p_\theta^2 + \frac{p_\phi^2}{\sin^2\theta} \right) \right] + V(r). \tag{2.111}$$

In order to see that this system satisfies the conditions of the Stäckel theorem, we introduce the matrix

$$(a_{ij}) = \begin{pmatrix} -r^{-2} & 1 & 0 \\ 0 & -\sin^{-2}\theta & 1 \\ 1 & 0 & 0 \end{pmatrix} \tag{2.112}$$

whose determinant is $\Delta = 1$ and the cofactors of the elements of the third row are:

$$\begin{aligned} A_1 &= 1 \\ A_2 &= r^{-2} \\ A_3 &= (r\sin\theta)^{-2}. \end{aligned} \tag{2.113}$$

Comparison to (2.97) shows that the functions f_i are

$$f_1 = \frac{1}{2m}p_r^2 + V(r)$$
$$f_2 = \frac{1}{2m}p_\theta^2 \qquad (2.114)$$
$$f_3 = \frac{1}{2m}p_\phi^2$$

and the proof is completed.

2.9 Angle–Action Variables of a Quadratic Hamiltonian

Let us consider the case of a Hamiltonian given by a quadratic form in q, p, with purely imaginary eigenvalues. Let it be

$$H_2(z) = \sum_{i,j=1}^{2N} \frac{1}{2} a_{ij} z_i z_j, \qquad (2.115)$$

where $z \equiv (q, p) \in \mathbf{R}^{2N}$. In this case, the techniques discussed in the previous sections to obtain angle–action variables cannot be used because the Hamiltonian does not have the form of the considered separable systems. However, the resulting differential equations are homogeneous and linear with constant coefficients and a few steps are enough to solve them. These equations are

$$\frac{dz}{dt} = -\mathsf{J}\frac{\partial H_2}{\partial z} = -\mathsf{J}\mathsf{S}z, \qquad (2.116)$$

where J is the symplectic unit matrix of order $2N$ and $\mathsf{S} = \left(\frac{\partial^2 H_2}{\partial z_i \partial z_j}\right) = (a_{ij})$ is the Hessian matrix of H_2. Let λ_i and $\tilde{A}_i$ be, respectively, the eigenvalues and eigenvectors of $-\mathsf{J}\mathsf{S}$. If we assume that all eigenvalues are distinct, the general solution of (2.116) is

$$z = \sum_{i=1}^{2N} c_i \tilde{A}_i \exp \lambda_i t, \qquad (2.117)$$

where c_i are arbitrary constants. The characteristic polynomial $P(\lambda) = \det(-\mathsf{J}\mathsf{S} - \lambda \mathsf{I})$ is even and, if λ is an eigenvalue of $-\mathsf{J}\mathsf{S}$, then so is $-\lambda$. The eigenvalues of $-\mathsf{J}\mathsf{S}$, which were assumed to be imaginary, may thus be written as

$$\lambda_k = -\mathrm{i}\omega_k, \qquad \lambda_{N+k} = \mathrm{i}\omega_k \qquad (k = 1, 2, \cdots, N). \qquad (2.118)$$

Let us now consider the matrix formed by the $2N$ eigenvectors, $\mathsf{A} \equiv (\tilde{A}_i)$, its transpose A' and let us form the matrix $\mathsf{R} = \mathsf{A}'\mathsf{J}\mathsf{A}$. A simple calculation shows that the elements of R are

$$\varrho_{ij} \stackrel{\text{def}}{=} \tilde{A}_i' \mathsf{J} \tilde{A}_j.$$

We have to prove the following lemma (see [71] Sect. II.C):

Lemma 2.9.1. *If $\tilde{A}_i$ and $\tilde{A}_j$ are eigenvectors of $-\mathsf{JS}$ corresponding to two eigenvalues λ_i, λ_j such that $\lambda_i + \lambda_j \neq 0$, then $\tilde{A}'_i \mathsf{J} \tilde{A}_j = 0$.*

The proof of this statement is very simple. We just have to recall that the eigenvalue λ_i and the eigenvector $\tilde{A}_i$ corresponding to it are related by $\mathsf{JS}\tilde{A}_i = -\lambda_i \tilde{A}_i$. It then follows that:

$$\lambda_i \tilde{A}'_i \mathsf{J} \tilde{A}_j = -\tilde{A}'_i \mathsf{S} \tilde{A}_j \quad \text{and}$$
$$\lambda_j \tilde{A}'_i \mathsf{J} \tilde{A}_j = \tilde{A}'_i \mathsf{S} \tilde{A}_j;$$

and so, $(\lambda_i + \lambda_j) \tilde{A}'_i \mathsf{J} \tilde{A}_j = 0$, that is, $\tilde{A}'_i \mathsf{J} \tilde{A}_j = 0$.

Corollary 2.9.1. $\tilde{A}'_i \mathsf{J} \tilde{A}_i = 0$ *for all* $i = 1, 2, \cdots, 2N$.

The following lemma is trivial.

Lemma 2.9.2. *For all $i, j = 1, 2, \cdots, 2N$, we have $\tilde{A}'_i \mathsf{J} \tilde{A}_j = -\tilde{A}'_j \mathsf{J} \tilde{A}_i$.*

A consequence of these lemmas is that the only terms of R that may be different from zero are those arising from eigenvectors corresponding to pairs of eigenvalues $\pm i\omega_k$. We assume $\varrho_{ij} \neq 0$ for the pairs i, j such that $|j - i| = N$. Otherwise, $\varrho_{ij} = 0$:

$$\mathsf{R} = \begin{pmatrix} 0 & 0 & \cdots & -\varrho_{N+1,1} & 0 & \cdots \\ 0 & 0 & \cdots & 0 & -\varrho_{N+2,2} & \cdots \\ \cdots & \cdots & \cdots & \cdots & \cdots & \cdots \\ \varrho_{N+1,1} & 0 & \cdots & 0 & 0 & \cdots \\ 0 & \varrho_{N+2,2} & \cdots & 0 & 0 & \cdots \\ \cdots & \cdots & \cdots & \cdots & \cdots & \cdots \end{pmatrix}. \quad (2.119)$$

Therefore, it is enough to rescale the eigenvectors (dividing the $\tilde{A}_k$ and $\tilde{A}_{N+k}$ by $\sqrt{\varrho_{N+k,k}}$ for all k) to obtain J instead of R. If D is the diagonal matrix

$$\mathsf{D} \stackrel{\text{def}}{=} \operatorname{diag}\left(\frac{1}{\sqrt{\varrho_{N+1,1}}}, \cdots, \frac{1}{\sqrt{\varrho_{2N,N}}}, \frac{1}{\sqrt{\varrho_{N+1,1}}}, \cdots, \frac{1}{\sqrt{\varrho_{2N,N}}}\right),$$

the matrix $\mathsf{M} = \mathsf{AD}$ is such that $\mathsf{M'JM} = \mathsf{J}$ and therefore, the linear transformation $\zeta \to z = \mathsf{AD}\zeta$ is canonical (see 1.36).

If we compare the equation of this transformation to (2.117), we obtain for the new canonical variables,

$$\zeta_k = c_k \sqrt{\varrho_{N+k,k}} \, e^{\lambda_k t} \qquad \zeta_{N+k} = c_{N+k} \sqrt{\varrho_{N+k,k}} \, e^{\lambda_{N+k} t}$$

$(k = 1, 2, \cdots, N)$. To complete the construction of the angle–action variables (w, J) of H_2, it is enough to introduce them through the Poincaré-like complex canonical variables $\sqrt{iJ_k} \, e^{-iw_k}$ and $\sqrt{iJ_k} \, e^{iw_k}$ and compare them to ζ. We get

2.9 Angle–Action Variables of a Quadratic Hamiltonian

$$w_k = \omega_k t - \alpha_k \tag{2.120}$$
$$J_k = -\mathrm{i}|c_k|^2 \varrho_{N+k,k} \quad (k = 1, 2, \cdots, N),$$

where α_k is the argument of c_k. Because of the rules of conjugation, it is enough to work with the equations giving the first N variables ζ_k. The other N equations repeat the same results. It is worth stressing some points: (i) the J_k are real since the $\varrho_{N+k,k}$ are imaginary; (ii) the J_k may be either positive or negative, according to the sign of $-\mathrm{i}\varrho_{N+k,k}$; (iii) the N complex integration constants c_k are changed into α_k, J_k; (iv) c_k and c_{N+k} are complex conjugates.

The direct comparison of equations (2.117) and (2.120) gives

$$z = \sum_{k=1}^{N} \sqrt{\frac{\mathrm{i}J_k}{\varrho_{N+k,k}}} \left(\tilde{A}_k \, \mathrm{e}^{-\mathrm{i}w_k} + \tilde{A}_{N+k} \, \mathrm{e}^{\mathrm{i}w_k} \right). \tag{2.121}$$

This equation is consistent with the fact that z is a real vector.

In terms of the angle–action variables, the new Hamiltonian follows straightforwardly from the equations $\partial H / \partial J_k = \dot{w}_k = \omega_k$, whose integration gives $H = \sum_k \omega_k J_k$, or, as a function of ζ, $H = -\sum_k \mathrm{i}\omega_k \zeta_k \zeta_{N+k}$. If we compare this result to

$$H = \frac{1}{2} z' S z = \frac{1}{2} \zeta' \mathsf{DA'SAD}\zeta,$$

we see that

$$\mathsf{DA'SAD} = \begin{pmatrix} 0 & 0 & \cdots & -\mathrm{i}\omega_1 & 0 & \cdots \\ 0 & 0 & \cdots & 0 & -\mathrm{i}\omega_2 & \cdots \\ \cdots & \cdots & \cdots & \cdots & \cdots & \cdots \\ -\mathrm{i}\omega_1 & 0 & \cdots & 0 & 0 & \cdots \\ 0 & -\mathrm{i}\omega_2 & \cdots & 0 & 0 & \cdots \\ \cdots & \cdots & \cdots & \cdots & \cdots & \cdots \end{pmatrix}.$$

This matrix is the Hessian of H calculated with respect to the new canonical variables ζ. It could be easily obtained from the properties of the matrices D, A and S, using the lemma given in Exercise 2.9.6, below.

Exercise 2.9.1. Show that the characteristic polynomial $P(\lambda) = \det(-\mathsf{JS} - \lambda \mathsf{I})$ is even.

Exercise 2.9.2. Show that the eigenvectors $\tilde{A}_k$ and $\tilde{A}_{N+k}$ of $-\mathsf{JS}$ corresponding to two complex conjugate eigenvalues are complex conjugate themselves.

Exercise 2.9.3. Show that, for $|i - j| = N$, the ϱ_{ij} are imaginary.

Exercise 2.9.4. Show that the transformation $(w, J) \to (\sqrt{\mathrm{i}J}\,\mathrm{e}^{-\mathrm{i}w}, \sqrt{\mathrm{i}J}\,\mathrm{e}^{\mathrm{i}w})$ is canonical.

Exercise 2.9.5. Show that ζ_k and ζ_{N+k} are *not* complex conjugates.

Exercise 2.9.6 (Lemma). Prove that for all $i, j = 1, 2, \cdots, 2N$, we have

$$\tilde{A}'_i \mathsf{S} \tilde{A}_j = -\lambda_j \varrho_{ji}.$$

Hint: Use the characteristic equation $\mathsf{JS}\tilde{A}_j = -\lambda_j \tilde{A}_j$.

2.9.1 Gyroscopic Systems

Let us consider the important particular case of the two-degrees-of-freedom gyroscopic system whose Hamiltonian is

$$H = \frac{p^2}{2} - [k, r, p] + W(r), \tag{2.122}$$

where $r \equiv (x, y), p \equiv (p_x, p_y)$, k is a unit vector perpendicular to the plane of motion and the potential energy is $W = \frac{1}{2}(ax^2 + by^2) + dxy$ (a, b, d are constants). (See Sect. 1.7; for the sake of simplicity, we have chosen units such that $m = 1$ and $|\Omega| = 1$.) The Hessian matrix is

$$S = \begin{pmatrix} a & d & 0 & -1 \\ d & b & 1 & 0 \\ 0 & 1 & 1 & 0 \\ -1 & 0 & 0 & 1 \end{pmatrix}. \tag{2.123}$$

Then,

$$-JS = \begin{pmatrix} 0 & 1 & 1 & 0 \\ -1 & 0 & 0 & 1 \\ -a & -d & 0 & 1 \\ -d & -b & -1 & 0 \end{pmatrix} \tag{2.124}$$

and the eigenvalues of $-JS$ are

$$\lambda_j = \pm \frac{1}{2}\sqrt{-2(a+b+2) \pm 2\sqrt{(a-b)^2 + 8(a+b) + 4d^2}}. \tag{2.125}$$

We assume that these eigenvalues are imaginary and write them as $\pm i\omega_1$ and $\pm i\omega_2$. This means that the parameters a, b, d of the given function W are such that $\phi \stackrel{\text{def}}{=} (a-b)^2 + 8(a+b) + 4d^2 \geq 0$ and $-(a+b+2) + \sqrt{\phi} < 0$.

The eigenvectors of $-JS$ are

$$\tilde{A}_j = \begin{pmatrix} -\lambda_j^3 - (b+1)\lambda_j + d \\ \lambda_j^2 + d\lambda_j - a + 1 \\ a\lambda_j^2 - b + ab - d^2 \\ d(\lambda_j^2 + 1) - (a+b)\lambda_j \end{pmatrix}. \tag{2.126}$$

The quantities $\varrho_{k+2,k}$ are immediate. We just point out the fact that, of the five parameters $a, b, d, \omega_1, \omega_2$, only three are independent. We use (2.125) to eliminate b, d and obtain

$$\begin{aligned}\varrho_{31} &= 2i\,\omega_1(\omega_1^2 - \omega_2^2)(1 - a + a\omega_1^2 - \omega_1^2\omega_2^2) \\ \varrho_{42} &= 2i\,\omega_2(\omega_2^2 - \omega_1^2)(1 - a + a\omega_2^2 - \omega_1^2\omega_2^2).\end{aligned} \tag{2.127}$$

The new angle–action variables are

$$\begin{aligned} w_k &= \omega_k t - \alpha_k \\ J_k &= -i|c_k|^2 \varrho_{N+k,k} \end{aligned} \quad (k = 1, 2, \cdots, N), \tag{2.128}$$

where $c_k = |c_k|e^{i\alpha_k}$ are the integration constants.

3
Classical Perturbation Theories

3.1 The Problem of Delaunay

Many general perturbation theories devised since the nineteenth century were founded on the powerful tools of Hamiltonian mechanics. They aimed at solving the specific problem of finding the solutions of the canonical system of $2N$ differential equations

$$\frac{dq_i}{dt} = \frac{\partial H}{\partial p_i}, \qquad \frac{dp_i}{dt} = -\frac{\partial H}{\partial q_i}, \tag{3.1}$$

$(i = 1, 2, \cdots, N)$, where the Hamiltonian H is the energy of the system and may be written as

$$H = \mathcal{H}_0(q, p) + \mathcal{R}(q, p, \varepsilon) \tag{3.2}$$

with $q \equiv (q_1, \cdots, q_N)$, $p \equiv (p_1, \cdots, p_N)$. $\mathcal{H}_0$ is the Hamiltonian of a separable system and $\mathcal{R}$ is a disturbing potential, analytical in some small parameter ε, and vanishing for $\varepsilon = 0$.

Since $\mathcal{H}_0$ is the Hamiltonian of a separable system, we may choose, as new variables, the angle–action variables (θ_i, J_i) associated with it[1]. The Hamiltonian system thus becomes

$$\frac{d\theta_i}{dt} = \frac{\partial H}{\partial J_i}, \qquad \frac{dJ_i}{dt} = -\frac{\partial H}{\partial \theta_i}, \tag{3.3}$$

where, now,

$$H = H_0(J) + R(\theta, J, \varepsilon) \tag{3.4}$$

is a smooth function in $\mathbf{T}^N \times \mathcal{O} \times I$ ($\mathcal{O}$ is an open set of $\mathbf{R}^N$ and $I \subset \mathbf{R}$).

---

[1] Generally, we use θ_i to denote the given angle variables (instead of w_i) to preserve the usual notations of perturbation theories. Also, to be consistent with the usual coordinate–momentum order, we adopt an angle–action order in all functions of these variables and in the equations of motion.

The characteristic properties of the new system are:

(a.) H_0 does not depend on the angle variables θ_i;
(b.) R is a 2π-periodic function of the angle variables;
(c.) H_0 depends only on actions. When H_0 has essential degeneracies, we may suppose that the corresponding actions were eliminated by means of a Schwarzschild transformation and that H_0 depends only on the non-degenerate actions J_μ ($\mu = 1, 2, \cdots, M$) ($M \leq N$).

Since introduced by Delaunay in his celebrated *Théorie de la Lune* [22], canonical perturbation theories have been constructed in agreement with the following scheme:

$$
\begin{array}{ccc}
& \phi & \\
\boxed{H(\theta, J)} & \longrightarrow & \boxed{H^*(J^*)} \\
\Downarrow\Downarrow & & \downarrow\downarrow \\
\boxed{\begin{array}{l}\theta = \theta(t) \\ J = J(t)\end{array}} & \xleftarrow{\phi^{-1}} & \boxed{\begin{array}{l}\theta^* = \nu^* t + \text{const} \\ J^* = \text{const}\end{array}}
\end{array}
$$

One seeks a canonical transformation $\phi : \mathbf{T}^N \times \mathbf{R}^N \Rightarrow \mathbf{T}^N \times \mathbf{R}^N$ such that the transformed Hamiltonian does not depend on the new angle variables θ^* and, thus, the canonical system that it defines may be trivially solved. After integration of the resulting equations, the inverse transformation allows us to change the solution thus obtained into the solution of the given problem.

In reality, the ideal scheme of Delaunay is extremely ambitious and Poincaré pointed out difficulties that makes its application generally impossible. We refer to his theorem on the non-convergence of the infinite series defining the transformation ϕ and the non-existence, in general, of analytic integrals of the equations of the motion, besides the energy integral. Poincaré's proof of this theorem is very illustrative in understanding how the set formed by the initial conditions leading to divergence becomes more densely filled at every step of the construction of H^* (see Sect. 3.10).

The general behavior of the dynamical systems defined by (3.1) is still largely unknown. The theorem of Kolmogorov guarantees the persistence, under small perturbations, of many of the invariant surfaces of the undisturbed integrable system, albeit in distorted form. However, many of them are destroyed, forming sets whose measure increases with the magnitude of the perturbations. Theories founded on this theorem have progressed enormously in the last half-century [19], but the behavior of systems with three or more degrees of freedom is yet, largely, a research subject. Notwithstanding their importance, the theories of Kolmogorov, Arnold and Moser (KAM) will not be treated in this book, except for some generalities concerning Kolmogorov's theorem. A full study of their consequences is beyond the scope of this book and the reader should look for them in the existing literature on Hamiltonian systems.

In spite of the difficulties pointed out by Poincaré and Kolmogorov, the schemes devised by Delaunay and Bohlin are of great practical utility to study the Hamiltonian systems resulting from the perturbation of an integrable system. The approximate solutions that they allow us to construct are valid, for limited time intervals, for initial conditions in relatively large sets. To avoid the above-mentioned convergence problems, perturbation theories are considered, in this book, under a finite formal point of view; the previous ideal scheme must be replaced by

$$\boxed{H(\theta, J)} \quad \xrightarrow{\phi_n} \quad \boxed{H^*(J^*) + \mathcal{R}_{n+1}(\theta^*, J^*, \varepsilon)}$$

$$\Downarrow \qquad\qquad\qquad \downarrow$$
$$\Downarrow \qquad\qquad\qquad \downarrow$$

$$\boxed{\begin{array}{l}\theta = \theta(t) + \mathcal{O}(\varepsilon^{n+1})\\ J = J(t) + \mathcal{O}(\varepsilon^{n+1})\end{array}} \xleftarrow{\phi_n^{-1}} \boxed{\begin{array}{l}\theta^*_{(n)} = \nu^* t + \text{const}\\ J^*_{(n)} = \text{const}\end{array}}$$

Now, the sought canonical transformation ϕ_n is such that the transformed Hamiltonian has a main part $H^*(J^*)$ independent of angle variables, and a remainder, $\mathcal{R}_{n+1}$, divisible by ε^{n+1}. The main part of the Hamiltonian defines an easily integrable dynamical system and the inverse transformation ϕ_n^{-1} transforms its solution into a formal solution *of order n* of the given problem.

It must, however, be emphasized that almost every theory presented in this book may be used to construct convergent solutions if accompanied by the frequency relocation algorithm presented in Sect. 3.11.1, provided that the conditions of Kolmogorov's theorem are satisfied. Frequency relocation is a powerful theoretical tool to avoid the uprise of small divisors, but is not used in the construction of formal low-order theories, because it is work-expensive and does not make low-order solutions significantly more accurate.

3.2 The Poincaré Theory

Under this title, we consider the construction of formal solutions of the canonical equations defined by the Hamiltonian

$$H = H_0(J) + \sum_{k=1}^{\infty} \varepsilon^k H_k(\theta, J), \qquad (3.5)$$

where we assume that H_0 is non-degenerate (see Sect. 2.7). Then, H_0 depends on all actions J_i. This case will often be referred to just as the case $M = N$ (see assumption (c.) in the previous section). In previous drafts of this book, it was called Lindstedt–Poincaré theory, since Poincaré, himself, called it Lindstedt theory ([80], Chap. IX). However, Lindstedt's name is being increasingly used to designate the direct calculation of the series (as done by Lindstedt [64]). The name now adopted avoids confusion with Lindstedt's direct method, and is justified by the fact that the use of Bohlin's ideas [8] to treat the general Delaunay problem is due to Poincaré.

Let us consider the canonical transformation $\phi_n : (\theta, J) \Rightarrow (\theta^*, J^*)$ defined by the equations

$$\theta_i^* = \frac{\partial S}{\partial J_i^*} \qquad J_i = \frac{\partial S}{\partial \theta_i}, \tag{3.6}$$

where the generating function $S = S(\theta, J^*, \varepsilon)$ is assumed to be a polynomial of degree n in ε and such that the transformation defined by (3.6) reduces to the identical transformation when $\varepsilon = 0$. Hence

$$S \stackrel{\text{def}}{=} \sum_{i=1}^{N} \theta_i J_i^* + \sum_{k=1}^{n} \varepsilon^k S_k(\theta, J^*) \tag{3.7}$$

and

$$\theta_i^* = \theta_i + \sum_{k=1}^{n} \varepsilon^k \frac{\partial S_k}{\partial J_i^*}, \qquad J_i = J_i^* + \sum_{k=1}^{n} \varepsilon^k \frac{\partial S_k}{\partial \theta_i}. \tag{3.8}$$

Since the time-dependent cases may be properly considered in the extended phase space, we may assume that $S_k(\theta, J^*)$, as well as $H_k(\theta, J)$ and the transformed Hamiltonian $H^*(J^*)$, does not depend on the independent variable t; the transformation defined by (3.6) is therefore conservative and we may write the conservation equation

$$H(\theta, J) = H^*(J^*) + \mathcal{R}_{n+1}(\theta^*, J^*, \varepsilon), \tag{3.9}$$

or, taking the transformation into consideration,

$$H\left(\theta, \frac{\partial S}{\partial \theta}\right) = H^*(J^*) + \mathcal{R}_{n+1}\left(\frac{\partial S}{\partial J^*}, J^*, \varepsilon\right). \tag{3.10}$$

The equations of Poincaré theory are obtained by substituting into (3.10) limited expansions of the functions H, H^* and S:

$$\begin{aligned} H^* &\stackrel{\text{def}}{=} H_0^* + \varepsilon H_1^* + \varepsilon^2 H_2^* + \cdots + \varepsilon^n H_n^* \\ S &= S_0 + \varepsilon S_1 + \varepsilon^2 S_2 + \cdots + \varepsilon^n S_n; \end{aligned} \tag{3.11}$$

the expansion of the given Hamiltonian H is more complicated because every $H_k(\theta, \partial S/\partial \theta)$ includes terms of several orders arising from S. The introduction of these terms is necessary since the identification in ε must be done using the variables θ, J^*. Those functions must be expanded beforehand, as described below.

3.2.1 Expansion of H_0

Following Taylor's theorem, we have

$$H_0 = H_0\left(\frac{\partial S}{\partial \theta}\right)_{\varepsilon=0} + \sum_{k=1}^{\infty} \frac{\varepsilon^k}{k!}\left[\frac{\mathrm{d}^k}{\mathrm{d}\varepsilon^k}H_0\left(\frac{\partial S}{\partial \theta}\right)\right]_{\varepsilon=0}, \qquad (3.12)$$

which may be written as

$$H_0 = G_{0,0} + \varepsilon G_{0,1} + \varepsilon^2 G_{0,2} + \cdots. \qquad (3.13)$$

The components $G_{0,k}$ are given by

$$G_{0,0} = H_0(J^*) \qquad (3.14)$$

and

$$G_{0,k} = \sum_{i=1}^{N} \nu_i^* \frac{\partial S_k}{\partial \theta_i} + \mathcal{E}_k, \qquad (3.15)$$

where

$$\nu_i^* \stackrel{\text{def}}{=} \frac{\partial H_0(J^*)}{\partial J_i^*}. \qquad (3.16)$$

The quantity J^* is introduced in these equations through

$$\Phi(J_i)_{\varepsilon=0} = \Phi\left(J_i^* + \sum_{k=1}^{n} \varepsilon^k \frac{\partial S_k}{\partial \theta_i}\right)_{\varepsilon=0} = \Phi(J_i^*)$$

for all considered functions $\Phi(J_i)$. The $\mathcal{E}_k$ are known functions of $S_1, S_2, \cdots, S_{k-1}$. In particular, we have

$$\mathcal{E}_1 = 0,$$

$$\mathcal{E}_2 = \frac{1}{2}\sum_{i=1}^{N}\sum_{j=1}^{N} \nu_{ij}^* \frac{\partial S_1}{\partial \theta_i}\frac{\partial S_1}{\partial \theta_j}, \qquad (3.17)$$

$$\mathcal{E}_3 = \sum_{i=1}^{N}\sum_{j=1}^{N} \nu_{ij}^* \frac{\partial S_1}{\partial \theta_i}\frac{\partial S_2}{\partial \theta_j} + \frac{1}{6}\sum_{i=1}^{N}\sum_{j=1}^{N}\sum_{\ell=1}^{N} \nu_{ij\ell}^* \frac{\partial S_1}{\partial \theta_i}\frac{\partial S_1}{\partial \theta_j}\frac{\partial S_1}{\partial \theta_\ell}, \qquad (3.18)$$

$$\mathcal{E}_4 = \sum_{i=1}^{N}\sum_{j=1}^{N} \nu_{ij}^* \frac{\partial S_1}{\partial \theta_i}\frac{\partial S_3}{\partial \theta_j} + \frac{1}{2}\sum_{i=1}^{N}\sum_{j=1}^{N} \nu_{ij}^* \frac{\partial S_2}{\partial \theta_i}\frac{\partial S_2}{\partial \theta_j}$$

$$+ \frac{1}{2}\sum_{i=1}^{N}\sum_{j=1}^{N}\sum_{\ell=1}^{N} \nu_{ij\ell}^* \frac{\partial S_1}{\partial \theta_i}\frac{\partial S_1}{\partial \theta_j}\frac{\partial S_2}{\partial \theta_\ell}$$

$$+ \frac{1}{24}\sum_{i=1}^{N}\sum_{j=1}^{N}\sum_{\ell=1}^{N}\sum_{m=1}^{N} \nu_{ij\ell m}^* \frac{\partial S_1}{\partial \theta_i}\frac{\partial S_1}{\partial \theta_j}\frac{\partial S_1}{\partial \theta_\ell}\frac{\partial S_1}{\partial \theta_m}, \qquad (3.19)$$

where we have introduced

$$\nu_{ij}^* \stackrel{\text{def}}{=} \frac{\partial^2 H_0(J^*)}{\partial J_i^* \partial J_j^*},$$

$$\nu_{ij\ell}^* \stackrel{\text{def}}{=} \frac{\partial^3 H_0(J^*)}{\partial J_i^* \partial J_j^* \partial J_\ell^*},$$

$$\nu_{ij\ell m}^* \stackrel{\text{def}}{=} \frac{\partial^4 H_0(J^*)}{\partial J_i^* \partial J_j^* \partial J_\ell^* \partial J_m^*}.$$

For the sake of future utilization of the above expansions, it is worth noting that, for $k \geq 3$, they may be written as

$$\mathcal{E}_k = \sum_{i=1}^{N} \sum_{j=1}^{N} \nu_{ij}^* \frac{\partial S_1}{\partial \theta_i} \frac{\partial S_{k-1}}{\partial \theta_j} + \mathcal{E}_k', \tag{3.20}$$

where $\mathcal{E}_k'$ represents a function of the derivatives of $S_1, S_2, \cdots, S_{k-2}$ (independent of S_{k-1} and S_k).

3.2.2 Expansion of H_k

In the same way as before, we write

$$H_k(\theta_i, \frac{\partial S}{\partial \theta_i}) = G_{k,k} + \varepsilon G_{k,k+1} + \varepsilon^2 G_{k,k+2} + \cdots, \tag{3.21}$$

where the components $G_{k,k+k'}$ are to be calculated by means of Taylor expansions. Following the same steps as for H_0, we obtain

$$G_{k,k} = H_k(\theta, J^*),$$

$$G_{k,k+1} = \sum_{i=1}^{N} \frac{\partial H_k(\theta, J^*)}{\partial J_i^*} \frac{\partial S_1}{\partial \theta_i}, \tag{3.22}$$

$$G_{k,k+2} = \sum_{i=1}^{N} \frac{\partial H_k(\theta, J^*)}{\partial J_i^*} \frac{\partial S_2}{\partial \theta_i} + \frac{1}{2} \sum_{i=1}^{N} \sum_{j=1}^{N} \frac{\partial^2 H_k(\theta, J^*)}{\partial J_i^* \partial J_j^*} \frac{\partial S_1}{\partial \theta_i} \frac{\partial S_1}{\partial \theta_j},$$

$$G_{k,k+3} = \sum_{i=1}^{N} \frac{\partial H_k(\theta, J^*)}{\partial J_i^*} \frac{\partial S_3}{\partial \theta_i} + \sum_{i=1}^{N} \sum_{j=1}^{N} \frac{\partial^2 H_k(\theta, J^*)}{\partial J_i^* \partial J_j^*} \frac{\partial S_1}{\partial \theta_i} \frac{\partial S_2}{\partial \theta_j}$$

$$+ \frac{1}{6} \sum_{i=1}^{N} \sum_{j=1}^{N} \sum_{\ell=1}^{N} \frac{\partial^2 H_k(\theta, J^*)}{\partial J_i^* \partial J_j^* \partial J_\ell^*} \frac{\partial S_1}{\partial \theta_i} \frac{\partial S_1}{\partial \theta_j} \frac{\partial S_1}{\partial \theta_\ell}.$$

3.2.3 Perturbation Equations

The next step is to introduce the above expansions in (3.10) and to identify all terms multiplying the same power of ε. For this task, we are helped by the following notation: the subscripts in $S_k, \mathcal{E}_k, H_k, H_k^*$ and the second subscript in the $G_{k',k}$ indicate the power of ε multiplying it in the complete equation, Thus, it follows that

$$H_0 = H_0^*,$$

$$\sum_{i=1}^{N} \nu_i^* \frac{\partial S_1}{\partial \theta_i} + H_1 = H_1^*,$$

$$\sum_{i=1}^{N} \nu_i^* \frac{\partial S_2}{\partial \theta_i} + G_{1,2} + H_2 + \mathcal{E}_2 = H_2^*, \quad (3.23)$$

$$\cdots\cdots$$

$$\sum_{i=1}^{N} \nu_i^* \frac{\partial S_k}{\partial \theta_i} + G_{1,k} + \cdots + G_{k-1,k} + H_k + \mathcal{E}_k = H_k^*,$$

$$\cdots\cdots$$

$$\sum_{i=1}^{N} \nu_i^* \frac{\partial S_n}{\partial \theta_i} + G_{1,n} + \cdots + G_{n-1,n} + H_n + \mathcal{E}_n = H_n^*;$$

the remaining terms have at least ε^{n+1} as a factor and are supposed to be grouped in the remainder $\mathcal{R}_{n+1}$. In these equations, all functions are calculated at the point (θ, J^*) [2].

The first of equations (3.23) gives H_0^* and means that H_0^* is the same function as H_0.

The other equations are first-order linear partial differential equations in the unknown functions $S_k(\theta)$. The generic or *homological* equation is

$$\sum_{i=1}^{N} \nu_i^* \frac{\partial S_k}{\partial \theta_i} = H_k^*(J^*) - \Psi_k(\theta, J^*), \quad (3.24)$$

---

[2] There is one question about notation that, although trivial, must be recalled to avoid possible misinterpretations. It is usual, in many chapters of Physics and Astronomy texts, to represent a given magnitude by the same notation no matter which independent variables are used in its definition. For instance, the energy of a perfect gas in a vessel is $\mathcal{U}$ no matter whether it is given as a function of the temperature or of the pressure. This was also done in previous chapters of this book. In the formulation of canonical perturbation theories, however, we must adopt strict rules: Every function symbol ϕ represents only one function: $\phi(x)$ and $\phi(y)$ indicate the same function ϕ calculated at the points x and y of its domain of definition.

where
$$\Psi_k = G_{1,k} + \cdots + G_{k-1,k} + H_k + \mathcal{E}_k.$$

We have to note that the functions $G_{k',k}$ are defined in such a way that they become known when the functions $S_1, \cdots, S_{k-k'}$ are known. The functions H_k are given functions (see 3.5) and the functions $\mathcal{E}_k$ are also known when $S_1, \cdots, S_{k-1}$ are known. Thus, the function Ψ_k is a completely known function provided that the equations corresponding to $S_1, \cdots, S_{k-1}$ were already solved; thus, the whole set of equations may be sequentially solved.

The homological equation is indeterminate since H_k^* is also unknown. The adopted choices for its solution are discussed in the next section. When it is solved for all $k \leq n$, we obtain the functions $S_k(\theta, J^*)$ and $H_k^*(J^*)$ and may perform the elementary operations leading to

$$\begin{aligned} \theta_{(n)}^* &= \nu_{(n)}^* t + \text{const} \\ J_{(n)}^* &= \text{const}, \end{aligned} \qquad (3.25)$$

which, through the inverse transformation ϕ_n^{-1}, lead to the formal solution of order n of the given Hamiltonian system.

3.3 Averaging Rule

To overcome the indetermination of the homological equation (3.24), we have to fix one of the two unknown functions. The main idea of canonical perturbation theories is that the canonical transformation performs an averaging and the resulting Hamiltonian has no periodic components. We thus adopt the following rule:

$$H_k^*(J^*) = <\Psi_k(\theta, J^*)> = \left(\frac{1}{2\pi}\right)^N \int_0^{2\pi} \cdots \int_0^{2\pi} \Psi_k d\theta_1 \cdots d\theta_N. \qquad (3.26)$$

Therefore, the homological partial differential equation becomes

$$\sum_{i=1}^N \nu_i^* \frac{\partial S_k}{\partial \theta_i} = <\Psi_k(\theta, J^*)> - \Psi_k(\theta, J^*) \qquad (3.27)$$

in which all terms on the right-hand side are periodic.

The averaging operation defined by (3.26) is such that all terms independent of θ_i ($i = 1, 2, \cdots, N$) are included in H_k^* and are absent from the right-hand side of (3.27). It is worth noting that, if non-periodic terms of this kind were allowed to remain in the right-hand side of the partial differential equation, they would appear in the solution S_k multiplied by a linear combination of the θ_i. As a consequence, the transformation ϕ_n would also include such linear combination as a factor and new and old variables would depart of each other with the speed of this combination.

It is of the utmost importance to emphasize that this averaging operation is not just the *scissors* averaging found in some applications, which consists of imposing

$$H_k^* = \left(\frac{1}{2\pi}\right)^N \int_0^{2\pi} \cdots \int_0^{2\pi} H_k d\theta_1 \cdots d\theta_N. \tag{3.28}$$

Scissors averaging and the one defined by (3.26) coincide only for $k = 1$. The so-called *averaging principle* defined by (3.28) has been critically considered by Arnold ([4], Chap. 10, Sect. 52B). I quote his comments: *"this principle is neither a theorem, an axiom, nor a definition; it is [$\cdots$] a vaguely formulated and, rigorously speaking, wrong proposition"*.

At variance with incomplete scissors techniques, the averaging defined by (3.26) is not based on any *principle* and aims only at giving a rule for the choice of the undetermined functions H_k^*. Such freedom of choice is allowed by the fact that the given recurrent partial differential equations are indeterminate and that it is necessary to fix one of the two unknown functions S_k and H_k^* to proceed.

3.3.1 Small Divisors. Non-Resonance Condition

The above functions Ψ_k generally have the form of truncated Fourier series:

$$\Psi_k = \sum_{h \in \mathbf{D}_k \subset \mathbf{Z}^N} A_{kh}(J^*) \exp\left(\mathrm{i} h \mid \theta\right). \tag{3.29}$$

The averaging operation leads to

$$H_k^* = A_{k0}(J^*)$$

and

$$\sum_{i=1}^N \nu_i^* \frac{\partial S_k}{\partial \theta_i} = -\sum_{h \in \mathbf{D}_k \setminus \{0\}} A_{kh} \exp\left(\mathrm{i} h \mid \theta\right). \tag{3.30}$$

The last equation admits the particular solution

$$S_k(\theta, J^*) = \sum_{h \in \mathbf{D}_k \setminus \{0\}} \frac{\mathrm{i} A_{kh} \exp\left(\mathrm{i} h \mid \theta\right)}{h \mid \nu^*}, \tag{3.31}$$

which introduces the divisors $(h \mid \nu^*)$. This is a common feature in perturbation theory and is the way in which small divisors may appear. Therefore, it is only valid if the following non-resonance condition is assumed.

Non-Resonance Condition. *The condition for the non-existence of small divisors is*

$$(h \mid \nu^*) \neq 0$$

for all vectors $h \in \mathbf{D}_k$ ($k \leq n$).

3.4 Degenerate Systems. The von Zeipel–Brouwer Theory

Poincaré theory cannot be used when, for $\varepsilon = 0$, the Hamiltonian is degenerate (in Schwarzschild's sense), that is, when H_0 does not depend on all actions J_i. Poincaré tried to overcome the difficulty represented, in this case, by the many identically null divisors of (3.31) through the sum of some arbitrary functions of θ to the solutions S_k. These arbitrary functions were later determined in such a way that the difficulties were transferred to higher orders and, in some cases, eventually eliminated. An improved theory, due to von Zeipel [96], was successfully used by Dirk Brouwer [14] to construct his solution of the equations of motion of an artificial Earth satellite.

Let us consider a Hamiltonian system with M non-degenerate and $N - M$ degenerate degrees of freedom. A Schwarzschild transformation allows it to be written as $H_0 = H_0(J_\mu)$, where J_μ ($\mu = 1, \cdots, M < N$) are the actions corresponding to the non-degenerate degrees of freedom. The actions J_ϱ ($\varrho = M+1, \cdots, N$) corresponding to degenerate degrees of freedom are absent from H_0; as a consequence, the undisturbed frequencies $\nu_\varrho = \partial H_0/\partial J_\varrho$ are identically equal to zero. The algorithm proposed by von Zeipel to deal with this case introduces an essential modification in the scheme of the Delaunay problem of Sect. 3.1. Now, a canonical transformation is sought such that the transformed Hamiltonian has a main part $H^*(\theta_\varrho^*, J^*)$ independent of the non-degenerate angles $\theta_1^*, \cdots, \theta_M^*$, but depending on the degenerate angles $\theta_{M+1}^*, \cdots, \theta_N^*$. That is,

$$\boxed{H(\theta, J)} \xrightarrow{\phi_n} \boxed{H^*(\theta_\varrho^*, J^*) + \mathcal{R}_{n+1}(\theta^*, J^*, \varepsilon).}$$

The main part of the new Hamiltonian defines a canonical system that may be reduced to M integrals, M quadratures, and a reduced Hamiltonian system with $N-M$ degrees of freedom. Indeed, the system defined by H^* is separated into two parts corresponding to the subscripts $\mu = 1, \cdots, M$ and $\varrho = M+1, \cdots, N$, respectively:

$$\dot{\theta}_\mu^* = \frac{\partial H^*}{\partial J_\mu^*} \qquad \dot{J}_\mu^* = 0$$
$$\dot{\theta}_\varrho^* = \frac{\partial H^*}{\partial J_\varrho^*} \qquad \dot{J}_\varrho^* = -\frac{\partial H^*}{\partial \theta_\varrho^*}. \tag{3.32}$$

The corresponding results are the M integrals

$$J_\mu^* = \text{const},$$

the canonical system of $N-M$ degrees of freedom given by (3.32), and M separated equations for $\dot{\theta}_\mu^*$ that may be solved by quadrature after the integration of the reduced canonical system.

3.4 Degenerate Systems. The von Zeipel–Brouwer Theory

Thus, the von Zeipel–Brouwer theory is not a theory seeking the formal solution of the given problem; it only leads to a reduction of the number of its degrees of freedom. In some favorable cases, the successive application of the theory may reduce the number of degrees of freedom to zero or to one, and thus the problem is solved (see Sect. 3.9). As an example, we mention Brouwer's original application of the theory [14]. There, a first operation reduced the number of degrees of freedom by only one unit, but a second one led to the complete solution of the problem.

The reduction of the number of degrees of freedom and the simplification of the equations due to the averaging often allows an easier analysis of problems for which a complete solution is not possible.

To obtain the implicit equations of the von Zeipel–Brouwer theory, we consider the canonical transformation $\phi_n : (\theta, J) \Rightarrow (\theta^*, J^*)$ defined by the equations

$$J_i = \frac{\partial S}{\partial \theta_i} \qquad \theta_i^* = \frac{\partial S}{\partial J_i^*} \tag{3.33}$$

with the generating function

$$S = \sum_{i=1}^{N} \theta_i J_i^* + \sum_{k=1}^{n} \varepsilon^k S_k(\theta, J^*). \tag{3.34}$$

Since the transformation is conservative, we have

$$H(\theta, J) = H^*(\theta_\varrho^*, J^*) + \mathcal{R}_{n+1}(\theta^*, J^*, \varepsilon), \tag{3.35}$$

or, taking (3.33) into account,

$$H\left(\theta, \frac{\partial S}{\partial \theta}\right) = H^*\left(\frac{\partial S}{\partial J_\varrho^*}, J^*\right) + \mathcal{R}_{n+1}\left(\frac{\partial S}{\partial J^*}, J^*, \varepsilon\right). \tag{3.36}$$

To identify both sides of (3.36), we use the same expansions of H and S already used in the Poincaré theory. However, we have to consider that H^* now depends also on some angles and, thus, assumptions similar to those made to obtain (3.23) are not sufficient. Indeed, when we assume

$$H^*(\theta_\varrho^*, J^*) = \sum_{k=0}^{n} \varepsilon^k H_k^*(\theta_\varrho^*, J^*) \tag{3.37}$$

we must take into account that every H_k^* ($k \neq 0$) depends also on ε through $\theta_\varrho^* = \partial S/\partial J_\varrho^*$ and, thus, we also need to consider the Taylor expansion of these terms.

It is worthwhile mentioning that the first accounts of this theory, following its successful application by Brouwer, missed the fact that, in more general situations, the functions H_k^* depend also on θ_ϱ and thus contribute to the formation of the terms in the von Zeipel–Brouwer perturbation equations of orders higher than k.

3.4.1 Expansion of H^*

Let us write

$$H^*\left(\frac{\partial S}{\partial J_\varrho^*}, J^*\right) \stackrel{\text{def}}{=} H_0^*(J^*) + \sum_{k=1}^n \varepsilon^k [H_k^*(\theta_\varrho, J^*) + G_k'^*(\theta_\varrho, J^*)], \quad (3.38)$$

where the functions $G_k'^*$ are easily obtained by writing down the Taylor expansion explicitly:

$$G_1'^* = 0,$$

$$G_2'^* = \sum_{\varrho=M+1}^{N} \frac{\partial H_1^*}{\partial \theta_\varrho} \frac{\partial S_1}{\partial J_\varrho^*}, \quad (3.39)$$

$$G_3'^* = \sum_{\varrho=M+1}^{N} \left(\frac{\partial H_1^*}{\partial \theta_\varrho} \frac{\partial S_2}{\partial J_\varrho^*} + \frac{1}{2} \sum_{\varrho'=M+1}^{N} \frac{\partial^2 H_1^*}{\partial \theta_\varrho \partial \theta_{\varrho'}} \frac{\partial S_1}{\partial J_\varrho^*} \frac{\partial S_1}{\partial J_{\varrho'}^*} + \frac{\partial H_2^*}{\partial \theta_\varrho} \frac{\partial S_1}{\partial J_\varrho^*} \right).$$

3.4.2 von Zeipel–Brouwer Perturbation Equations

When the functions in (3.36) are replaced by their expansions, and all terms that multiply the same power of ε are identified, we obtain

$$H_0 = H_0^*,$$

$$\sum_{\mu=1}^M \nu_\mu^* \frac{\partial S_1}{\partial \theta_\mu} + H_1 = H_1^*,$$

$$\sum_{\mu=1}^M \nu_\mu^* \frac{\partial S_2}{\partial \theta_\mu} + G_{1,2} + H_2 + \mathcal{E}_2 = H_2^* + G_2'^*, \quad (3.40)$$

$$\sum_{\mu=1}^M \nu_\mu^* \frac{\partial S_3}{\partial \theta_\mu} + G_{1,3} + G_{2,3} + H_3 + \mathcal{E}_3 = H_3^* + G_3'^*,$$

$$\cdots\cdots$$

$$\sum_{\mu=1}^M \nu_\mu^* \frac{\partial S_k}{\partial \theta_\mu} + G_{1,k} + \cdots + G_{k-1,k} + H_k + \mathcal{E}_k = H_k^* + G_k'^*.$$

$$\cdots\cdots$$

$$\sum_{\mu=1}^M \nu_\mu^* \frac{\partial S_n}{\partial \theta_\mu} + G_{1,n} + \cdots + G_{n-1,n} + H_n + \mathcal{E}_n = H_n^* + G_n'^*.$$

The remaining terms have at least ε^{n+1} as a factor and are supposed to be grouped in the remainder $\mathcal{R}_{n+1}$.

3.4 Degenerate Systems. The von Zeipel–Brouwer Theory   73

The functions $\mathcal{E}_k$ and $G_{k,k'}$ are the same as those defined in Sect. 3.2. However, since H_0 now depends only on $J_1, J_2, \cdots, J_M$, the summations in (3.17)–(3.22) are restricted to the subscripts $1, 2, \cdots, M$. In particular, (3.20) becomes

$$\mathcal{E}_k = \sum_{\mu=1}^{M} \sum_{\mu'=1}^{M} \nu^*_{\mu\mu'} \frac{\partial S_1}{\partial \theta_\mu} \frac{\partial S_{k-1}}{\partial \theta_{\mu'}} + \mathcal{E}'_k \quad (k \geq 3), \qquad (3.41)$$

where $\mathcal{E}'_k$ is a function of the derivatives of the functions $S_1, \cdots, S_{k-2}$ with respect to $\theta_1, \cdots, \theta_M$ (note that $\mathcal{E}'_2 = 0$).

As before, the first von Zeipel–Brouwer perturbation equation gives H_0^* and means that H_0^* is the same function as H_0, where we have just replaced the J_μ by J_μ^*. The other equations are the homological first-order linear partial differential equations giving $S_k(\theta)$:

$$\sum_{\mu=1}^{M} \nu^*_\mu \frac{\partial S_k}{\partial \theta_\mu} = -\Psi_k(\theta, J^*) + H_k^*(\theta_\varrho, J^*), \qquad (3.42)$$

where the functions Ψ_k are, now,

$$\Psi_k = G_{1,k} + \cdots + G_{k-1,k} + H_k + \mathcal{E}_k - G'^*_k, \qquad (3.43)$$

and are completely known if the functions $S_1, \cdots, S_{k-1}$ and $H_1^*, \cdots, H_{k-1}^*$ are known.

3.4.3 The von Zeipel Averaging Rule

To overcome the indetermination of (3.40), we have to fix H_k^*. The averaging rule used in Poincaré theory needs a modification to be applied in this case. Indeed, if we intend to avoid identically null divisors in (3.31), we need to exclude from the summation all degenerate terms, that is, terms for which the first M components of $h \in \mathbf{Z}^N$ are zero. In such terms, $(h \mid \nu^*) \equiv 0$ because $\nu^*_\varrho \equiv 0$ for $\varrho = M+1, \cdots, N$. The von Zeipel averaging rule is, then,

$$H_k^*(\theta_\varrho, J^*) = <\Psi_k(\theta, J^*)>, \qquad (3.44)$$

where $< \cdots >$ stands, now, for the average over the angles θ_μ ($\mu = 1, \cdots, M$) only, on $[0, 2\pi]$. The angles θ_ϱ ($\varrho = M+1, \cdots, N$) are not included in the averaging. Therefore, we have

$$H_k^*(\theta_\varrho, J^*) = \Psi_{k(S)}(J^*) + \Psi_{k(LP)}(\theta_\varrho, J^*) \qquad (3.45)$$

and

$$\sum_{\mu=1}^{M} \nu^*_\mu \frac{\partial S_k}{\partial \theta_\mu} = -\Psi_{k(SP)}(\theta, J^*), \qquad (3.46)$$

where the subscripts S, LP, SP stand for different parts in the Fourier expansion of Ψ_k, as follows:

- $\Psi_{k(S)}(J^*)$ – *Secular* part of Ψ_k. This is the average of Ψ_k over *all* angles.
- $\Psi_{k(LP)}(\theta_\varrho, J^*)$ – *Long-period* part of Ψ_k. This is the collection of all periodic terms of Ψ_k independent of the fast angles θ_μ ($\mu = 1, \cdots, M$).
- $\Psi_{k(SP)}(\theta, J^*)$ – *Short-period* part of Ψ_k. This is the collection of all periodic terms of Ψ_k dependent on at least one of the fast angles θ_μ ($\mu = 1, \cdots, M$).

The solution of the equations follows closely the same steps as the solution of the equations of the Poincaré theory.

3.5 Small Divisors and Resonance

When $\Psi_{k(SP)}$ is replaced by its Fourier expansion, (3.46) becomes

$$\sum_{\mu=1}^{M} \nu_\mu^* \frac{\partial S_k}{\partial \theta_\mu} = - \sum_{h \in \mathbf{D}_{k(SP)}} A_{kh}(J^*) \cos(h \mid \theta), \qquad (3.47)$$

where $\mathbf{D}_{k(SP)} \subset \mathbf{Z}^N$ is a set of vectors of N integer components with at least one of the M first components different from zero. Equation (3.47) has the particular solution

$$S_k(\theta, J^*) = - \sum_{h \in \mathbf{D}_{k(SP)}} \frac{A_{kh}(J^*) \sin(h \mid \theta)}{(h \mid \nu^*)}. \qquad (3.48)$$

This solution introduces the divisors $(h \mid \nu^*)$. This is a common feature in perturbation theory and is the way in which small divisors, which impair the convergence of the solution, appear in the process of its construction. When some of the $(h \mid \nu^*)$ become nearly zero, the von Zeipel–Brouwer theory fails and, in such cases, different procedures must be adopted. (See Chap. 4.)

The non-resonance condition in this case is:

Non-Resonance Condition. *The condition for the non-existence of small divisors is*

$$(h \mid \nu^*) \neq 0$$

for all vectors $h \in \mathbf{D}_{k(SP)}$ ($k \leq n$).

3.5.1 Elimination of the Non-Critical Short-Period Angles

When the non-resonance condition is not satisfied, the terms with angle combinations leading to small divisors (*critical terms*) can no longer be eliminated using the theories of the previous sections. To study these cases, let us assume that there are L ($L \leq M$) independent commensurability relations

$$(h_\ell \mid \nu^*) = 0 \qquad (3.49)$$

nearly satisfied, simultaneously, by the frequencies ν_μ^*.

The von Zeipel–Brouwer theory may still be used to eliminate the non-critical periodic terms, but the scheme and the averaging operation need to be modified. Instead of those previously adopted, we adopt, in this case, the scheme

$$\boxed{H(\theta, J)} \xrightarrow{\phi_n} \boxed{H^*(h_\ell|\theta^*, \theta_\varrho^*, J^*) + \mathcal{R}_{n+1}(\theta^*, J^*, \varepsilon)}$$

and the averaging rule

$$H_k^*(h_\ell|\theta, \theta_\varrho, J^*) = <\Psi_k(\theta, J^*)>, \qquad (3.50)$$

where $< \cdots >$ stands for the average over the angles θ_μ ($\mu = 1, \cdots, M$) on $[0, 2\pi]$ but, now, only when they are not in a critical combination. In this case, the canonical transformation is sought in such a way that the transformed Hamiltonian has a main part $H^*(h_\ell|\theta^*, \theta_\varrho^*, J^*)$, independent only of the angles conjugate to the actions $J_1^*, \cdots, J_M^*$ which do not reduce themselves to one of the critical combinations $(h_\ell \mid \theta^*)$.

This problem may be treated in a simple way if we perform, beforehand, a Lagrange point transformation. We introduce a set of N new angles defined by:

$$\begin{aligned} \phi_\ell &= (h_\ell \mid \theta) & \ell &= 1, \cdots, L \\ \phi_{\ell'} &= (h_{\ell'} \mid \theta) & \ell' &= L+1, \cdots, M \\ \phi_\varrho &= \theta_\varrho & \varrho &= M+1, \cdots, N, \end{aligned} \qquad (3.51)$$

where $(h_\ell \mid \theta)$ are the given L critical angles and $(h_{\ell'} \mid \theta)$ are $M - L$ arbitrary linear combinations of the θ_μ independent of the critical angles $(h_\ell, h_{\ell'} \in \mathbf{Z}^M)$. The change in the actions may be easily obtained by means of the Jacobian canonical condition in the particular form:

$$\sum_{i=1}^{N} J_i \, \delta\theta_i = \sum_{i=1}^{N} I_i \, \delta\phi_i$$

or

$$\sum_{i=1}^{N} J_i \, \delta\theta_i = \sum_{\ell=1}^{L} I_\ell(h_\ell \mid \delta\theta^*) + \sum_{\ell'=L+1}^{M} I_{\ell'}(h_{\ell'} \mid \delta\theta^*) + \sum_{\varrho=M+1}^{N} I_\varrho \, \delta\theta_\varrho^*.$$

The identification of both sides of this equation leads to the linear relations defining the actions J_i as functions of the I_i:

$$\begin{aligned} J_\mu &= \sum_{\lambda=1}^{M} I_\lambda h_{\lambda,\mu} & \mu &= 1, \cdots, M \\ J_\varrho &= I_\varrho, \end{aligned} \qquad (3.52)$$

where $h_{\lambda,\mu}$ are the M integer components of h_λ (for an example, see [48]).

There are $N - M + L$ degenerate degrees of freedom. L of them are accidentally degenerate and $N - M$ are essentially degenerate (see Sect. 2.7). The von Zeipel–Brouwer theory is applied in exactly the same way as before except that, now, the degenerate angles are both the ϕ_ℓ (resonant) and the ϕ_ϱ. Now, the canonical transformation $(\phi, I) \to (\phi^*, I^*)$ eliminates the non-critical short-period angles $\phi_{\ell'}$, but the critical and degenerate angles remain in the transformed Hamiltonian H^*. Every function $\Psi_k(\phi, I^*)$ appearing in the homological equations is now decomposed as

$$\Psi_k^*(\phi, I^*) = \Psi_{k(S)}(I^*) + \Psi_{k(LP)}(\phi_\varrho, I^*) + \Psi_{k(K)}(\phi_\ell, \phi_\varrho, I^*) + \Psi_{k(SP)}(\phi, I^*),$$

where the subscripts S, LP, K, SP stand for different parts in the Fourier expansion of Ψ_k, as follows:

- $\Psi_{k(S)}(I^*)$ – *Secular* part. This is the average of Ψ_k over *all* angles.
- $\Psi_{k(LP)}(\phi_\varrho, I^*)$ – *Long-period* part of Ψ_k. This is the collection of all periodic terms of Ψ_k independent of the angles ϕ_μ ($\mu = 1, \cdots, M$).
- $\Psi_{k(K)}(\phi_\ell, \phi_\varrho, I^*)$ – *Critical* part. This is the collection of all periodic terms of Ψ_k independent of the fast angles $\phi_{\ell'}$ ($\ell' = L+1, \cdots, M$), but depending on at least one of the critical angles ϕ_ℓ ($\ell = 1, \cdots, L$).
- $\Psi_{k(SP)}(\phi, I^*)$ – *Short-period* part of Ψ_k. This is the collection of all terms of Ψ_k dependent on at least one of the fast angles $\phi_{\ell'}$ ($\ell' = L+1, \cdots, M$).

With the new averaging rule and the above decomposition of the functions, the homological equation of von Zeipel–Brouwer theory gives

$$\mathcal{H}_k^*(\phi_\ell, \phi_\varrho, I^*) = \Psi_{k(S)}(I^*) + \Psi_{k(LP)}(\phi_\varrho, I^*) + \Psi_{k(K)}(\phi_\ell, \phi_\varrho, I^*) \qquad (3.53)$$

and

$$\sum_{\mu=1}^{M} \nu_\mu^* \frac{\partial S_k}{\partial \phi_\mu} = -\Psi_{k(SP)}(\phi, I^*). \qquad (3.54)$$

The transformed Hamiltonian is

$$\mathcal{H}^* = \mathcal{H}_0^*(I_\mu^*) + \sum_{k=1}^{n} \varepsilon^k \mathcal{H}_k^*(\phi_\ell^*, \phi_\varrho^*, I^*),$$

independent of the angles $\phi_{\ell'}^*$. Therefore, the $I_{\ell'}^*$ are constants and $\mathcal{H}^*$ is the Hamiltonian of a canonical system with $N - M + L$ degrees of freedom. In the new variables, the commensurability relations given by (3.49) are simply written

$$\nu_\ell^* = 0 \qquad (\ell = 1, 2, \cdots, L),$$

where, now,

$$\nu_\ell^* = \frac{\partial \mathcal{H}_0^*}{\partial I_\ell^*}.$$

This reduced form will be adopted in the study of resonant problems in the forthcoming chapters.

3.6 An Example – Part I

The application of the von Zeipel–Brouwer theory follows straightforwardly from the principles and formulas stated in the preceding sections. An example will serve to summarize the ideas and to make clear the directions for other applications.

We consider the Hamiltonian

$$H = -\frac{1}{2J_1^2} + \nu_2 J_2 + \varepsilon R(\theta, J) \tag{3.55}$$

with

$$R(\theta, J) = \sum_{s=-\infty}^{+\infty} \Big(A_s \cos s(\theta_1 - \theta_2) + B_s \cos[s\theta_1 - (s+1)\theta_2] \Big)$$

$$+ \sum_{s=-\infty}^{+\infty} M_s \sqrt{-J_3} \cos[s\theta_1 - (s+1)\theta_2 + \theta_3]$$

$$+ \sum_{s=-\infty}^{+\infty} L_s \sqrt{-J_3} \cos[s(\theta_1 - \theta_2) + \theta_3],$$

where $A_0 = a(J_1) + b(J_1) J_3$, and $A_s(s \neq 0), B_s, M_s, L_s$ are known functions of J_1 with $A_s = A_{-s}$. ε is a small parameter.

This example is suggested by a classical problem of the Mechanics of the Solar System. H is the Hamiltonian of the elliptic restricted problem of three bodies and governs the motion of an asteroid under the joint action of the Sun and Jupiter, when Jupiter is assumed to move on a fixed Keplerian ellipse around the Sun. θ_1 is the mean longitude of the asteroid, θ_2 is the mean longitude of Jupiter, $\nu_2 = \dot{\theta}_2$, and θ_3 is the longitude of the asteroid perihelion. In terms of the Keplerian elements of the asteroid, the actions are

$$\begin{aligned} J_1 &= L = \sqrt{a}, \\ J_3 &= G - L = J_1(\sqrt{1-e^2} - 1) \end{aligned} \tag{3.56}$$

(see Sect. 1.1; note that $J_3 < 0$). J_2 is the momentum conjugate to the mean longitude of Jupiter. The main axis of the reference system was taken directed to Jupiter's perihelion.

In the function R, we have kept the main parts and some less important ones necessary to make this example more illustrative. The units are the length of the Sun–Jupiter distance, the universal gravitational constant, and the solar mass. ε is the mass of Jupiter[3].

---

[3] Some authors use signs opposite to those in (3.55). However, in this book, H is the energy itself and not its opposite. In the same way, R is not the so-called disturbing function, but the potential of the disturbing force.

To apply the results of the preceding sections, let us first note that, in this example, we have

$$H_0 = -\frac{1}{2J_1^2} + \nu_2 J_2 \tag{3.57}$$

$$\begin{aligned} N &= 3 \\ M &= 2 \\ H_1 &= R(\theta, J) \\ H_k &= 0 \qquad (k \geq 2); \end{aligned} \tag{3.58}$$

also

$$\nu_1^* = \frac{\partial H_0(J^*)}{\partial J_1^*} = \frac{1}{J_1^{*3}}, \qquad \nu_2^* = \frac{\partial H_0(J^*)}{\partial J_2^*} = \nu_2$$

and

$$\nu_{11}^* = \frac{-3}{J_1^{*4}}, \qquad \nu_{12}^* = \nu_{22}^* = 0.$$

(because of the adopted units, $\nu_2 \approx 1$).

We then obtain the two first von Zeipel–Brouwer perturbation equations, which are

$$H_0^* = H_0(J^*) = -\frac{1}{2J_1^{*2}} + \nu_2 J_2^*$$

and

$$\sum_{\mu=1}^{2} \nu_\mu^* \frac{\partial S_1}{\partial \theta_\mu} = H_1^* - R(\theta, J^*).$$

(It is not superfluous to emphasize again that all functions in the Poincaré and in the von Zeipel–Brouwer perturbation equations have θ, J^* as independent variables.) The function $R(\theta, J^*)$ may be decomposed into its secular, long-period and short-period parts:

$$R_{(S)}(J^*) = A_0^* = a(J_1^*) + b(J_1^*)J_3^* \tag{3.59}$$

$$R_{(LP)}(\theta_\varrho, J^*) = L_0^* \sqrt{-J_3^*} \cos\theta_3 \tag{3.60}$$

and

$$\begin{aligned} R_{(SP)}(\theta, J^*) =& \sum_{s \in \mathbf{Z} \setminus \{0\}} \left(A_s^* \cos s(\theta_1 - \theta_2) + L_s^* \sqrt{-J_3^*} \cos\left[s(\theta_1 - \theta_2) + \theta_3\right] \right) \\ &+ \sum_{s \in \mathbf{Z}} B_s^* \cos\left[s\theta_1 - (s+1)\theta_2\right] \\ &+ \sum_{s \in \mathbf{Z}} M_s^* \sqrt{-J_3^*} \cos\left[s\theta_1 - (s+1)\theta_2 + \theta_3\right], \end{aligned} \tag{3.61}$$

where $A_{s(s\neq 0)}^* = A_s(J_1^*), B_s^* = B_s(J_1^*), M_s^* = M_s(J_1^*)$ and $L_s^* = L_s(J_1^*)$. The application of the averaging operation defined by (3.44) gives

$$H_1^*(\theta_3, J^*) = R_{(S)}(J^*) + R_{(LP)}(\theta_3, J^*) \tag{3.62}$$

and
$$\sum_{\mu=1}^{2} \nu_\mu^* \frac{\partial S_1}{\partial \theta_\mu} = -R_{(SP)}(\theta, J^*) \tag{3.63}$$

or, after integration,

$$\begin{aligned}
S_1(\theta, J^*) = &- \sum_{s \in \mathbf{Z} \setminus \{0\}} \left(\frac{A_s^* \sin s(\theta_1 - \theta_2)}{s(\nu_1^* - \nu_2^*)} + \frac{L_s^* \sqrt{-J_3^*} \sin [s(\theta_1 - \theta_2) + \theta_3]}{s(\nu_1^* - \nu_2^*)} \right) \\
&- \sum_{s \in \mathbf{Z}} \frac{B_s^* \sin [s\theta_1 - (s+1)\theta_2]}{s\nu_1^* - (s+1)\nu_2^*} \\
&- \sum_{s \in \mathbf{Z}} \frac{M_s^* \sqrt{-J_3^*} \sin [s\theta_1 - (s+1)\theta_2 + \theta_3]}{s\nu_1^* - (s+1)\nu_2^*}
\end{aligned} \tag{3.64}$$

when we assume that $\nu_1^* - \nu_2^* \neq 0$ and $s\nu_1^* - (s+1)\nu_2^* \neq 0$ for all $s \in \mathbf{Z}$.

The next von Zeipel–Brouwer perturbation equation is

$$\sum_{\mu=1}^{2} \nu_\mu^* \frac{\partial S_2}{\partial \theta_\mu} = -\Psi_2(\theta, J^*) + H_2^*(\theta_3, J^*), \tag{3.65}$$

where

$$\Psi_2 = \sum_{i=1}^{3} \frac{\partial R}{\partial J_i^*} \frac{\partial S_1}{\partial \theta_i} + \frac{1}{2} \nu_{11}^* \left(\frac{\partial S_1}{\partial \theta_i} \right)^2 - \frac{\partial H_1^*}{\partial \theta_3} \frac{\partial S_1}{\partial J_3^*}. \tag{3.66}$$

See (3.43), (3.17), (3.22) and (3.39).

Ψ_2 may be decomposed into its secular, long-period and short-period parts. The first summation in (3.66) gives

$$\sum_{i=1}^{3} \left(\frac{\partial R_{(S)}}{\partial J_i^*} \frac{\partial S_1}{\partial \theta_i} + \frac{\partial R_{(LP)}}{\partial J_i^*} \frac{\partial S_1}{\partial \theta_i} + \frac{\partial R_{(SP)}}{\partial J_i^*} \frac{\partial S_1}{\partial \theta_i} \right).$$

Because of the elementary properties of the product of trigonometric functions, the terms

$$\sum_{i=1}^{3} \left(\frac{\partial R_{(S)}}{\partial J_i^*} \frac{\partial S_1}{\partial \theta_i} + \frac{\partial R_{(LP)}}{\partial J_i^*} \frac{\partial S_1}{\partial \theta_i} \right)$$

are short-periodic, while the summation

$$\sum_{i=1}^{3} \frac{\partial R_{(SP)}}{\partial J_i^*} \frac{\partial S_1}{\partial \theta_i}$$

will contribute secular, long-period and short-period terms. The secular terms arise from the terms of the same (or opposite) arguments in the derivatives of $R_{(SP)}$ and S_1:

$$\left[\sum_{i=1}^{3}\frac{\partial R_{(SP)}}{\partial J_i^*}\frac{\partial S_1}{\partial \theta_i}\right]_{(S)} = -\frac{1}{2}\sum_{s\in\mathbf{Z}\setminus\{0\}}\frac{1}{\nu_1^* - \nu_2^*}\left(2\frac{\partial A_s^*}{\partial J_1^*}A_s^* - \frac{\partial L_s^*}{\partial J_1^*}L_s^*J_3^* - \frac{L_s^{*2}}{2s}\right)$$

$$-\frac{1}{2}\sum_{s\in\mathbf{Z}}\frac{\left(\frac{\partial B_s^*}{\partial J_1^*}sB_s^* - \frac{\partial M_s^*}{\partial J_1^*}sM_s^*J_3^* - \frac{M_s^{*2}}{2}\right)}{s\nu_1^* - (s+1)\nu_2^*}; \quad (3.67)$$

in the same way, the long-period terms arise from the products of terms whose arguments differ by a multiple of θ_3:

$$\left[\sum_{i=1}^{3}\frac{\partial R_{(SP)}}{\partial J_i^*}\frac{\partial S_1}{\partial \theta_i}\right]_{(LP)} = -\frac{1}{2}\sum_{s\in\mathbf{Z}\setminus\{0\}}\frac{2\sqrt{-J_3^*}}{\nu_1^* - \nu_2^*}\frac{\partial(A_s^*L_s^*)}{\partial J_1^*}\cos\theta_3$$

$$-\frac{1}{2}\sum_{s\in\mathbf{Z}\setminus\{0\}}\frac{s\sqrt{-J_3^*}}{s\nu_1^* - (s+1)\nu_2^*}\frac{\partial(B_s^*M_s^*)}{\partial J_1^*}\cos\theta_3$$

$$+\frac{1}{2}\sum_{s\in\mathbf{Z}\setminus\{0\}}\frac{\partial L_s^*}{\partial J_1^*}\frac{L_{-s}^*J_3^*}{(\nu_1 - \nu_2)}\cos 2\theta_3 \quad (3.68)$$

We may calculate, in a similar way, the secular and long-period parts of the second term of (3.66) obtaining

$$\left[\frac{1}{2}\nu_{11}^*\left(\frac{\partial S_1}{\partial \theta_i}\right)^2\right]_{(S)} = -\frac{3}{4J_1^{*4}}\sum_{s\in\mathbf{Z}\setminus\{0\}}\left(\frac{2A_s^{*2} - L_s^{*2}J_3^*}{(\nu_1^* - \nu_2^*)^2} + \frac{s^2(B_s^{*2} - M_s^{*2}J_3^*)}{[s\nu_1^* - (s+1)\nu_2^*]^2}\right) \quad (3.69)$$

and

$$\left[\frac{1}{2}\nu_{11}^*\left(\frac{\partial S_1}{\partial \theta_i}\right)^2\right]_{(LP)} = -\frac{3}{4J_1^{*4}}\sum_{s\in\mathbf{Z}\setminus\{0\}}\left(\frac{2A_s^*L_s^*\sqrt{-J_3^*}}{(\nu_1^* - \nu_2^*)^2}\cos\theta_3 \quad (3.70)\right.$$

$$\left. + \frac{s^2 B_s^* M_s^*\sqrt{-J_3^*}}{[s\nu_1^* - (s+1)\nu_2^*]^2}\cos\theta_3 - \frac{L_s^*L_{-s}^*J_3^*}{(\nu_1^* - \nu_2^*)^2}\cos 2\theta_3\right).$$

Finally, we would have to consider the contributions from $\frac{\partial H_1^*}{\partial \theta_3}\frac{\partial S_1}{\partial J_3^*}$, but all terms arising from this part of Ψ_2 are short-periodic. The short-period contributions of this term and of the other two considered before will not be written. They are of no importance in the context of a low-order example. In practical problems, they need to be calculated and the use of one algebraic manipulator is appropriate.

The application of the von Zeipel averaging rule to (3.65) gives the second-order term of the averaged (or secular) Hamiltonian:

$$H_2^*(\theta_3^*, J^*) = \Psi_{2(S)}(J^*) + \Psi_{2(LP)}(\theta_3^*, J^*) \quad (3.71)$$

and the equation

$$\sum_{\mu=1}^{2} \nu_\mu^* \frac{\partial S_2}{\partial \theta_\mu} = -\Psi_{2(SP)}(\theta, J^*), \tag{3.72}$$

the integration of which gives the second-order term of the generating function that formally reduces the Hamiltonian given by (3.55) to one degree of freedom with a remainder of order $\mathcal{O}(\varepsilon^3)$.

3.7 Linear Secular Theory

By secular theory, we mean the study of the solutions of the system defined by the Hamiltonian obtained after the elimination of all fast angles.

Let us consider the averaged Hamiltonian resulting from the previous example

$$H^* = \frac{1}{2J_1^{*2}} + \nu_2 J_2^* + \varepsilon \left(R_{(S)}(J^*) + R_{(LP)}(\theta_3^*, J^*) \right) + \varepsilon^2 \left(\Psi_{2(S)}(J^*) + \Psi_{2(LP)}(\theta_3^*, J^*) \right).$$

By construction, it is independent of θ_1^* and θ_2^*. The corresponding system of equations is

$$\begin{aligned}
\frac{d\theta_1^*}{dt} &= \frac{\partial H^*}{\partial J_1^*} & \frac{dJ_1^*}{dt} &= -\frac{\partial H^*}{\partial \theta_1^*} = 0 \\
\frac{d\theta_2^*}{dt} &= \frac{\partial H^*}{\partial J_2^*} = \nu_2 & \frac{dJ_2^*}{dt} &= -\frac{\partial H^*}{\partial \theta_2^*} = 0 \\
\frac{d\theta_3^*}{dt} &= \frac{\partial H^*}{\partial J_3^*} & \frac{dJ_3^*}{dt} &= -\frac{\partial H^*}{\partial \theta_3^*}.
\end{aligned} \tag{3.73}$$

From these equations, it follows that J_1^* and J_2^* are constants. In the original asteroidal problem, J_1^* constant means a^* constant, or, in a first approximation, that the "average" semi-major axis of the asteroid's orbit, $< a >$, is constant. This fact is sometimes mentioned by saying that there are no long-period terms in the asteroid's semi-major axis. However, this is true only in this approximation. Indeed, instantaneous and proper values are related through (3.33)–(3.34), that is,

$$\begin{aligned}
\theta_i &= \theta_i^* + \varepsilon \phi_i^{(a)}(\theta, J^*, \varepsilon) \\
J_i &= J_i^* + \varepsilon \phi_i^{(b)}(\theta, J^*, \varepsilon) \qquad (i = 1, 2).
\end{aligned} \tag{3.74}$$

The iterations to solve this system introduce, at second and higher orders, products of terms whose arguments are different but have equal short-period parts. The results are long-period terms (that is, terms whose argument is a multiple of θ_3).

82   3 Classical Perturbation Theories

The third pair of equations is independent of the other two and may be separated from the system. It is more easily studied if we introduce the Poincaré variables
$$x = \sqrt{-2J_3^*}\cos\theta_3^* \qquad y = \sqrt{-2J_3^*}\sin\theta_3^*. \qquad (3.75)$$

If we discard the terms independent of J_3^* and θ_3^*, H^* becomes
$$\begin{aligned}\hat{H}^*(\theta_3^*, J^*) &= C_0 J_3^* + C_1\sqrt{-J_3^*}\cos\theta_3^* + C_2 J_3^* \cos 2\theta_3^* \qquad (3.76)\\ &= \frac{-C_0}{2}(x^2+y^2) + \frac{C_1}{\sqrt{2}}x - \frac{C_2}{2}(x^2-y^2),\end{aligned}$$

where
$$\begin{aligned}C_0 = \varepsilon b &+ \frac{\varepsilon^2}{2}\sum_{s\in\mathbb{Z}\setminus\{0\}}\left(\frac{L_s^*}{\nu_1^*-\nu_2^*}\frac{\partial L_s^*}{\partial J_1^*} + \frac{sM_s^*}{s\nu_1^*-(s+1)\nu_2^*}\frac{\partial M_s^*}{\partial J_1^*}\right)\\ &+ \frac{3\varepsilon^2}{4J_1^{*4}}\sum_{s\in\mathbb{Z}\setminus\{0\}}\left(\frac{L_s^{*2}}{(\nu_1^*-\nu_2^*)^2} + \frac{s^2 M_s^{*2}}{[s\nu_1^*-(s+1)\nu_2^*]^2}\right), \qquad (3.77)\end{aligned}$$

$$\begin{aligned}C_1 = \varepsilon L_0^* &- \frac{\varepsilon^2}{2}\sum_{s\in\mathbb{Z}\setminus\{0\}}\left(\frac{2}{\nu_1^*-\nu_2^*}\frac{\partial(A_s^* L_s^*)}{\partial J_1^*} + \frac{s}{s\nu_1^*-(s+1)\nu_2^*}\frac{\partial(B_s^* M_s^*)}{\partial J_1^*}\right)\\ &- \frac{3\varepsilon^2}{4J_1^{*4}}\sum_{s\in\mathbb{Z}\setminus\{0\}}\left(\frac{2 A_s^* L_s^*}{(\nu_1^*-\nu_2^*)^2} + \frac{s^2 B_s^* M_s^*}{[s\nu_1^*-(s+1)\nu_2^*]^2}\right), \qquad (3.78)\end{aligned}$$

$$C_2 = \frac{\varepsilon^2}{2}\sum_{s\in\mathbb{Z}\setminus\{0\}}\left(\frac{\partial L_s^*}{\partial J_1^*}\frac{L_{-s}^*}{(\nu_1^*-\nu_2^*)} + \frac{3}{2J_1^{*4}}\frac{L_s^* L_{-s}^*}{(\nu_1^*-\nu_2^*)^2}\right). \qquad (3.79)$$

When high-order terms are neglected, the corresponding differential equations are linear. Secular theories of this kind are called *linear secular theories*. Actually, linear equations arise generally, at this order, when all functions in the given problem are assumed to be analytical in x, y, and J_3 is a small quantity. This is the case in many Celestial Mechanics problems when θ_3, J_3 have the definitions given by (3.56) (note that, in the asteroidal problem, J_3 is of the order of the square of the orbital eccentricity).

The resulting linear system of differential equations is
$$\frac{dx}{dt} = \frac{\partial H^*}{\partial y} = -(C_0 - C_2)y \qquad (3.80)$$
$$\frac{dy}{dt} = -\frac{\partial H^*}{\partial x} = (C_0 + C_2)x - \frac{C_1}{\sqrt{2}} \qquad (3.81)$$

(to fix the signs in these equations, one may note that the Poisson bracket of the new variables defined by (3.75) is $\{x,y\} = \{\theta_3^*, J_3^*\} = 1$). These equations have the general solution

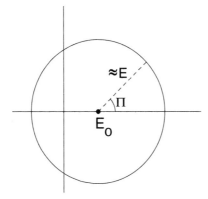

Fig. 3.1. Secular motion in the (x, y) plane

$$x = E\sqrt{1 - \frac{C_2}{C_0}} \cos(\omega t + \Pi_0) + E_0$$
$$y = E\sqrt{1 + \frac{C_2}{C_0}} \sin(\omega t + \Pi_0),$$
(3.82)

where

$$\omega = \sqrt{C_0^2 - C_2^2} \qquad E_0 = \frac{C_1}{\sqrt{2}(C_0 + C_2)}$$

and E, Π_0 are integration constants. This solution is shown in Fig. 3.1. It has two components:

- One free component with amplitudes $E\sqrt{1 - \frac{C_2}{C_0}}$ in x and $E\sqrt{1 + \frac{C_2}{C_0}}$ in y, and polar angle $\Pi = \omega t + \Pi_0$; in terms of the given example, E is the so-called *proper eccentricity* and Π the longitude of the *proper perihelion*. Note that, since $C_2/C_0 = \mathcal{O}(\varepsilon)$, the proper frequency is $\omega \simeq |C_0|$.
- One forced component of amplitude E_0 (the *forced eccentricity*) directed along the x-axis. Since C_0 and C_1 are both of order $\mathcal{O}(\varepsilon)$, this quantity is not controlled by the size of ε. ($E_0 \simeq L_0^*/\sqrt{2}b$.)

If the integration constant E is smaller than E_0 and if the quantity C_2/C_0 can be neglected, the trajectory in the (x, y) plane does not include the origin and the angle θ_3^* oscillates about 0. (The asteroid perihelion oscillates about the direction of Jupiter's perihelion.) In the other case, when $E > E_0$, the angle θ_3 circulates with period $2\pi/\omega$.

3.8 An Example – Part II

In the example of the application of the von Zeipel–Brouwer theory considered in Sect. 3.6, the divisors $\nu_1^* - \nu_2^*$ and $s\nu_1^* - (s+1)\nu_2^*$ appeared. Let us reconsider

that example, but assuming, now, that the values of the actions J_1 and J_2 are such that
$$\nu_1 - 2\nu_2 \simeq 0. \tag{3.83}$$

In this case, the non-resonance condition is no longer satisfied by the previous solution and the von Zeipel–Brouwer theory may not be used to eliminate the short-period terms dependent on the critical combination $\theta_1 - 2\theta_2$. Therefore, instead of the decomposition of $R(\theta, J^*)$, given by (3.62), we must consider in $R_{(SP)}$ only the non-critical terms. That is:

$$R_{(SP)}(\theta, J^*) = \sum_{s \in \mathbf{Z} \setminus \{0\}} \left(A_s^* \cos s(\theta_1 - \theta_2) + L_s^* \sqrt{-J_3^*} \cos[s(\theta_1 - \theta_2) + \theta_3] \right)$$
$$+ \sum_{s \in \mathbf{Z} \setminus \{1\}} B_s^* \cos[s\theta_1 - (s+1)\theta_2] \tag{3.84}$$
$$+ \sum_{s \in \mathbf{Z} \setminus \{1\}} M_s^* \sqrt{-J_3^*} \cos[s\theta_1 - (s+1)\theta_2 + \theta_3]$$

and the critical terms are included separately in

$$R_{(K)}(\theta, J^*) = B_1^* \cos(\theta_1 - 2\theta_2) + M_1^* \sqrt{-J_3^*} \cos(\theta_1 - 2\theta_2 + \theta_3). \tag{3.85}$$

The averaging rule fixed by (3.50) leads to

$$H_1^*(\theta_1 - 2\theta_2, \theta_3, J^*) = R_{(S)}(J^*) + R_{(LP)}(\theta_3, J^*) + R_{(K)}(\theta_1 - 2\theta_2, \theta_3, J^*) \tag{3.86}$$

and

$$\sum_{\mu=1}^{2} \nu_\mu^* \frac{\partial S_1}{\partial \theta_\mu} = -R_{(SP)}(\theta, J^*) \tag{3.87}$$

or, after integration,

$$S_1(\theta, J^*) = -\sum_{s \in \mathbf{Z} \setminus \{0\}} \left(\frac{A_s^* \sin s(\theta_1 - \theta_2)}{s(\nu_1^* - \nu_2^*)} + \frac{L_s^* \sqrt{-J_3^*} \sin[s(\theta_1 - \theta_2) + \theta_3]}{s(\nu_1^* - \nu_2^*)} \right)$$
$$- \sum_{s \in \mathbf{Z} \setminus \{1\}} \frac{B_s^* \sin[s\theta_1 - (s+1)\theta_2]}{s\nu_1^* - (s+1)\nu_2^*}$$
$$- \sum_{s \in \mathbf{Z} \setminus \{1\}} \frac{M_s^* \sqrt{-J_3^*} \sin[s\theta_1 - (s+1)\theta_2 + \theta_3]}{s\nu_1^* - (s+1)\nu_2^*}.$$

We still assume that $\nu_1^* - \nu_2^* \neq 0$ and $s\nu_1^* - (s+1)\nu_2^* \neq 0$ for all $s \neq 1$. It is worth noting that we have not yet done the Lagrangian point transformation indicated in Sect. 3.5.1 because, in this case, we have only one resonance and it is easy to trace the critical terms and to separate them from the others.

The next von Zeipel–Brouwer perturbation equation is

$$\sum_{\mu=1}^{2} \nu_\mu^* \frac{\partial S_2}{\partial \theta_\mu} = -\Psi_2(\theta, J^*) + H_2^*(\theta_1 - 2\theta_2, \theta_2, J^*), \qquad (3.88)$$

where Ψ_2 is similar to the one found in Sect. 3.5.1 and must be decomposed following the rules given there. The secular and long-period parts of Ψ_2 are the same as given in (3.67)–(3.70), except for those terms having $(\nu_1^* - 2\nu_2^*)$ as divisor, which will no longer appear because the corresponding argument was excluded from $R_{(SP)}$ and S_1. The critical terms in Ψ_2 arise from two combinations of short-period angles:

(a.) $s'(\theta_1 - \theta_2)$ and $s\theta_1 - (s+1)\theta_2$ when $s' = 1 - s$;
(b.) $s'(\theta_1 - \theta_2)$ and $s\theta_1 - (s+1)\theta_2$ when $s' = s - 1$.

They have arguments $\theta_1 - 2\theta_2 - \theta_3$, $\theta_1 - 2\theta_2$, $\theta_1 - 2\theta_2 + \theta_3$, and $\theta_1 - 2\theta_2 + 2\theta_3$. Thus, we have

$$\begin{aligned} \Psi_{2(K)} &= (K_1^* + K_2^* J_3^*) \cos(\theta_1 - 2\theta_2) + K_3^* \sqrt{-J_3^*} \cos(\theta_1 - 2\theta_2 - \theta_3) \\ &\quad + K_4^* \sqrt{-J_3^*} \cos(\theta_1 - 2\theta_2 + \theta_3) + K_5^* J_3^* \cos(\theta_1 - 2\theta_2 + 2\theta_3), \end{aligned}$$

where the coefficients K_i^* are functions of J_1^*. The calculation of the coefficients is elementary, but cumbersome. The short periodic part is, also, lengthy. They will not be given here.

The application of the von Zeipel averaging rule now gives

$$H_2^* = \Psi_{2(S)}(J^*) + \Psi_{2(LP)}(\theta_3, J^*) + \Psi_{2(K)}(\theta_1 - 2\theta_2, \theta_3, J^*) \qquad (3.89)$$

and

$$\sum_{\mu=1}^{2} \nu_\mu^* \frac{\partial S_2}{\partial \theta_\mu} = -\Psi_{2(SP)}(\theta, J^*). \qquad (3.90)$$

When the calculations are done, we obtain a new canonical system whose Hamiltonian is

$$\begin{aligned} H^*(\theta^*, J^*) &= -\frac{1}{2J_1^{*2}} + \nu_2 J_2^* + \varepsilon \left(A_0^* + L_0^* \sqrt{-J_3^*} \cos\theta_3^* + B_1^* \cos(\theta_1^* - 2\theta_2^*) \right. \\ &\quad \left. + M_1^* \sqrt{-J_3^*} \cos(\theta_1^* - 2\theta_2^* + \theta_3^*) \right) \qquad (3.91) \\ &\quad + \varepsilon^2 \left(\Psi_{2(S)}(J^*) + \Psi_{2(LP)}(\theta_3^*, J^*) \right. \\ &\quad + (K_1^* + K_2^* J_3^*) \cos(\theta_1^* - 2\theta_2^*) + K_3^* \sqrt{-J_3^*} \cos(\theta_1^* - 2\theta_2^* - \theta_3^*) \\ &\quad \left. + K_4^* \sqrt{-J_3^*} \cos(\theta_1^* - 2\theta_2^* + \theta_3^*) + K_5^* J_3^* \cos(\theta_1^* - 2\theta_2^* + 2\theta_3^*) \right). \end{aligned}$$

It differs from the given Hamiltonian H by a remainder $\mathcal{R}_3$ divisible by ε^3. The variables θ^*, J^* are related to the original variables θ, J by means of the transformation

$$J_i = \frac{\partial S}{\partial \theta_i}, \qquad \theta_i^* = \frac{\partial S}{\partial J_i^*}, \qquad (3.92)$$

where
$$S = (\theta \mid J^*) + \varepsilon S_1(\theta, J^*) + \varepsilon^2 S_2(\theta, J^*). \tag{3.93}$$

The Lagrange point transformation defined by

$$\begin{aligned}
\phi_1 &= \theta_1^* - 2\theta_2^* & I_1 &= J_1^* \\
\phi_2 &= \theta_2^* & I_2 &= J_2^* + 2J_1^* \\
\phi_3 &= \theta_3^* & I_3 &= J_3^*
\end{aligned} \tag{3.94}$$

leads to a Hamiltonian where the angle ϕ_2 is absent. Thus, $I_2 = J_2^* + 2J_1^*$ is a constant and the averaged Hamiltonian may be written

$$\begin{aligned}
\widehat{H}^*(\phi, I) =\ & -\frac{1}{2I_1^2} + \nu_2(I_2 - 2I_1) + \varepsilon \left(A_0^* + L_0^*\sqrt{-I_3}\cos\phi_3 + B_1^*\cos\phi_1 \right. \\
& + M_1^*\sqrt{-I_3}\cos(\phi_1 + \phi_3) \bigg) + \varepsilon^2 \left(\widehat{\Psi}_{2(S)}(I) + \widehat{\Psi}_{2(LP)}(\phi_3, I) \right. \\
& + (K_1^* + K_2^* I_3)\cos\phi_1 + K_3^*\sqrt{-I_3}\cos(\phi_1 - \phi_3) \\
& + K_4^*\sqrt{-I_3}\cos(\phi_1 + \phi_3) + K_5^* I_3 \cos(\phi_1 + 2\phi_3) \bigg),
\end{aligned} \tag{3.95}$$

where, for simplicity, we have kept the same symbol for the functions appearing in the coefficients notwithstanding the fact that they are, now, expressed with the new variables.

The transformed system has, now, two degrees of freedom (one degree of freedom more than in the non-resonant case). The theories allowing for the formal elimination of the critical angles from the Hamiltonian are the subject of Chaps. 4 and 8.

3.9 Iterative Use of von Zeipel–Brouwer Operations

After one application of the basic operation of the von Zeipel–Brouwer theory, the given system is split into two parts. One of them is a $N - M$ degrees-of-freedom canonical system whose Hamiltonian is

$$H^*(\theta_\varrho^*, J_\mu^*, J_\varrho^*, \varepsilon) = \sum_{k=1}^{n} \varepsilon^k H_k^*(\theta_\varrho^*, J_\mu^*, J_\varrho^*) \tag{3.96}$$

(we have dropped the term $H_0^*(J_\mu^*)$ that gives no contribution to the new differential equations). The actions were separated into J_μ^* and J_ϱ^*. J_μ^* are constants and enter in the canonical equations as mere parameters.

In the favorable case where H_1^* does not depend on the θ_ϱ^*, a new application of the von Zeipel–Brouwer theory may, again, reduce the number of degrees of freedom. Eventually, successive applications of the operation may provide the solution of the Delaunay problem.

The best-known example of a problem solved by two successive operations is the main problem of the theory of the Earth's artificial satellites [14]. In that problem, the undisturbed motion is a Kepler motion, that is, $H_0 = -\frac{1}{2}J_1^{-2}$, where J_1 is the only action appearing in H_0. The perturbation is

$$H_1 = -\frac{C}{r^3}(3\sin^2\varphi - 1), \tag{3.97}$$

where C is a constant factor determined by the physical parameters of the Earth, and r and φ are the satellite radius vector and latitude over the Earth's equator. After substitution of the Delaunay angle–action variables, this becomes

$$H_1 = \sum_{h \in \mathbf{D}_1 \subset \mathbf{Z}^3} A_h(J) \cos(h|\theta), \tag{3.98}$$

where the elements of the set $\mathbf{D}_1$ have the following properties:

- $h_3 = 0$ (that is, H_1 does not depend on the third Delaunay angle, the longitude of the ascending node);
- $h_2 = 0$ when $h_1 = 0$ (that is, the second Delaunay angle, the argument of the perigee, never appears alone in the cosine arguments of H_1).

We then have the decomposition

$$\begin{aligned} H_{1(S)} &= A_0(J), \\ H_{1(SP)} &= \sum_{h \in \mathbf{D}_1 \setminus \{0\}} A_h(J) \cos(h|\theta), \\ H_{1(LP)} &= 0. \end{aligned} \tag{3.99}$$

Then, $H_1^* = <H_1> = H_{1(S)} + H_{1(LP)}$ is a function of the actions only. $H_1^* = A_0(J^*)$ may play the role of the "undisturbed" Hamiltonian in a new application of the von Zeipel–Brouwer theory, which allows us to eliminate the angle θ_2^* present in the higher-order terms H_k^* ($k \geq 2$).

Theories in Celestial Mechanics are often classified as lunar or planetary according to $H_{1(LP)} = 0$ or $H_{1(LP)} \neq 0$, respectively. The different behavior arises from the fact that the adopted small parameter is not the same in these theories. In lunar theories, the small parameter is the inverse distance to the disturbing body (the Sun). In planetary theories, it is the mass of the disturbing body (another planet) in units of the solar mass. The different hierarchy of the terms in the expansion of the disturbing potential in lunar theories is such that $H_{1(LP)} = 0$. Then, the new "undisturbed" Hamiltonian $H_1^*(J_\varrho^*)$ is not degenerate (in Schwarzschild's sense), and the system defined by H^* may be formally solved through an application of the Poincaré theory. If $H_1^*(J_\varrho^*)$ is degenerate, the von Zeipel–Brouwer theory can be used again to eliminate the non-degenerate degrees of freedom. The possibility of a further reduction will now depend on whether or not the resulting H^{**} depend on the remaining angles $\theta_{\varrho'}^{**}$.

In the case of planetary theories H_1^* depend on the longitudes of the perihelion and ascending node (see [58], Sect. 83). Thus, a second application of the von Zeipel–Brouwer theory to obtain a formal solution in terms of pure trigonometric series is not possible in this case.

3.10 Divergence of the Series. Poincaré's Theorem

The Poincaré theory leads to the solution of a given Hamiltonian system by means of a Jacobian canonical transformation whose generating function is (see 3.31):

$$S_k(\theta, J^*) = \sum_{h \in \mathbf{D}_k \setminus \{0\}} \frac{iA_{kh} \exp(ih \mid \theta)}{h \mid \nu^*}, \qquad (3.100)$$

for whose existence it is necessary to assume that $(h \mid \nu^*) \neq 0$ for all $h \in \mathbf{D}_k \setminus \{0\}$ (*non-resonance condition*). Poincaré noted that, since the product of two Fourier polynomials introduces new combinations of the angles and increases the number of terms in the resulting polynomial, the products appearing in the construction of Ψ_k make any non-trivial $\mathbf{D}_k$ grow with k and, as k grows, values of $(h \mid \nu^*)$ smaller than any arbitrarily small given limit may be formed. The series are then divergent in any open set of the phase space.

For this reason the canonical perturbation theories discussed in this book are always considered as finite processes. However, infinite processes may be considered if some more stringent conditions are adopted. This is the case in Kolmogorov's theorem.

3.11 Kolmogorov's Theorem

Let us consider the same Hamiltonian system of Sect. 3.2:

$$H = H_0(J) + \sum_{k=1}^{\infty} \varepsilon^k H_k(\theta, J), \qquad (3.101)$$

where H_0 is the Hamiltonian of an integrable system satisfying a non-degeneracy condition more restrictive than Schwarzschild's non-degeneracy condition assumed in the previous sections. Kolmogorov's non-degeneracy condition is

$$\det\left(\frac{\partial^2 H_0}{\partial J_i \partial J_j}\right) \neq 0. \qquad (3.102)$$

The unperturbed system defined by H_0 admits non-degenerate quasiperiodic solutions

$$\begin{aligned} \theta_i &= \nu_i t + \text{const} \\ J_i &= \text{const} \end{aligned} \qquad (i = 1, 2, \cdots, N), \qquad (3.103)$$

where $\theta \in \mathbf{T}^N$ are angles conjugate to the actions J and

$$\nu_i \stackrel{\text{def}}{=} \frac{\partial H_0(J)}{\partial J_i}. \qquad (3.104)$$

Let us consider one of the above solutions, say

3.11 Kolmogorov's Theorem

$$\theta_i = \theta_i^o = \nu_i^o t + \text{const}$$
$$J_i = J_i^o = \text{const} \qquad (i = 1, 2, \cdots, N) \tag{3.105}$$

and let us assume that the frequencies ν_i^o satisfy the Diophantine condition

$$|(h|\nu^o)| \geq K||h||^{-(N+1)} \tag{3.106}$$

for all $h \in \mathbf{Z}^N \backslash \{0\}$ and a certain $K(\nu^o) > 0$.

The theorem of Kolmogorov [57] states that this solution persists when the system is perturbed, provided only that the perturbation is sufficiently small[4].

To understand the nature of such preserved solutions, that is, of the so-called Kolmogorov or KAM tori, one may recall that in any neighborhood $\mathcal{V}(J^o)$, there are infinitely many points J^* for which $(h \mid \nu^*) = 0$ for some $h \in \mathbf{Z}^N \backslash \{0\}$ and the series are divergent.

3.11.1 Frequency Relocation

To explain the procedure followed by Kolmogorov to obtain quasiperiodic solutions, let us present the construction of the canonical transformation as an extension of the Poincaré theory.

In Poincaré's theory, for given initial conditions, we look for a quasiperiodic solution starting at them. The frequencies of the solution of order n are given by the derivatives of $H^*_{(n)}$. To each order n of approximation, there corresponds a different set of proper frequencies ν^*. Even if the solution were to converge to an actual quasiperiodic solution of the given Hamiltonian, the frequencies of that solution would not be precisely known, being improved as the order of the approximation grows.

To guarantee that, at $J^* = J^o$, the frequencies of the disturbed and undisturbed systems are the same, we have to introduce a slight modification in Poincaré theory, relocating the frequencies in such a way that we have the same proper frequencies at every order of approximation. To do this, we split the Jacobian generating function into two parts:

$$S_k = S'_k + S''_k \qquad (k \in \mathbf{Z}). \tag{3.107}$$

S'_k is determined using the averaging rule of Poincaré theory:

$$\sum_{i=1}^{N} \nu_i^* \frac{\partial S'_k}{\partial \theta_i} = <\Psi_k> - \Psi_k(\theta, J^*). \tag{3.108}$$

The remaining part of the homological equation is

---

[4] For simplicity, all functions in this section are considered to be analytical in the angle–action variables and in the small parameter ε.

$$\sum_{i=1}^{N} \nu_i^* \frac{\partial S_k''}{\partial \theta_i} = H_k^*(J^*) - <\Psi_k>. \qquad (3.109)$$

Let us now determine S_k'' in such a way that

$$\left. \frac{\partial H_k^*(J^*)}{\partial J^*} \right|_{J^*=J^o} = 0. \qquad (3.110)$$

This definition allows us to eliminate from H_k^* the linear terms in $(J^* - J^o)$. To achieve this elimination, we expand $<\Psi_k>$ and ν_i^* in powers of $(J^* - J^o)$. Let $\sum_j B_{kj}(J_j^* - J_j^o)$ be the linear term of the expansion of $<\Psi_k>$. The linear term of ν_i^* is

$$\sum_{j=1}^{N} \left. \frac{\partial \nu_i^*}{\partial J_j^*} \right|_{J^*=J^o} (J_j^* - J_j^o)$$

or

$$\sum_{j=1}^{N} \frac{\partial^2 H_0(J^o)}{\partial J_i^o \partial J_j^o} (J_j^* - J_j^o).$$

Equating the linear terms of both sides of (3.109), it follows that

$$\sum_{i=1}^{N} \frac{\partial^2 H_0(J^o)}{\partial J_i^o \partial J_j^o} \frac{\partial S_k''}{\partial \theta_i} = -B_{kj} \qquad (3.111)$$

(H_k^* has no linear term, by construction). The solution of this trivial equation is

$$S_k'' = \xi_k \mid \theta, \qquad (3.112)$$

where $\xi_k \in \mathbf{R}^N$ is a constant vector given by the solutions of the linear equations

$$\sum_{i=1}^{N} \frac{\partial^2 H_0(J^o)}{\partial J_i^o \partial J_j^o} \xi_{ki} = -B_{kj}. \qquad (3.113)$$

We recall that, by hypothesis,

$$\det \left(\frac{\partial^2 H_0(J^o)}{\partial J_i^o \partial J_j^o} \right) \neq 0.$$

The kth component of the transformed Hamiltonian then becomes

$$H_k^*(J^*) = <\Psi_k> + \sum_{i=1}^{N} \nu_i^* \xi_{ki},$$

which, because of the definition of the ξ_{ki}, has no linear term in $J^* - J^o$ and, thus, satisfies the given hypothesis.

The function S_k'' does not obey the conditions fixed in Sect. 3.3. It is proportional to an angle and goes to infinity as t increases. However, it will not generate terms of this kind in the explicit equations of the transformation, because S only appears in the Poincaré algorithm through derivatives with respect to angles.

When $J^* = J^o$, the solutions of the dynamical system whose Hamiltonian is H^* are

$$\begin{aligned} \theta_i^* &= \nu_i^o t + \text{const} \\ J_i^* &= J_i^o = \text{const}, \end{aligned} \quad (3.114)$$

that is, the quasiperiodic solution of the undisturbed Hamiltonian is transformed into a quasiperiodic solution of the disturbed system, with the same frequencies ν_i^o. The equal frequencies of the undisturbed and disturbed solutions is achieved because of the adequate choice of the functions S_k''.

The procedure described above is very similar to that of Poincaré's theory, differing from it only in the averaging rule adopted to solve the problem of the indetermination of the homological equation.

3.11.2 Convergence

The crucial part of Kolmogorov's theorem is the proof of the convergence of the infinite series

$$S = \sum_{i=1}^{N} \theta_i J_i^* + \sum_{k=1}^{\infty} \varepsilon^k S_k(\theta, J^*)$$

and

$$H^* = H_0^* + \sum_{k=1}^{\infty} \varepsilon^k H_k^*$$

at the point $J^* = J^o$. At this point we have $\nu^* = \nu^o$, and the proof follows from the Diophantine condition and the rules of decrease of the coefficients of a Fourier series. These two properties enable the determination of a limiting ε^* such that, for $\varepsilon < \varepsilon^*$, the given series converge (see [3], [40]).

In order to have easier control of the small divisors and simplify the proof, Kolmogorov adopted an approach different from that described above. Instead of just looking for one canonical transformation (generated by S), Kolmogorov sought a succession of canonical transformations, each generated by a first-order Poincaré algorithm.

The first canonical transformation, $\phi : (\theta, J) \Rightarrow (\theta^*, J^*)$, is defined by the generating function

$$S = \sum_{i=1}^{N} \theta_i J_i^* + \varepsilon S_1(\theta, J^*),$$

where S_1 is determined by

$$\sum_{i=1}^{N} \nu_i^* \frac{\partial S_1}{\partial \theta_i} = H_1^* - H_1, \quad (3.115)$$

with the additional split of S_1 into $S_1' + S_1''$ as discussed in the previous section to eliminate the linear terms in $J^* - J^o$ from H_1^*. If the same hypotheses of the previous section are adopted, the Fourier series giving S_1 is convergent.

Since this algorithm is limited to first order, there is a remainder $\mathcal{R}_2$ divisible by ε^2, and the Hamiltonian of the resulting system is

$$H^* = H_0^*(J^*) + \varepsilon H_1^*(J^*) + \mathcal{R}_2(\theta^*, J^*, \varepsilon).$$

So, we have a new perturbed system whose integrable "undisturbed" part is $\widehat{H}_0^* = H_0^* + \varepsilon H_1^*$, and the perturbation $\mathcal{R}_2(\theta^*, J^*, \varepsilon)$ is of the order of ε^2.

The second canonical transformation $\phi^* : (\theta^*, J^*) \Rightarrow (\theta^{**}, J^{**})$ is defined by the generating function

$$S^* = \sum_{i=1}^{N} \theta_i^* J_i^{**} + \varepsilon^2 S_2^*(\theta^*, J^{**}),$$

with S_2^* determined by

$$\sum_{i=1}^{N} \nu_i^{**} \frac{\partial S_2^*}{\partial \theta_i^*} = \widehat{H}_1^{**} - \widehat{H}_1^*, \tag{3.116}$$

where we include in $\widehat{H}_1^*$ all terms of $\mathcal{R}_2$ of orders $\mathcal{O}(\varepsilon^2)$ and $\mathcal{O}(\varepsilon^3)$. As before, the Fourier series giving S_2^* is convergent.

Again, since the algorithm is limited to first order, there is a remainder $\mathcal{R}_4$. Since the small parameter is ε^2, the remainder is divisible by ε^4, and the Hamiltonian of the resulting system is

$$H^{**} = \widehat{H}_0^{**}(J^{**}) + \varepsilon^2 \widehat{H}_1^{**}(J^{**}) + \mathcal{R}_4(\theta^{**}, J^{**}, \varepsilon).$$

So, we have a new perturbed system whose integrable "undisturbed" part is $\widehat{\widehat{H}}_0^{**} = \widehat{H}_0^{**}(J^{**}) + \varepsilon^2 \widehat{H}_1^{**}(J^{**})$, and the perturbation $\mathcal{R}_4(\theta^{**}, J^{**}, \varepsilon)$ is of order ε^4.

The next step is again the canonical transformation defined by a first-order algorithm with ε^4 as small parameter and considering the perturbation $\varepsilon^4 \widehat{\widehat{H}}_1^{**}$ which includes all terms of orders $\mathcal{O}(\varepsilon^4)$ to $\mathcal{O}(\varepsilon^7)$ of $\mathcal{R}_4$. And so on.

This algorithm is sometimes called super-convergent because it resembles the super-convergent Newton's method for finding the root of an equation following an approximation scheme in which the order of the "error" is squared at each step. That is, we have the sequence $\varepsilon, \varepsilon^2, \varepsilon^4, \varepsilon^8, \cdots$ instead of $\varepsilon, \varepsilon^2, \varepsilon^3, \varepsilon^4, \cdots$.

The simplicity of the super-convergent approach is striking and, certainly, a key point in the proof of the theorem. However, this approach and that of Sect. 3.11.1 should lead to the same solution.

3.11.3 Degenerate Systems

Kolmogorov's theorem has been extended by Arnold [3], [5] to the case where H_0 depends only on M ($M < N$) actions and

$$\det\left(\frac{\partial^2 H_0}{\partial J_\mu \partial J_{\mu'}}\right)_{\mu,\mu'=1,\cdots,M} \neq 0 \qquad (3.117)$$

(*proper degeneracy*). The proof is a combination of the proof of Kolmogorov's theorem and the rules for iterative use of von Zeipel–Brouwer theory. The procedure followed in Kolmogorv's theorem is initially used to construct a canonical transformation that reduces the given Hamiltonian to $N - M$ degrees of freedom. If the frequencies ν_μ^o ($\mu = 1, \cdots, M$) satisfy the Diophantine condition, the series giving the transformation is convergent and the reduction is not merely formal as in the cases studied in this book. The resulting system is written

$$H^* = H_0^*(J_\mu^*) + \sum_{k=1}^{\infty} \varepsilon^k H_k^*(\theta_\varrho^*, J^*),$$

($\varrho = M+1, \cdots, N$). The angles θ_μ^* ($\mu = 1, \cdots, M$) are ignorable and the actions J_μ^* are constants. The system is then reduced to $N - M$ degrees of freedom. We may rescale the independent variable to εt and delete the constant term $H_0^*(J_\mu^*)$. The Hamiltonian of the reduced system is

$$\mathcal{H} = H_1^*(\theta_\varrho^*, J^*) + \sum_{k=1}^{\infty} \varepsilon^k H_{k+1}^*(\theta_\varrho^*, J^*),$$

where, by construction,

$$H_1^*(\theta_\varrho^*, J^*) = H_{1(S)}(J^*) + H_{1(LP)}(\theta_\varrho^*, J^*).$$

In the particular case where $H_{1(LP)}^*(\theta_\varrho^*, J^*) = 0$, the reduced Hamiltonian becomes

$$\mathcal{H} = H_{1(S)}(J^*) + \sum_{k=1}^{\infty} \varepsilon^k H_{k+1}^*(\theta_\varrho^*, J^*),$$

to which Kolmogorov's theorem may be applied once more provided that

$$\det\left(\frac{\partial^2 H_{1(S)}(J)}{\partial J_\varrho \partial J_{\varrho'}}\right)_{\varrho,\varrho'=M+1,\cdots,N} \neq 0. \qquad (3.118)$$

The combination of the two operations leads to convergent series describing quasiperiodic motions with the given frequencies ν_j^o. The additional conditions concern the first-order perturbation H_1: they are the absence there of long-period terms, i.e., $H_{1(LP)}(\theta_\varrho, J) = 0$, and the non-degeneracy (in Kolmogorov's sense) of the secular perturbation $H_{1(S)}(J_\varrho)$.

3.11.4 Degeneracy in the Extended Phase Space

A particular kind of degeneracy occurs when the given system results from an extension of the phase space. In that case, one linear term in the additional momentum is added to the Hamiltonian (see Sect. 1.6) so that the resulting Hessian has a row of zeros and is identically equal to zero. Let us assume, for the sake of simplicity, that the system is non-degenerate in Schwarzschild's sense, that the action J_N is the only one for which we have

$$\frac{\partial \nu_N}{\partial J_i} \equiv 0 \qquad (i = 1, \cdots, N) \tag{3.119}$$

and that the Hessian matrix of H_0 has rank $N-1$ and is such that

$$\det \left(\frac{\partial^2 H_0}{\partial J_\mu \partial J_{\mu'}} \right)_{\mu,\mu'=1,\cdots,N-1} \neq 0. \tag{3.120}$$

The construction of S'_k is not affected by the degeneracy in Kolmogorov's sense as long as the system is non-degenerate in Schwarzschild's sense. The condition of non-degeneracy in Kolmogorov's sense is however necessary in the construction of S''_k to establish the one-to-one correspondence between actions and frequencies. However if the degeneracy in Kolmogorov's sense results from an extension of the phase space, the given Hamiltonian is such that

$$\frac{\partial H_0}{\partial J_N} = \nu_N \qquad \frac{\partial H_k}{\partial J_N} = 0 \qquad (k \geq 1);$$

as a consequence, the functions Ψ_k are independent of J_N (the functions $\mathcal{E}'$ and $G_{k',k}$ of Sects. 3.2.1 and 3.2.2 are independent of J_N). Therefore, we have to eliminate from Ψ_k only the linear terms in $(J^*_\mu - J^o_\mu)$ ($\mu = 1, \cdots, N-1$) and the subscript in (3.113) may be restricted to $1, \cdots, N-1$. The other operations are not affected by this particular kind of degeneracy and a quasiperiodic solution exist in this case for frequency sets satisfying the Diophantine condition (see [5], Chapt. 5, Sect. 3).

In this case, H_0 is said to be isoenergetically non-degenerate. The condition for *isoenergetic non-degeneracy* has the general form

$$\det \begin{pmatrix} \left(\frac{\partial^2 H_0}{\partial J_i \partial J_j}\right) & \left(\frac{\partial H_0}{\partial J_i}\right) \\ \left(\frac{\partial H_0}{\partial J_j}\right)' & 0 \end{pmatrix}_{i,j=1,\cdots,N} \neq 0 \tag{3.121}$$

3.12 Inversion of a Jacobian Transformation

One successful application of the theories discussed in this chapter results in a new Hamiltonian $H^*(J^*)$ (or $H^*(\theta^*_\varrho, J^*)$), and a canonical transformation

3.12 Inversion of a Jacobian Transformation

defined by a Jacobian generating function $S(\theta, J^*)$. From the definitions (Sect. 3.2), we have

$$\theta_i = \theta_i^* - \sum_{k=1}^{n} \varepsilon^k \frac{\partial S_k(\theta, J^*)}{\partial J_i^*} = \theta_i^* + \varepsilon \phi_i^{(a)}(\theta, J^*; \varepsilon) \qquad (3.122)$$

$$J_i = J_i^* + \sum_{k=1}^{n} \varepsilon^k \frac{\partial S_k(\theta, J^*)}{\partial \theta_i^*} = J_i^* + \varepsilon \phi_i^{(b)}(\theta, J^*; \varepsilon). \qquad (3.123)$$

The transformation is not explicitly given because θ is still present in the right-hand sides. Thus, an inversion is necessary to reach the solution of the given problem. In low-order theories, the inversion can be achieved through a straightforward iterative procedure. In more general cases, we may use an extension of the well-known Lagrange formula (see Sect. 3.12.1).

Only the N equations (3.122) are implicit. Once they are solved, a mere substitution of the results into (3.123) is enough to complete the inversion. However, we will prefer to combine both into only one equation:

$$z = z^* + \varepsilon \phi(z; J^*, \varepsilon), \qquad (3.124)$$

where $z, z^* \in \mathbf{T}^N \times \mathcal{O}$, and $\phi : \mathbf{T}^N \times \mathcal{O} \to \mathbf{T}^N \times \mathcal{O}$ is analytical. ($\mathcal{O}$ is an open set of $\mathbf{R}^N$.) J^* appears in the function ϕ as an external parameter and is ignored during the inversion.

According to Lagrange's theorem, the solution of (3.124) is

$$z = z^* + \varepsilon \phi(z^*; J^*, \varepsilon) + \sum_{k \in \mathbf{N}} \frac{\varepsilon^{k+1}}{k!} G_k(z^*), \qquad (3.125)$$

where the functions $G_k(z)$ are given by

$$G_0(z) = \phi(z; J^*, \varepsilon),$$

$$G_k(z) = \sum_{\ell=0}^{k-1} \binom{k-1}{\ell} \left(\frac{\partial G_{k-\ell-1}(z)}{\partial z} \right) G_\ell(z), \qquad (k \geq 1). \qquad (3.126)$$

One may note that we have, on the right-hand side of the last equation, the product of a matrix by a vector. Because of the mixed form of the Jacobian transformation, $\phi(z; J^*, \varepsilon)$ does not depend on J, but only on θ. As a consequence, all $G_k(z)$ ($k \geq 0$) depend only on θ and the right-hand half of the matrix ($\partial G_{k-\ell-1}/\partial z$) is formed by zeros. Hence, the above equation may be written

$$G_k(z) = \sum_{\ell=0}^{k-1} \binom{k-1}{\ell} \left(\frac{\partial G_{k-\ell-1}(z)}{\partial \theta} \right) \widehat{G}_\ell(z), \qquad (3.127)$$

where $\widehat{G}_\ell$ is the restriction of G_ℓ to its first N components.

3.12.1 Lagrange Implicit Function Theorem

The one-dimensional Lagrange implicit function theorem is expressed by a well-known formula allowing the root of an implicit equation involving an analytic function to be obtained. It was extended to N dimensions by several authors. We reproduce here the extension due to Feagin and Gottlieb [29].

Theorem 3.12.1. *Consider the equation*

$$z = z^* + \varepsilon \phi(z),$$

where $z \in \mathbf{C}^N$, $\phi : \mathbf{C}^N \to \mathbf{C}^N$ *is analytic in a neighborhood of* $z = z^*$ *and* ε *is a (small) real parameter. Let* $f : \mathbf{C}^N \to \mathbf{C}^N$ *be a given analytical function in the neighborhood of* $z = z^*$. *Then*

$$f(z) = f(z^*) + \sum_{k=1}^{\infty} \frac{\varepsilon^k}{k!} F_k(z^*),$$

where the functions F_k *are given by*

$$F_0(z^*) = f(z^*),$$

$$F_k(z^*) = \sum_{\ell=0}^{k-1} \binom{k-1}{\ell} \left(\frac{\partial F_{k-\ell-1}(z^*)}{\partial z^*} \right) G_\ell(z^*) \quad (k \geq 1),$$

$$G_0(z^*) = \phi(z^*),$$

$$G_\ell(z^*) = \sum_{m=0}^{\ell-1} \binom{\ell-1}{m} \left(\frac{\partial G_{\ell-m-1}(z^*)}{\partial z^*} \right) G_m(z^*) \quad (k \geq 1).$$

Exercise 3.12.1. Show that, in the particular case $f(z) = z$, $F_k = k G_{k-1}$.

3.12.2 Practical Considerations

The inversion given in this chapter is not often used. In many applications, only $H^*(\theta^*, J^*)$ matters. For instance, in many applications in Mathematics, it is enough to know that $H^*(\theta^*, J^*)$ and $H(\theta, J)$ are related through a smooth transformation. In Astronomy, we may devise several kinds of applications, ranging from the qualitative study of the evolution of given systems of bodies, to the construction of ephemerides. In qualitative studies, generally, only $H^*(\theta^*, J^*)$ matters. We know that the solutions of the given Hamiltonian $H(\theta, J)$ are in the neighborhood of the solutions of $H^*(\theta^*, J^*)$, and this is often enough for our purposes. In the construction of ephemerides, on the contrary, the purpose is to predict the actual position of one body at a given time, and we have to obtain $\theta = \theta(t); J = J(t)$. Thus, the transformation has to be done. In fact, precise ephemerides are nowadays constructed mainly with numerical integrations. However, in the case of fast-moving objects, like close planetary satellites, the propagation of numerical errors limits the validity of numerical integrations to spans of time that are not enough large. In this case, long-term ephemerides are obtained by means of formal solutions.

Another example where formal solutions are often used is the computation of proper elements. Proper elements may be defined in several different ways, but the classical and more rigorous definition is to define them as the J^* (see [61]). Indeed, if the corresponding Delaunay problem has been solved (to a given order), and led to a Hamiltonian $H(J^*)$, these quantities are constant (in fact, only almost constant because the remainder $\mathcal{R}_{n+1}$ still depends on θ^*). To know the proper elements of a given body, it is necessary to know how to relate J^* to the actually observed quantities θ, J. Thus, knowledge of the function $J^* = J^*(\theta, J)$ is necessary. Again, the whole determination can be done by means of numerical integrations, and is often done. However, numerical procedures may present drawbacks. Consider the case of the asteroids. There are more than 100 000 of them. They are affected by very-long-period perturbations, and precise numerical integrations over some millions of years are necessary to derive good proper elements. This must be done for every asteroid, and repeated each time our knowledge of the actual orbit is improved, making the numerical approach inefficient.

Finally, we should note that even when the construction of a formal solution is more convenient than numerical integrations, the cumbersome step represented by the inversion of the Jacobian transformation can be done numerically. The solution of a system of a few algebraic equations close to an identity is a very inexpensive one-time numerical operation, which is not impaired by error's propagation like numerical integrations.

3.13 Lindstedt's Direct Calculation of the Series

By the end of the ninetenth century, the methods of Celestial Mechanics were translated into the language of Hamiltonian mechanics, using Jacobian generating functions to span the canonical transformations. Those methods aimed, generally, at dealing with perturbed oscillators given by equations of the form

$$\frac{d^2 x}{dt^2} + \omega^2 x = \varepsilon f(x, t). \tag{3.128}$$

The essential feature, kept unaltered since then, is the search for solutions in which the angles and the momenta (or actions) are given by

$$\theta = \nu t + \sum_{h \in \mathbf{Z}^N} A_h \exp\left(ih|\nu t\right) \tag{3.129}$$

and

$$J = \sum_{h \in \mathbf{Z}^N} B_h \exp\left(ih|\nu t\right), \tag{3.130}$$

respectively.

Solutions of this kind are often called *Lindstedt series* [64]. Excepted for the linear terms νt, they are in "pure trigonometric form" (cf. Charlier [20]), that is, quasiperiodic functions with N fundamental frequencies.

In Poincaré theory, we seek an integrable Hamiltonian $H^*(J^*)$, whose solutions are
$$\begin{aligned} \theta_i^* &= \nu_i^* t + \text{const} \\ J_i^* &= \text{const} \end{aligned} \tag{3.131}$$
and then we transform these solutions by means of the inverse of the canonical transformation defined by $S(\theta, J^*)$, to obtain the solutions of the given system. The averaging operation defined in Sect. 3.3 guarantees that all S_k are Lindstedt series. Therefore, the final solution has the required form.

It is worth recalling that most of the earlier planetary theories, founded on Lagrange's variation of the elements, included terms in powers of t and the so-called "Poisson terms", $t^a \exp(\mathrm{i}h|\nu t)$, mixing powers of t and trigonometric functions. These terms, describing the solution with high precision during a short interval of time, generally deteriorate rapidly as t increases. As the motion of the planets is slow, approximations of this kind are generally good for predicting purposes and remain valid for decades or even centuries. Nevertheless, it is not possible to use them in the study of fast motions. It is noteworthy that the revival of von Zeipel's theory, in modern Celestial Mechanics, happened when a good theory became necessary to describe the motion of Earth's artificial satellites [14].

If we know that a quasiperiodic solution (formal or exact) exists, we can construct it directly without the need for any of the previous methods. It is, in principle, possible to substitute the solutions given by (3.129) and (3.130) directly into the equations and to solve the resulting infinite set of equations resulting from the identification of both sides to obtain the unknown ν, A_h, B_h. The solutions are constructed order by order, in ε.

The direct calculation of the series has some drawbacks. One, obvious, is the extra amount of work resulting from the separate consideration of each of the $2N$ equations. It can be done only when the given equations are very simple. The second drawback has been pointed out by Giorgilli [39]: the direct calculation of the series leads to an increase in the number of terms with small divisors. However, the solutions with both techniques cannot be different and cancellations occur (many huge contributions that compensate among themselves).

4

Resonance

4.1 The Method of Delaunay's Lunar Theory

Delaunay was the first astronomer to use the mechanics of Hamilton and Jacobi to obtain the approximated solution of the equations of motion of a celestial body. His lunar theory [22] is a pioneer work in many respects. We credit Delaunay with the introduction of the set of angle–action variables ℓ, g, h, L, G, H in which the Lagrange equations for the variation of the orbital elements under a perturbation are canonical. His theory of the motion of the Moon is not a collection of clever tricks, as other theories in the old Celestial Mechanics. Having obtained the variation equations in canonical form, his problem was to find the solutions of the differential equations defined by the Hamiltonian

$$H = H_0(J) + \varepsilon \sum_{h \in \mathbf{D}} A_h(J) \cos(h|\theta), \tag{4.1}$$

where the canonical variables are $J \equiv (J_1, \ldots, J_N)$ and $\theta \equiv (\theta_1, \ldots, \theta_N)$, ε is a small parameter and $\mathbf{D} \subset \mathbf{Z}^N$. The technique adopted by Delaunay is methodologically very clear. He defined an *operation* and performed it, successively, almost 500 times. This operation starts with the choice of one argument $(h_1|\theta)$ in (4.1) and the consideration of the dynamical system defined by the abridged Hamiltonian

$$\mathcal{F}_1 = H_0(J) + \varepsilon A_{h_1}(J) \cos(h_1|\theta). \tag{4.2}$$

This system is integrable, since the angles θ_i are present only through the linear combination $(h_1|\theta)$. The main step of one Delaunay operation is to obtain a particular solution of this selected system and to use this solution to derive a canonical transformation leading to the elimination of the term $A_{h_1}(J)\cos(h_1|\theta)$ from the given Hamiltonian. (In fact, it is a transformation leading to the substitution of this term by others with much smaller coefficients.) To obtain the solution of the dynamical system defined by $\mathcal{F}$, we introduce the Jacobian generating function

$$S(\theta, J^*) \stackrel{\text{def}}{=} (\theta|J^*) + \Sigma(\theta, J^*), \tag{4.3}$$

where Σ is a function of order $\mathcal{O}(\varepsilon)$, and consider the Hamilton–Jacobi equation

$$E_1 = H_0\left(\frac{\partial S}{\partial \theta}\right) + \varepsilon A_{h_1}\left(\frac{\partial S}{\partial \theta}\right) \cos(h_1|\theta). \tag{4.4}$$

The functions of $\partial S/\partial \theta_i$, on the right-hand side of this equation, may be expanded about $\partial S/\partial \theta_i = J_i^*$ and (4.4) becomes

$$E_1 = H_0(J^*) + \sum_{i=1}^{N} \frac{\partial H_0(J^*)}{\partial J_i^*} \frac{\partial \Sigma}{\partial \theta_i} + \varepsilon A_{h_1}(J^*) \cos(h_1|\theta) + \mathcal{O}(\varepsilon^2). \tag{4.5}$$

At variance with the standard Hamilton–Jacobi theory, we do not look for a complete solution of the equation. We assume $E_1 = H_0(J^*)$, and seek a suitable particular solution of the partial differential equation for Σ:

$$0 = \sum_{i=1}^{N} \frac{\partial H_0(J^*)}{\partial J_i^*} \frac{\partial \Sigma}{\partial \theta_i} + \varepsilon A_{h_1}(J^*) \cos(h_1|\theta) + \mathcal{O}(\varepsilon^2). \tag{4.6}$$

If the higher-order terms are neglected, we have the immediate particular solution

$$\Sigma = -\frac{\varepsilon A_{h_1}(J^*) \sin(h_1|\theta)}{(h_1|\nu^*)}, \tag{4.7}$$

where $\nu^* \equiv (\nu_1^*, \nu_2^*, \cdots, \nu_N^*)$ and

$$\nu_i^* = \frac{\partial H_0(J^*)}{\partial J_i^*}. \tag{4.8}$$

Once we have obtained a first-order solution of the dynamical system spanned by $\mathcal{F}_1$, we go back to the given Hamiltonian H and perform the transformation of the variables generated by the function S:

$$\theta_i^* = \frac{\partial S}{\partial J_i^*} = \theta_i + \frac{\partial \Sigma}{\partial J_i^*}, \qquad J_i = \frac{\partial S}{\partial \theta_i} = J_i^* + \frac{\partial \Sigma}{\partial \theta_i}. \tag{4.9}$$

To complete the exposition of a Delaunay operation, we write the full Hamiltonian as

$$H = \mathcal{F}_1 + \Delta \mathcal{F}. \tag{4.10}$$

Hence, according to (4.5)–(4.7), when the above variable change is done, $\mathcal{F}_1$ becomes

$$\mathcal{F}_1^*(\theta^*, J^*) = H_0(J^*) + \mathcal{O}(\varepsilon^2), \tag{4.11}$$

that is, E_1 plus the higher-order terms of (4.6), which were neglected when (4.7) was obtained. With the same change, the additional part $\Delta \mathcal{F}(\theta, J)$ is transformed into $\Delta \mathcal{F}(\theta^*, J^*) + \mathcal{O}(\varepsilon^2)$. (The function $\Delta \mathcal{F}$ is the same as before.)

The result of the Delaunay operation is, then, a new Hamiltonian

$$H^* = H_0(J^*) + \Delta \mathcal{F}(\theta^*, J^*) + \mathcal{O}(\varepsilon^2) \qquad (4.12)$$

differing formally from the given one, in only two respects:

(a.) the term $\varepsilon A_{h_1} \cos(h_1|\theta)$ has disappeared;
(b.) new terms of order $\mathcal{O}(\varepsilon^2)$ were added.

In this way, performing as many operations as necessary, we may expect to eliminate from H all periodic terms of order $\mathcal{O}(\varepsilon)$. Indeed, as shown in the previous chapter, all these operations can be performed at one stroke, by finding the function S generating a transformation that eliminates all periodic terms of order $\mathcal{O}(\varepsilon)$.

We may also expect to eliminate, with a second sequence of operations, those terms of order $\mathcal{O}(\varepsilon^2)$, after that, the terms of order $\mathcal{O}(\varepsilon^3)$, and so on. In reality, as discussed in Sect. 3.12, this is not so. The combination of the arguments $(h|\theta)$ in the transformation of H tends to enlarge the set of values of h (the maximum of $|h|$ increases). Thus, values of h for which $(h|\nu^*)$ is too small can be reached (Poincaré Theorem) and the Delaunay theory, as well as the theories of the previous chapter (with the exception of Kolmogorov's) cannot be extended indefinitely. Only a finite number of operations can be done and the non-resonance condition $(h|\nu^*) \neq 0$ must be verified for all $h \in \mathbf{D}$, and for all h generated in the calculations. Otherwise, the theory needs to be modified as discussed thereafter.

We may also consider the case where one or more values $h \in \mathbf{D}$ are already such that $(h|\nu^*) \equiv 0$. This case happens when $H_0(J)$ is degenerate, that is, when H_0 does not depend on all components of J. One essential degeneracy of this kind appears in Celestial Mechanics where H_0 depends only on the Delaunay variable L and on the variable Λ, the canonical conjugate to the time t:

$$H_0 = -\frac{\mu^2}{2L^2} + \Lambda. \qquad (4.13)$$

In this case, the Delaunay theory does not allow one to get rid of the terms independent of both the time t and the mean anomaly ℓ (conjugate to L). In the particular problem of the motion of the Moon, periodic terms of this kind do not exist in the given perturbation (see the discussion in Sect. 3.9) and the theory developed by Delaunay allowed all periodic terms of order $\mathcal{O}(\varepsilon)$ to be eliminated.

4.2 Introduction of the Square Root of the Small Parameter

Let us consider, in this section, the equations of the Delaunay theory in the case where one resonance exists. Let us assume that

$$(h_1|\nu^*) = 0 \tag{4.14}$$

for some $h_1 \in \mathbf{D}$ and some point $J^* \in \mathcal{O}$ ($\mathcal{O}$ is the open set of $\mathbf{R}^N$ under study). We may continue as in the previous section up to equation (4.7). However, in this case, the resonance $(h_1|\nu^*) = 0$ happens at one point of $\mathcal{O}$. At such a point, the first term in the right-hand side of (4.6) vanishes and the equation becomes singular. If we do not get rid of this singularity and continue calculating as before, the divisor appearing in the result will become null when the exact resonance is reached. To study this problem, we will perform the same sequence of calculations as in the previous section, but keeping in explicit form some second-order terms.

For the sake of simplicity, we will only consider, here, the simplest case of only one degree of freedom, in which case the resonance assumption given by (4.14) becomes, simply,

$$\nu_1^* = 0. \tag{4.15}$$

Let us introduce again the generating function as

$$S(\theta_1, J_1^*) = \theta_1 J_1^* + \Sigma(\theta_1, J_1^*)$$

and let us expand the function

$$H_0(J_1) = H_0\left(\frac{\partial S}{\partial \theta_1}\right) = H_0\left(J_1^* + \frac{\partial \Sigma}{\partial \theta_1}\right).$$

Then

$$H_0(J_1) = H_0(J_1^*) + \nu_1^* \frac{\partial \Sigma}{\partial \theta_1} + \frac{1}{2}\nu_{11}^* \left(\frac{\partial \Sigma}{\partial \theta_1}\right)^2 + \cdots,$$

where we have introduced

$$\nu_1^* = \frac{\mathrm{d}H_0(J_1^*)}{\mathrm{d}J_1^*}, \qquad \nu_{11}^* = \frac{\mathrm{d}^2 H_0(J_1^*)}{\mathrm{d}J_1^{*2}}. \tag{4.16}$$

In the same way, we expand

$$R_1(\theta_1, J_1) \stackrel{\mathrm{def}}{=} A_{h_1}(J_1)\cos\theta_1$$

to obtain

$$R_1(\theta_1, J_1) = R_1(\theta_1, J_1^*) + \frac{\partial R_1(\theta_1, J_1^*)}{\partial J_1^*}\frac{\partial \Sigma}{\partial \theta_1} + \cdots.$$

When these expansions are substituted into the Hamilton–Jacobi equation (4.4), we obtain

$$E_1 = H_0(J_1^*) + \nu_1^* \frac{\partial \Sigma}{\partial \theta_1} + \frac{1}{2}\nu_{11}^* \left(\frac{\partial \Sigma}{\partial \theta_1}\right)^2 + \cdots + \varepsilon R_1(\theta_1, J_1^*) + \varepsilon \frac{\partial R_1}{\partial J_1^*}\frac{\partial \Sigma}{\partial \theta_1} + \cdots \tag{4.17}$$

and (4.6), correspondingly, becomes

$$\nu_1^* \frac{\partial \Sigma}{\partial \theta_1} + \frac{1}{2} \nu_{11}^* \left(\frac{\partial \Sigma}{\partial \theta_1}\right)^2 + \cdots + \varepsilon R_1(J_1^*, \theta_1) + \varepsilon \frac{\partial R_1}{\partial J_1^*} \frac{\partial \Sigma}{\partial \theta_1} + \cdots = 0. \quad (4.18)$$

Let us, now, investigate the algebraic inversion of this equation. This is done with the help of some classical results of Weierstrass' implicit functions theory. However, instead of making an application of the theory itself, we prefer, here, to adapt it to the present problem.

Equation (4.18) may be written in a more compact form as

$$\mathcal{F}(\sigma, \varepsilon) = a_{01}\varepsilon + a_{10}\sigma + a_{20}\sigma^2 + \sum_i \sum_j a_{ij} \sigma^i \varepsilon^j = 0, \quad (4.19)$$

where

$$\sigma = \frac{\partial \Sigma}{\partial \theta_1} \quad (4.20)$$

and the a_{ij} have obvious meanings. When the resonance condition

$$a_{10} = \nu_1^* = 0$$

holds, the leading terms in the expansion of $\mathcal{F}(\sigma, \varepsilon)$ are $a_{01}\varepsilon$ and $a_{20}\sigma^2$. Therefore, the only possibility of having $\mathcal{F}(\sigma, \varepsilon) = 0$, identically, with $a_{01} \neq 0$ and $a_{20} \neq 0$, is that the solution $\sigma(\varepsilon)$ has, at the origin, an algebraic critical point of order 2. Then, we may write

$$\sigma = b_1 \sqrt{\varepsilon} + b_2 \varepsilon + b_3 \varepsilon \sqrt{\varepsilon} + \cdots. \quad (4.21)$$

Since $\sqrt{\varepsilon}$ has two branches, we have two solutions forming a system of two algebraic functions, each corresponding to one branch of $\sqrt{\varepsilon}$. It is worth emphasizing that, when the series written in (4.19) is convergent in a neighborhood of the origin, the fundamental theorem on algebraic functions can be used to prove the convergence of the solutions given by (4.21).

4.2.1 Garfinkel's Abnormal Resonance

One hypothesis implicitly considered above and in this whole chapter is $\nu_{11}^* \neq 0$. The case $\nu_{11}^* = 0$ was called, by Garfinkel, *abnormal*. In such a case, $a_{10} = a_{20} = 0$ and the leading terms of the expansion of $\mathcal{F}(\sigma, \varepsilon)$ are $a_{01}\varepsilon$ and $a_{30}\sigma^3$. Therefore, the origin is an algebraic critical point of order 3 and we have to use the cube root of ε instead of the square root in the series expansion of $\sigma(\varepsilon)$.

4.3 Delaunay Theory According to Poincaré

Poincaré considered Delaunay theory in the first part of his chapter on Bohlin's theory ([80], Chap. XIX). He considered the one-degree-of-freedom problem

4 Resonance

with a disturbing potential formed by the term $\varepsilon R_1 = \varepsilon A_1 \cos\theta_1$ only. In this section, we present the complete Delaunay theory for the canonical equations defined by the Hamiltonian

$$H = H_0(J_1) + \sum_{k=1}^{\infty} \varepsilon^k H_{2k}(\theta_1, J_1). \tag{4.22}$$

One may note that the subscripts were modified to indicate the order of the terms in $\sqrt{\varepsilon}$.

The initial calculations are the same as in the previous section. Since we know that, in the neighborhood of the resonance, Σ may be expanded in a power series in $\sqrt{\varepsilon}$, we consider the canonical transformation

$$(\theta_1, J_1) \Rightarrow (\alpha, E)$$

defined by the Jacobian generating function

$$S = \theta_1 J_1^* + \sum_{k=1}^{n} \varepsilon^{k/2} S_k(\theta_1, E), \tag{4.23}$$

where J_1^* is the solution of the equation giving the exact resonance:

$$\nu_1(J_1^*) = \left(\frac{dH_0}{dJ_1}\right)_{J_1=J_1^*} = 0. \tag{4.24}$$

Poincaré considered, separately, the case $\nu_1^* = 0$ and the general case $\nu_1^* \neq 0$ (but close to zero). The consideration of the case $\nu_1^* \neq 0$ is, however, not necessary and is not done here.

The equations of the canonical transformation are

$$\alpha = \frac{\partial S}{\partial E}, \qquad J_1 = \frac{\partial S}{\partial \theta_1} \tag{4.25}$$

and the transformed Hamiltonian is assumed to have a main part

$$\varepsilon E + H^*(E)$$

independent of α, and a remainder $\mathcal{R}_{n+1}$ divisible by $\varepsilon^{(n+1)/2}$.

The solution is given by the integral

$$E = \text{const} \tag{4.26}$$

and the quadrature

$$\alpha = \int \frac{\partial}{\partial E}(H^* + \varepsilon E)\, dt. \tag{4.27}$$

Since the transformation is conservative, we have

$$H(\theta_1, J_1) = \varepsilon E + H^*(E) + \mathcal{R}_{n+1}(\alpha, E). \tag{4.28}$$

Taking into account the canonical transformation generated by S, this equation becomes

$$H\left(\theta_1, \frac{\partial S}{\partial \theta_1}\right) = \varepsilon E + H^*(E) + \mathcal{R}_{n+1}. \tag{4.29}$$

To identify both sides of (4.29) according to the powers of $\sqrt{\varepsilon}$, we need the power-series expansions of H_k and H^*. These expansions are identical to those performed in Poincaré theory (see Sects. 3.2.1 and 3.2.2). We have

$$H_0 = G_{0,0} + \varepsilon G_{0,2} + \varepsilon^{3/2} G_{0,3} + \cdots + \varepsilon^{n/2} G_{0,n} + \cdots \tag{4.30}$$

$$H_k = G_{k,k} + \varepsilon^{1/2} G_{k,k+1} + \varepsilon G_{k,k+2} + \cdots + \varepsilon^{n/2} G_{k,n} + \cdots \tag{4.31}$$

and

$$H^*(E) = \sum_{k=0}^{n} \varepsilon^{k/2} H_k^*(E). \tag{4.32}$$

All remaining terms are at least of order $\varepsilon^{(n+1)/2}$. Since $\nu_1^* = 0$, then $G_{0,1} = 0$ and $G_{0,k} = \mathcal{E}_k$ (see 3.15). The functions $G_{k,j}$ are defined by (3.22). In particular, $G_{k,k} = H_k(\theta_1, J_1^*)$.

The identification in the powers of the small parameter is made simple by the fact that ε is always explicit in the formulas and that all other quantities are finite. Thus, we have

$$H_0(J_1^*) = H_0^*,$$

$$0 = H_1^*,$$

$$\frac{1}{2}\nu_{11}^* \left(\frac{\partial S_1}{\partial \theta_1}\right)^2 + H_2(\theta_1, J_1^*) = H_2^* + E,$$

$$\nu_{11}^* \frac{\partial S_1}{\partial \theta_1} \frac{\partial S_2}{\partial \theta_1} + G_{2,3} + \mathcal{E}_3' = H_3^*, \tag{4.33}$$

$$\cdots\cdots$$

$$\nu_{11}^* \frac{\partial S_1}{\partial \theta_1} \frac{\partial S_k}{\partial \theta_1} + G_{2,k+1} + G_{4,k+1} + \cdots + \mathcal{E}_{k+1}' = H_{k+1}^*,$$

$$\cdots\cdots$$

$$\nu_{11}^* \frac{\partial S_1}{\partial \theta_1} \frac{\partial S_{n-1}}{\partial \theta_1} + G_{2,n} + G_{4,n} + \cdots + \mathcal{E}_n' = H_n^*.$$

(The functions $\mathcal{E}_k'$ are those defined implicitly by (3.20).) All remaining terms have at least $\varepsilon^{(n+1)/2}$ as a factor and are supposed to be grouped with the remainder $\mathcal{R}_{n+1}$.

As in the theories of the previous chapter, the first equation gives H_0^* and says that it is the value of the function H_0 at $J_1 = J_1^*$. Thus H_0^* is, now, just a number (it does not depend on the new variables α, E). The second equation says that $H_1^* = 0$. The third equation is the fundamental equation of Delaunay theory (the *Delaunay* or *Delaunay–Poincaré equation*):

$$\frac{1}{2}\nu_{11}^*\left(\frac{\partial S_1}{\partial \theta_1}\right)^2 + H_2(\theta_1, J_1^*) - H_2^* = E. \tag{4.34}$$

This equation is indeterminate while H_2^* is not fixed. This indetermination is overcome by introducing the averaging rule

$$H_2^* = <H_2(\theta_1, J_1^*)>, \tag{4.35}$$

where $<\cdots>$ stands for the average over the angle θ_1. Therefore, we have

$$\frac{1}{2}\nu_{11}^*\left(\frac{\partial S_1}{\partial \theta_1}\right)^2 + H_{2(K)} = E, \tag{4.36}$$

where

$$H_{2(K)}(J_1^*) = H_2(\theta_1, J_1^*) - <H_2(\theta_1, J_1^*)>. \tag{4.37}$$

Taking into account that the functions $G_{2,k+1}, G_{4,k+1}, \cdots$ and $\mathcal{E}'_{k+1}$ are completely known when the functions $S_1, S_2, \cdots, S_{k-1}$ are known, the generic or *homological* form of (4.33) (for $k \geq 2$) is

$$\nu_{11}^* \frac{\partial S_1}{\partial \theta_1}\frac{\partial S_k}{\partial \theta_1} + \Psi_{k+1}^*(\theta_1, E) = H_{k+1}^*(E), \tag{4.38}$$

where Ψ_{k+1} represents known functions. At variance with the fundamental Delaunay–Poincaré equation, the homological equation is linear and it is sufficient to obtain particular solutions of it.

4.3.1 First-Approximation Solution

When a complete integral of the fundamental equation is known, the generating function

$$S_{(1)} = \theta_1 J_1^* + \sqrt{\varepsilon}\, S_1(\theta_1, E)$$

defines a canonical transformation leading to a transformed Hamiltonian independent of α, except for terms factored by, at least, $\varepsilon^{3/2}$.

From the equations of the canonical transformation we have

$$J_1 = J_1^* + \sqrt{\varepsilon}\frac{\partial S_1}{\partial \theta_1} + \mathcal{O}(\varepsilon) = J_1^* \pm \sqrt{\frac{2\varepsilon}{\nu_{11}^*}(E - H_{2(K)})} + \mathcal{O}(\varepsilon), \tag{4.39}$$

$$\alpha = \sqrt{\varepsilon}\frac{\partial S_1}{\partial E} + \mathcal{O}(\varepsilon) = \pm\frac{\partial}{\partial E}\int\sqrt{\frac{2\varepsilon}{\nu_{11}}(E - H_{2(K)})}\, d\theta_1 + \mathcal{O}(\varepsilon). \tag{4.40}$$

The last equation, combined with (4.27) (which is reduced, at this order, to $\alpha = \int \varepsilon\, dt$), gives

$$t - t_0 = \int \frac{\pm d\theta_1}{\sqrt{2\varepsilon\nu_{11}(E - H_{2(K)})}} + \mathcal{O}(\varepsilon) \tag{4.41}$$

showing that the time scale of resonant phenomena is inversely proportional to $\sqrt{\varepsilon}$, that is, the frequencies associated with the resonance are proportional to $\sqrt{\varepsilon}$.

Equations (4.39) and (4.41) are the formal solutions of order $\mathcal{O}(\sqrt{\varepsilon})$ of the problem of Delaunay, in the presence of one resonance, in one degree of freedom.

4.4 Garfinkel's Ideal Resonance Problem

Let us use the Delaunay theory to obtain a complete solution of the Ideal Resonance Problem. This problem, thoroughly studied by Garfinkel [37], is defined as the problem of obtaining a formal solution of order $\varepsilon^{n/2}$ of the canonical equations defined by the Hamiltonian

$$H = H_0(J_1) - \varepsilon A(J_1) \cos \theta_1 \tag{4.42}$$

in the neighborhood of the value J_1^* for which $\nu_1 = dH_0/dJ_1 = 0$. The disturbing term has not, here, the same form $2\varepsilon A(J_1)\sin^2(\theta_1/2)$ considered in Garfinkel's work, but the two forms are equivalent.

This Hamiltonian system has two equilibrium solutions, viz. $\theta_1 = 0$ and $\theta_1 = \pi$ whose stability depends on the sign of $A^*\nu_{11}^*$ ($A^* = A(J_1^*)$). Without loss of generality, we assume that $A^*\nu_{11}^* > 0$ and the stable equilibrium is at $\theta_1 = 0$; otherwise, it is enough to change θ_1 into $\theta_1' + \pi$ so that the system satisfies this assumption.

The fundamental equation corresponding to the Hamiltonian of (4.42) is

$$\frac{1}{2}\nu_{11}^* \left(\frac{\partial S_1}{\partial \theta_1}\right)^2 - A^* \cos \theta_1 = E \tag{4.43}$$

or

$$\frac{\partial S_1}{\partial \theta_1} = \pm\sqrt{\frac{2}{\nu_{11}^*}(E + A^* \cos \theta_1)}, \tag{4.44}$$

where we take into account that $H_2^* = < -A^* \cos \theta_1 > = 0$.

We may note that this fundamental equation is nothing but the Hamilton–Jacobi equation of the simple pendulum. However, at variance with the conventional simple pendulum, the "inverse mass" ν_{11}^* may be either positive or negative. The solutions of the simple pendulum given in Sect. B.1 apply without modification. We just have to take care of the sign differences between the cases $\nu_{11}^* < 0$ and $\nu_{11}^* > 0$.

The homological equation is (see 4.38)

$$\frac{\partial S_k}{\partial \theta_1} = \frac{1}{\nu_{11}^*} \left(\frac{\partial S_1}{\partial \theta_1}\right)^{-1} (H_{k+1}^* - \Psi_{k+1}^*) \qquad (k = 2, \cdots, n), \tag{4.45}$$

where Ψ_{k+1}^* is a polynomial in the derivatives of $S_1, S_2, \cdots, S_{k-1}$ whose coefficients are constants or derivatives of H_2. For instance

$$\Psi_3^* = \frac{\partial S_1}{\partial \theta_1} \left[\frac{1}{6} \nu_{111}^* \left(\frac{\partial S_1}{\partial \theta_1} \right)^2 + \frac{\partial H_2}{\partial J_1^*} \right] \tag{4.46}$$

and

$$\Psi_4^* = \frac{1}{2} \nu_{11}^* \left(\frac{\partial S_2}{\partial \theta_1} \right)^2 + \frac{1}{2} \nu_{111}^* \left(\frac{\partial S_1}{\partial \theta_1} \right)^2 \frac{\partial S_2}{\partial \theta_1} + \frac{1}{24} \nu_{1111}^* \left(\frac{\partial S_1}{\partial \theta_1} \right)^4$$
$$+ \frac{\partial H_2}{\partial J_1^*} \frac{\partial S_2}{\partial \theta_1} + \frac{1}{2} \frac{\partial^2 H_2}{\partial J_1^{*2}} \left(\frac{\partial S_1}{\partial \theta_1} \right)^2.$$

From the previous equations, we may write

$$H_2(\theta_1, J_1^*) = -A^* \cos \theta_1 = E - \frac{1}{2} \nu_{11}^* \left(\frac{\partial S_1}{\partial \theta_1} \right)^2 \tag{4.47}$$

and

$$\frac{\partial^k H_2}{\partial J_1^{*k}} = -\frac{\mathrm{d}^k A^*}{\mathrm{d} J_1^{*k}} \cos \theta_1 = \frac{1}{A^*} \frac{\mathrm{d}^k A^*}{\mathrm{d} J_1^{*k}} \left[E - \frac{1}{2} \nu_{11}^* \left(\frac{\partial S_1}{\partial \theta_1} \right)^2 \right], \tag{4.48}$$

that is, H_2 and its derivatives may be written as polynomials in the first derivative of S_1. Therefore, $\Psi_3^*, \Psi_4^*, \cdots, \Psi_{n-1}^*$ may be, successively, written as polynomials in the first derivative of S_1:

$$\Psi_{k+1}^* = \sum_{k'=0}^{k+1} C_{k,k'} \left(\frac{\partial S_1}{\partial \theta_1} \right)^{k'}, \tag{4.49}$$

where $C_{k,k'} = 0$ when k and k' have the same parity; then, (4.45) may be written as

$$\frac{\partial S_k}{\partial \theta_1} = \frac{1}{\nu_{11}^*} \left(\frac{\partial S_1}{\partial \theta_1} \right)^{-1} \left[H_{k+1}^* - \sum_{k'=0}^{k+1} C_{k,k'} \left(\frac{\partial S_1}{\partial \theta_1} \right)^{k'} \right]. \tag{4.50}$$

To avoid the singularity at the libration boundaries, where $\partial S_1/\partial \theta_1 = 0$, H_{k+1}^* may be chosen to be such that the coefficient of $(\partial S_1/\partial \theta_1)^{-1}$ in (4.50) vanishes:

$$H_{k+1}^* = C_{k,0}.$$

(One may note that $H_k^* = 0$ for all k odd because of the parity rule of the coefficients $C_{k,k'}$.) The homological equation then becomes

$$\frac{\partial S_k}{\partial \theta_1} = -\frac{1}{\nu_{11}^*} \sum_{k'=1}^{k+1} C_{k,k'} \left(\frac{\partial S_1}{\partial \theta_1} \right)^{k'-1}. \tag{4.51}$$

In particular, for $k = 2$, we have

$$\frac{\partial S_2}{\partial \theta_1} = -\frac{E}{A^* \nu_{11}^*} \frac{dA^*}{dJ_1^*} - \left(\frac{\nu_{111}^*}{6\nu_{11}^*} - \frac{1}{2A^*} \frac{dA^*}{dJ_1^*}\right) \left(\frac{\partial S_1}{\partial \theta_1}\right)^2. \qquad (4.52)$$

Once S is known, we may construct the formal solutions of the Ideal Resonance Problem. To order $\mathcal{O}(\varepsilon)$, they are:

$$\begin{aligned}
J_1 &= J_1^* + \sqrt{\varepsilon}\,\frac{\partial S_1}{\partial \theta_1} + \varepsilon \frac{\partial S_2}{\partial \theta_1} \\
\alpha &= \varepsilon(t - t_0) = \sqrt{\varepsilon}\,\frac{\partial S_1}{\partial E} + \varepsilon \frac{\partial S_2}{\partial E}.
\end{aligned} \qquad (4.53)$$

4.4.1 Garfinkel–Jupp–Williams Integrals

The integration of (4.51), for all k, involves the integrals

$$\mathcal{I}_k = \int \left(\frac{\partial S_1}{\partial \theta_1}\right)^k d\theta_1 = \int \left(\frac{2}{\nu_{11}^*}(E + A^* \cos \theta_1)\right)^{k/2} d\theta_1,$$

which can be calculated by means of recurrence formulas [36]. Differentiating (4.43) with respect to θ_1, we obtain

$$\frac{\partial S_1}{\partial \theta_1} \frac{\partial}{\partial \theta_1}\left(\frac{\partial S_1}{\partial \theta_1}\right) = -\frac{A^*}{\nu_{11}^*} \sin \theta_1. \qquad (4.54)$$

Hence,

$$\frac{\partial}{\partial \theta_1}\left(\frac{\partial S_1}{\partial \theta_1}\right)^k = -k\left(\frac{\partial S_1}{\partial \theta_1}\right)^{k-2} \frac{A^*}{\nu_{11}^*} \sin \theta_1, \qquad (4.55)$$

$$\frac{\partial^2}{\partial \theta_1^2}\left(\frac{\partial S_1}{\partial \theta_1}\right)^k = k(k-2)\left(\frac{\partial S_1}{\partial \theta_1}\right)^{k-4}\left(\frac{A^*}{\nu_{11}^*}\right)^2 \sin^2 \theta_1 - k\left(\frac{\partial S_1}{\partial \theta_1}\right)^{k-2} \frac{A^*}{\nu_{11}^*} \cos \theta_1.$$

The trigonometric functions may be eliminated with the help of (4.43) giving

$$\frac{\partial^2}{\partial \theta_1^2}\left(\frac{\partial S_1}{\partial \theta_1}\right)^k = -\frac{k^2}{4}\left(\frac{\partial S_1}{\partial \theta_1}\right)^k + k(k-1)\frac{E}{\nu_{11}^*}\left(\frac{\partial S_1}{\partial \theta_1}\right)^{k-2}$$

$$+ k(k-2)\frac{A^{*2} - E^2}{\nu_{11}^{*2}}\left(\frac{\partial S_1}{\partial \theta_1}\right)^{k-4}$$

whose integration, with respect to θ_1, followed by the use of (4.55) and the definition of $\mathcal{I}_k$, yields

$$\frac{k}{4}\mathcal{I}_k = \left(\frac{\partial S_1}{\partial \theta_1}\right)^{k-2} \frac{A^*}{\nu_{11}^*} \sin \theta_1 + (k-1)\frac{E}{\nu_{11}^*}\mathcal{I}_{k-2} + (k-2)\frac{A^{*2} - E^2}{\nu_{11}^{*2}}\mathcal{I}_{k-4}. \qquad (4.56)$$

(The integration constant is chosen to be such that $\mathcal{I}_k = 0$ at $\theta_1 = 0$.) Thus, all integrals are known when we know a sequence of four of them.

For even k, the integrals are elementary and are the same no matter whether the motion is a libration or a circulation:

$$\mathcal{I}_0 = \int d\theta_1 = \theta_1,$$

$$\mathcal{I}_2 = \int \left(\frac{\partial S_1}{\partial \theta_1}\right)^2 d\theta_1 = \frac{2E}{\nu_{11}^*}\theta_1 + \frac{2A^*}{\nu_{11}^*}\sin\theta_1. \quad (4.57)$$

For odd k, the integrals are elliptic and we have to consider separately the cases where $E\nu_{11}^* > A^*\nu_{11}^*$ (circulation), $|E| < |A^*|$ (libration) and $E = A^*$ (asymptotic motion). This will be done in the forthcoming sections.

4.4.2 Circulation ($E\nu_{11}^* > A^*\nu_{11}^* > 0$)

Let us calculate the solutions of the Ideal Resonance Problem in the case of circulations. The first step is to calculate the Garfinkel–Jupp–Williams integrals necessary to generate the solutions at all orders. To complete the set of four integrals necessary to span the whole set, we need two of them with odd values of k. They are

$$\mathcal{I}_{-1} = \pm\sqrt{\frac{2\nu_{11}^*}{E+A^*}}\, \mathcal{F}\left(\frac{\theta_1}{2},\kappa\right),$$

and

$$\mathcal{I}_1 = \pm\sqrt{\frac{8}{\nu_{11}^*}(E+A^*)}\, \mathcal{E}\left(\frac{\theta_1}{2},\kappa\right),$$

where $\mathcal{F}(\frac{\theta_1}{2},\kappa)$ and $\mathcal{E}(\frac{\theta_1}{2},\kappa)$ are incomplete elliptic integrals[1] of the first and second kind, respectively, of modulus

$$\kappa = \sqrt{\frac{2A^*}{E+A^*}} \qquad (0 < \kappa < 1).$$

Double signs were used in front of the square roots to stress that these functions have two branches each corresponding to a distinct family of circulations.

The solutions of the equations for S_k ($k=1$ and $k=2$) are

$$S_1 = \mathcal{I}_1 = \pm\frac{4}{\kappa}\sqrt{\frac{A^*}{\nu_{11}^*}}\, \mathcal{E}\left(\frac{\theta_1}{2},\kappa\right) \quad (4.58)$$

and

$$S_2 = -\frac{E}{A^*\nu_{11}^*}\frac{dA^*}{dJ_1^*}\mathcal{I}_0 - \frac{\nu_{111}^*}{6\nu_{11}^*}\mathcal{I}_2 + \frac{1}{2A^*}\frac{dA^*}{dJ_1^*}\mathcal{I}_2,$$

---

[1] The slight change in the usual notation for the elliptic integrals made here ($\mathcal{F}$ and $\mathcal{E}$ instead of F and E) is necessary to avoid confusion with other functions in the book. $\mathbb{K}$ and $\mathbb{E}$ are the corresponding complete elliptic integrals.

or
$$S_2 = -\frac{E\nu_{111}^*}{3\nu_{11}^{*2}}\theta_1 + \left(\frac{dA^*}{dJ_1^*} - \frac{A^*\nu_{111}^*}{3\nu_{11}^*}\right)\frac{\sin\theta_1}{\nu_{11}^*}. \tag{4.59}$$

From (4.58) and (4.59) we have, respectively,
$$\frac{\partial S_1}{\partial E} = \pm\frac{\kappa}{\sqrt{A^*\nu_{11}^*}}\mathcal{F}\left(\frac{\theta_1}{2},\kappa\right), \qquad \frac{\partial S_2}{\partial E} = -\frac{\nu_{111}^*\theta_1}{3\nu_{11}^{*2}}.$$

Therefore,
$$t = t_0 \pm \frac{\kappa}{\sqrt{\varepsilon A^*\nu_{11}^*}}\mathcal{F}\left(\frac{\theta_1}{2},\kappa\right) - \frac{\nu_{111}^*}{3\nu_{11}^{*2}}\theta_1 + \mathcal{O}(\sqrt{\varepsilon}), \tag{4.60}$$

where the upper sign corresponds to prograde circulations and the lower one to retrograde circulations. $\theta_1 = 0$ when $t = t_0$.

The period of the circulations is given, to this order, by
$$T = \frac{2\kappa}{\sqrt{\varepsilon A^*\nu_{11}^*}}\mathbb{K}(\kappa) \mp \frac{2\pi\nu_{111}^*}{3\nu_{11}^{*2}} + \mathcal{O}(\sqrt{\varepsilon}), \tag{4.61}$$

where $\mathbb{K}(\kappa)$ is the complete elliptic integral of the first kind of modulus κ.

The use of Jacobian functions is necessary to write the complete solution, as well as, for instance, to give the explicit form of the time law $\theta = \theta(t)$. Inverting the elliptic integral in (4.60) we obtain
$$\theta_1 = \pm 2\,\mathrm{am}\left(\tau + \frac{\sqrt{\varepsilon A^*\nu_{11}^*}}{\kappa}\frac{\nu_{111}^*}{3\nu_{11}^{*2}}\theta_1\right) + \mathcal{O}(\varepsilon),$$

where am is the Jacobian *amplitude*, and
$$\tau = \frac{\sqrt{\varepsilon A^*\nu_{11}^*}}{\kappa}(t - t_0). \tag{4.62}$$

To the given order of approximation, we may still write
$$\theta_1 = \pm 2\,\mathrm{am}\,\tau + 4\frac{\sqrt{\varepsilon A^*\nu_{11}^*}}{\kappa}\frac{\nu_{111}^*}{3\nu_{11}^{*2}}\,\mathrm{am}\,\tau\,\mathrm{dn}\,\tau + \mathcal{O}(\varepsilon), \tag{4.63}$$

where dn is the Jacobian *delta amplitude* elliptic function.

The variation of the action J_1, to the same order of approximation, is
$$J_1 = J_1^* \pm \frac{2}{\kappa}\sqrt{\frac{\varepsilon A^*}{\nu_{11}^*}}\sqrt{1-\kappa^2\sin^2\frac{\theta_1}{2}} - \frac{\varepsilon E\nu_{111}^*}{3\nu_{11}^{*2}}$$
$$+ \varepsilon\left(\frac{dA^*}{dJ_1^*} - \frac{A^*\nu_{111}^*}{3\nu_{11}^*}\right)\frac{\cos\theta_1}{\nu_{11}^*} + \mathcal{O}(\varepsilon\sqrt{\varepsilon}), \tag{4.64}$$

where the upper sign corresponds to motions above the libration zone ($J_1 > J_1^*$) and the lower sign to motions below the libration zone ($J_1 < J_1^*$). One should be aware that the relationship between the double signs in (4.63) and (4.64) is not always the same. When $\nu_{11}^* > 0$, the upper (resp. lower) sign in one of them corresponds to the upper (resp. lower) sign in the other (the circulations above the libration zone are prograde and the circulations below the libration zone are retrograde). When $\nu_{11}^* < 0$, we have to consider that the second of them carries the sign of $\partial S_1/\partial \theta_1$ (which is proportional to $\sqrt{A^*/\nu_{11}^*}$), while the first of them carries the sign of $\partial S_1/\partial E$ (which is proportional to $\nu_{11}^* \sqrt{A^*/\nu_{11}^*}$; written as $\sqrt{A^*\nu_{11}^*}$). Then, when $\nu_{11}^* < 0$, these two partial derivatives have opposite signs and, to the upper sign in one of the equations, corresponds the lower sign in the other (the circulations above the libration zone are retrograde and the circulations below the libration zone are prograde).

In the inner limit $\kappa \to 1$, we have $\mathbb{K} \to \infty$ and, thus, $T \to \infty$. The outer limit $\kappa \to 0$ corresponds to $E \to \infty$. From (4.56) and (4.57), it is evident that, for k even, $\mathcal{I}_k$ has a leading term in $E^{k/2}$; thus, for $\kappa \to 0$, the series giving the function S is divergent, meaning that this theory does not allow us to study the motion far of the resonance; it is only valid in the region of *deep resonance* where $\kappa > \mathcal{O}(\sqrt{\varepsilon})$ and where the general theories of the previous chapter would fail because of the small divisor ν_1^*.

4.4.3 Libration ($|E| < |A^*|$)

The basic equations for librations and circulations are the same. However, elliptic integrals must be treated in a different way since, now, $\kappa > 1$. We need to use the reciprocal modulus transformation

$$\sin \zeta = \kappa \sin \frac{\theta_1}{2} \qquad (4.65)$$

(see Sect. B.1.2) and the solutions describing the librations are obtained from those describing the circulations by means of the well-known relations

$$\kappa \mathcal{F}\left(\frac{\theta_1}{2}, \kappa\right) = \mathcal{F}\left(\zeta, \frac{1}{\kappa}\right) \qquad (4.66)$$

and

$$\kappa \, \mathcal{E}\left(\frac{\theta_1}{2}, \kappa\right) = \kappa^2 \, \mathcal{E}\left(\zeta, \frac{1}{\kappa}\right) - (\kappa^2 - 1) \, \mathcal{F}\left(\zeta, \frac{1}{\kappa}\right). \qquad (4.67)$$

We thus have

$$\mathcal{I}_{-1} = \sqrt{\frac{\nu_{11}^*}{A^*}} \, \mathcal{F}\left(\zeta, \frac{1}{\kappa}\right)$$

and

$$\mathcal{I}_1 = 4\sqrt{\frac{A^*}{\nu_{11}^*}} \left[\mathcal{E}\left(\zeta, \frac{1}{\kappa}\right) + \beta \mathcal{F}\left(\zeta, \frac{1}{\kappa}\right)\right],$$

4.4 Garfinkel's Ideal Resonance Problem

where
$$\beta = \frac{1-\kappa^2}{\kappa^2} \tag{4.68}$$

and
$$\zeta = \arcsin\left(\kappa \sin\frac{\theta_1}{2}\right) = \arcsin\sqrt{\frac{A^*(1-\cos\theta_1)}{E+A^*}}. \tag{4.69}$$

The integrals $\mathcal{I}_0$ and $\mathcal{I}_2$ are the same as before.

The solution of the equation for $k=1$ is again
$$S_1 = \mathcal{I}_1,$$

(with the new value of $\mathcal{I}_1$); for $k=2$, the solution is the same as for circulations. We also have
$$\frac{\partial S_1}{\partial E} = \frac{1}{\sqrt{A^*\nu_{11}^*}} \mathcal{F}\left(\zeta, \frac{1}{\kappa}\right).$$

Substitution of these results into (4.53) gives, now, the time law
$$t = t_0 + \frac{1}{\sqrt{\varepsilon A^* \nu_{11}^*}} \mathcal{F}\left(\zeta, \frac{1}{\kappa}\right) - \frac{\nu_{111}^*}{3\nu_{11}^{*2}}\theta_1 + \mathcal{O}(\sqrt{\varepsilon}), \tag{4.70}$$

where we assume $\theta_1 = 0$ and $\dot\theta_1 > 0$ (or $\zeta = 0$ and $\dot\zeta > 0$) at $t = t_0$.

The period of the librations is the time for θ_1 to perform a complete oscillation between the boundaries of the libration. We may first note that the term proportional to θ_1 does not contribute to the period since the angle θ_1 will be brought back to the initial value without completing one revolution; this term only says that θ_1 is faster in one direction than in another (if $\nu_{111}^* > 0$, it is faster when θ_1 grows). We have to consider, then, only the contribution of the term involving the elliptic integral, whose calculation is the same as for the simple pendulum:
$$T = \frac{4}{\sqrt{\varepsilon A^* \nu_{11}^*}} \mathbb{K}\left(\frac{1}{\kappa}\right) + \mathcal{O}(\sqrt{\varepsilon}). \tag{4.71}$$

The inversion of the elliptic integral in (4.70) gives, now,
$$\kappa \sin\frac{\theta_1}{2} = \sin\zeta = \mathrm{sn}\left\{\sqrt{\varepsilon A^* \nu_{11}^*}\left[(t-t_0) + \left(\frac{\nu_{111}^*}{3\nu_{11}^{*2}}\right)\theta_1\right]\right\} + \mathcal{O}(\varepsilon), \tag{4.72}$$

where sn is the Jacobian *sine amplitude* elliptic function with modulus $1/\kappa$. An iteration over θ_1 is necessary to complete the inversion of (4.70).

The analog of (4.64), in this case, is
$$J_1 = J_1^* \pm \frac{2}{\kappa}\sqrt{\frac{\varepsilon A^*}{\nu_{11}^*}}\cos\zeta - \frac{\varepsilon E\nu_{111}^*}{3\nu_{11}^{*2}} + \varepsilon\left(\frac{\mathrm{d}A^*}{\mathrm{d}J_1^*} - \frac{A^*\nu_{111}^*}{3\nu_{11}^*}\right)\frac{\cos\theta_1}{\nu_{11}^*} + \mathcal{O}(\varepsilon\sqrt{\varepsilon}). \tag{4.73}$$

In all equations before the last one, we have not used double signs since the two branches of the square roots meet at the boundary of the libration and are parts of the same solution. However, as in the case of circulations, when $\nu_{11}^* < 0$, $\sqrt{A^*/\nu_{11}^*}$ and $\sqrt{A^*\nu_{11}^*}$ must be considered with different signs. Hence, a double sign was included in the last equation, the positive sign holding when $\nu_{11}^* > 0$ and the negative one when $\nu_{11}^* < 0$.

The inner limit $\kappa \to \infty$ ($\frac{1}{\kappa} \to 0$) corresponds to $E \to -A^*$, that is, to the stable equilibrium point. The outer limit $\kappa \to 1$ corresponds to the separatrix (see below).

4.4.4 Asymptotic Motions ($E = A^*$)

When $\kappa = 1$ ($E = A^*$), (4.44) becomes, simply,

$$\frac{\partial S_1}{\partial \theta_1} = \pm \sqrt{\frac{2A^*}{\nu_{11}^*}(1 + \cos\theta_1)} = \pm \sqrt{\frac{4A^*}{\nu_{11}^*}} \cos\frac{\theta_1}{2};$$

the corresponding integral is only pseudo-elliptic and gives

$$S_1 = \pm\sqrt{\frac{8A^*}{\nu_{11}^*}(1 - \cos\theta_1)} = \pm\sqrt{\frac{16A^*}{\nu_{11}^*}}\sin\frac{\theta_1}{2}.$$

The derivative $\partial S_1/\partial E$ needs some special consideration since, now, E is a constant. This derivative may be obtained by calculating $\partial^2 S_1/\partial\theta_1\partial E$ from (4.44), then making $E = A^*$, and integrating with respect to θ_1. Then

$$\frac{\partial S_1}{\partial E} = \pm\int\sqrt{\frac{1}{4A^*\nu_{11}^*}}\sec\frac{\theta_1}{2}d\theta_1 = \pm\sqrt{\frac{1}{A^*\nu_{11}^*}}\ln\tan\left(\frac{\pi}{4} + \frac{\theta_1}{4}\right)$$

($-\pi < \theta_1 < \pi$). The formal solution of order $\mathcal{O}(\varepsilon)$, for this particular choice of the integration constant, is

$$J_1 = J_1^* \pm \sqrt{\frac{4A^*\varepsilon}{\nu_{11}^*}}\cos\frac{\theta_1}{2} - \frac{A^*\varepsilon\nu_{111}^*}{3\nu_{11}^{*2}}(1 + \cos\theta_1) + \frac{\varepsilon}{\nu_{11}^*}\frac{dA^*}{dJ_1^*}\cos\theta_1 + \mathcal{O}(\varepsilon\sqrt{\varepsilon}),$$
(4.74)

$$t = t_0 \pm \sqrt{\frac{1}{\varepsilon A^*\nu_{11}^*}}\ln\tan\left(\frac{\pi}{4} + \frac{\theta_1}{4}\right) - \frac{\nu_{111}^*}{3\nu_{11}^{*2}}\theta_1 + \mathcal{O}(\sqrt{\varepsilon}). \quad (4.75)$$

In these two equations, each choice in the double signs corresponds to one of the separatrices. They are to be chosen in accordance with the same rules used for circulations: upper or lower separatrix in the double sign of (4.74) and prograde or retrograde motion in the double sign of (4.75). The terms coming from the derivatives of S_2 introduce an asymmetric correction to the height of the pendulum separatrices and on the asymptotic motions on them.

4.5 Angle–Action Variables of the Ideal Resonance Problem

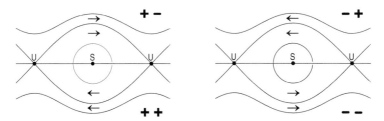

Fig. 4.1. Solutions of the Ideal Resonance Problem for diverse sign choices. *Left:* $\nu_{11}^* > 0$. *Right:* $\nu_{11}^* < 0$

4.5 Angle–Action Variables of the Ideal Resonance Problem

The angle–action variables of the Ideal Resonance Problem may be easily calculated using the one-degree-of freedom formulas of Sect. 2.1.1. We just have to pay attention to the need of some notation changes, since J_1 was already used to denote the actions in the undisturbed ($\varepsilon = 0$) problem. We will calculate the new angle w_1 and the new action

$$\Lambda_1 = \pm \frac{1}{2\pi} \oint (J_1 - J_1^*) d\theta_1, \tag{4.76}$$

in the two regimes of periodic motion: circulation and libration. The introduction of J_1^* in the function under the integral sign has the effect of adding a constant to the definition given by (2.6); this can always be done, since actions are defined except for an arbitrary additive constant.

4.5.1 Circulation

From (4.76) and (4.64), we have

$$\Lambda_1 = \pm \frac{4}{\kappa \pi} \sqrt{\frac{\varepsilon A^*}{\nu_{11}^*}}\, \mathbb{E}(\kappa) \mp \frac{\varepsilon E \nu_{111}^*}{3\nu_{11}^{*2}} + \mathcal{O}(\varepsilon \sqrt{\varepsilon}), \tag{4.77}$$

where the sign in front of the integral is to be fixed in accordance with the rules stated in Sect. 2.1.2. It is positive when $\dot\theta_1 > 0$ and negative when $\dot\theta_1 < 0$. Combining this rule with the double sign of S_1, there are four possible sign combinations: As a rule of thumb, the first of the double signs is $+$ when $\nu_{11}^* > 0$ and $-$ when $\nu_{11}^* < 0$ and the second one is $+$ for retrograde motions and $-$ for prograde motions. (See Fig. 4.1.)

The calculation of w_1 gives

$$w_1 = \pm \frac{\pi \mathcal{F}(\theta_1/2, \kappa)}{\mathbb{K}(\kappa)} - \frac{\sqrt{\varepsilon A^* \nu_{11}^*}\, \nu_{111}^* \pi}{3\kappa \nu_{11}^{*2} \mathbb{K}(\kappa)} \left(\theta_1 - \frac{\pi \mathcal{F}(\theta_1/2, \kappa)}{\mathbb{K}(\kappa)} \right) + \mathcal{O}(\varepsilon). \tag{4.78}$$

or $w_1 = \dot{w}_1(t - t_0)$, where

$$\dot{w}_1 = \frac{2\pi}{T} = \frac{\pi\sqrt{\varepsilon A^* \nu_{11}^*}}{\kappa \mathbb{K}(\kappa)} \left(1 \pm \frac{\pi \nu_{111}^* \sqrt{\varepsilon A^* \nu_{11}^*}}{3\nu_{11}^{*2} \kappa \mathbb{K}(\kappa)}\right) + \mathcal{O}(\varepsilon\sqrt{\varepsilon}), \qquad (4.79)$$

the double signs corresponding to prograde or retrograde circulations as in (4.60).

4.5.2 Libration

We continue as before, just taking into account that, in the libration regime, $\mathcal{I}_1$ is not the same as for a circulation. In this case, the contribution of some terms of $J_1 - J_1^*$ vanishes, since θ_1 oscillates in a bounded interval returning to the initial value after one libration period, without performing a complete rotation. The first approximation of the angle–action variables of the libration is, thus, the same as in the simple pendulum (with just a different constant factor and a double sign in Λ_1):

$$\begin{aligned}
\Lambda_1 &= \pm \frac{8}{\pi}\sqrt{\frac{\varepsilon A^*}{\nu_{11}^*}}\left[\mathbb{E}\left(\frac{1}{\kappa}\right) + \beta\mathbb{K}\left(\frac{1}{\kappa}\right)\right] + \mathcal{O}(\varepsilon^{3/2}) \\
&= \pm \frac{2}{\kappa^2}\sqrt{\frac{\varepsilon A^*}{\nu_{11}^*}}\left(1 + \frac{1}{8\kappa^2} + \cdots\right) + \mathcal{O}(\varepsilon^{3/2})
\end{aligned} \qquad (4.80)$$

and

$$w_1 = \frac{\pi \mathcal{F}(\zeta, \kappa^{-1})}{2\mathbb{K}(\kappa^{-1})} - \frac{\pi \nu_{111}^* \sqrt{\varepsilon A^* \nu_{11}^*}}{6\nu_{11}^{*2} \mathbb{K}(\kappa^{-1})} \theta_1 + \mathcal{O}(\varepsilon) \qquad (4.81)$$

or, $w_1 = \dot{w}_1(t - t_0)$, where

$$\dot{w}_1 = \frac{2\pi}{T} = \frac{\pi\sqrt{\varepsilon A^* \nu_{11}^*}}{2\mathbb{K}(\kappa^{-1})} + \mathcal{O}(\varepsilon\sqrt{\varepsilon}). \qquad (4.82)$$

The inversion of (4.81) gives

$$\sin\zeta = \operatorname{sn}\left(\frac{2\mathbb{K}}{\pi}w_1 + \sqrt{\varepsilon A^* \nu_{11}^*}\,\frac{\nu_{111}^* \theta_1}{3\nu_{11}^{*2}} + \mathcal{O}(\varepsilon)\right) \qquad (4.83)$$

or

$$\sin\zeta = \operatorname{sn}\left(\frac{2\mathbb{K}}{\pi}w_1\right) + \frac{\pi}{2\mathbb{K}}\frac{d}{dw_1}\operatorname{sn}\left(\frac{2\mathbb{K}}{\pi}w_1\right)\sqrt{\varepsilon A^* \nu_{11}^*}\,\frac{\nu_{111}^* \theta_1}{3\nu_{11}^{*2}} + \mathcal{O}(\varepsilon). \qquad (4.84)$$

All elliptic functions and integrals have modulus κ^{-1}. The elliptic function may be replaced by its Fourier expansion[2]

---
[2] See [17], Sect. 908.

4.5 Angle–Action Variables of the Ideal Resonance Problem

$$\operatorname{sn}\left(\frac{2\mathbb{K}}{\pi}w_1\right) = \frac{\pi\kappa}{\mathbb{K}}\sum_{j=0}^{\infty}\operatorname{csch}\left[\left(j+\frac{1}{2}\right)\chi(\kappa^{-1})\right]\sin(2j+1)w_1,$$

where $\chi(\kappa^{-1}) = \dfrac{\pi\mathbb{K}(\sqrt{1-\kappa^{-2}})}{\mathbb{K}(\kappa^{-1})}$ (see B.31). We also know that

$$\theta_1 = 2\arcsin\left(\frac{1}{\kappa}\sin\zeta\right)$$

and some iterations are needed to obtain the expansion of θ_1 at a given order. Here, it is useful to recall that

$$\operatorname{csch}\left[\left(j+\frac{1}{2}\right)\chi\right] = 2(e^{-\chi})^{j+\frac{1}{2}}\left\{1-(e^{-\chi})^{2j+1}\right\}^{-1}$$

and that $\lim_{\kappa^{-1}\to 0}\chi(\kappa^{-1}) = \infty$.

In an analogous way, we may use (4.84) to obtain similar expansions for $\cos\zeta$:

$$\cos\zeta = \operatorname{cn}\left(\frac{2\mathbb{K}}{\pi}w_1\right) + \frac{\pi}{2\mathbb{K}}\frac{d}{dw_1}\operatorname{cn}\left(\frac{2\mathbb{K}}{\pi}w_1\right)\sqrt{\varepsilon A^*\nu_{11}^*}\frac{\nu_{111}^*\theta_1}{3\nu_{11}^{*2}} + \mathcal{O}(\varepsilon) \quad (4.85)$$

and

$$\operatorname{cn}\left(\frac{2\mathbb{K}}{\pi}w_1\right) = \frac{\pi\kappa}{\mathbb{K}}\sum_{j=0}^{\infty}\operatorname{sech}\left[\left(j+\frac{1}{2}\right)\chi(\kappa^{-1})\right]\cos(2j+1)w_1$$

where cn is the Jacobian *cosine amplitude* elliptic function with modulus $1/\kappa$. We also recall that

$$\operatorname{sech}\left[\left(j+\frac{1}{2}\right)\chi\right] = 2\left(e^{-\chi}\right)^{j+\frac{1}{2}}\left\{1+(e^{-\chi})^{2j+1}\right\}^{-1}.$$

This series may be substituted into (4.73) to obtain J_1.

4.5.3 Small-Amplitude Librations

When the amplitude of the librations is small, that is, when $\kappa^{-1} \sim 0$, we may consider only the leading terms of the Taylor expansions of the elliptic integrals in powers of κ^{-1} and, thus, obtain

$$\theta_1 = \frac{2}{\kappa}\sin w_1 + \mathcal{O}(\kappa^{-3}),$$

$$J_1 = J_1^* \pm \frac{2}{\kappa}\sqrt{\frac{\varepsilon A^*}{\nu_{11}^*}}\cos w_1 + \mathcal{O}(\sqrt{\varepsilon}\kappa^{-3}).$$

To obtain θ_1 and J_1 as functions of the action Λ_1, we need to invert (4.80) with respect to κ^{-1}:

$$\frac{1}{\kappa} = \sqrt{\frac{|\Lambda_1|}{2}} \left(\frac{\nu_{11}^*}{\varepsilon A^*}\right)^{\frac{1}{4}} \left(1 - \frac{|\Lambda_1|}{16}\sqrt{\frac{\nu_{11}^*}{8A^*}} + \cdots \right) + \mathcal{O}(\varepsilon^{\sim 1}). \qquad (4.86)$$

It is also useful to introduce the libration frequency

$$\dot{w}_1 = \frac{2\pi}{T} = \sqrt{\varepsilon A^* \nu_{11}^*}\left(1 - \frac{1}{4\kappa^2} + \cdots\right) = \sqrt{\varepsilon A^* \nu_{11}^*} - \frac{1}{8}\nu_{11}^* \Lambda_1 + \cdots. \qquad (4.87)$$

An easy calculation allows us to obtain

$$\theta_1 = \sqrt{\frac{2\Lambda_1 \nu_{11}^*}{\dot{w}_1}} \sin w_1 + \mathcal{O}\left(\kappa^{-3}\right); \qquad (4.88)$$

$$J_1 = J_1^* \pm \sqrt{\frac{2\Lambda_1 \dot{w}_1}{\nu_{11}^*}} \cos w_1 + \mathcal{O}\left(\sqrt{\varepsilon}\kappa^{-3}\right). \qquad (4.89)$$

These equations give, at the lower order of approximation, θ_1, J_1 as functions of the angle–action variables of Garfinkel's Ideal Resonance Problem. We recall that Λ_1 and ν_{11}^* can be either positive or negative, but their product or quotient is always positive. $\dot{w}_1$ is always positive. The sign in front of the square root of (4.89) is positive or negative according to the sign of ν_{11}^*. The calculation of terms of higher orders requires more work, but it does not present any difficulty. (See Sect. 8.8.1.)

4.6 Morbidelli's Successive Elimination of Harmonics

The central idea of Delaunay's lunar theory has been explored by Morbidelli [76] and used to study the overlap of resonances in the phase space of the dynamical system defined by the Hamiltonian

$$H = H_0(J) + \varepsilon \sum_{h \in \mathbf{D}} A_h(J) \cos(h|\theta). \qquad (4.90)$$

Morbidelli's successive elimination of harmonics starts with the choice of an argument $(h_1|\theta)$ of H and the consideration of the system defined by the abridged Hamiltonian

$$\mathcal{F}_1 = H_0(J) + \varepsilon A_{h_1}(J) \cos(h_1|\theta), \qquad (4.91)$$

where $h_1 \equiv (h_{1(1)}, h_{1(2)}, \cdots, h_{1(N)}) \in \mathbf{Z}^N$ This system is integrable. However, at variance with Delaunay theory, the non-resonance condition $(h_1|\nu) \neq 0$ is not assumed; on the contrary, the term to start the procedure is selected from among the most important resonant terms in the domain of the phase

4.6 Morbidelli's Successive Elimination of Harmonics

space under study. It is chosen in the set of resonant terms, by its topological consequences. For instance, we may define the resonance strength of a term by its width – defined as the maximum separation between the two branches of the separatrices. From the equations of Sect. 4.4.4, we have

$$\Delta J_{\text{sep}} = 4\sqrt{\left|\frac{\varepsilon A_{h_1}}{\tilde{\nu}_{11}}\right|}, \tag{4.92}$$

where $\tilde{\nu}_{11}$ is the second derivative of $H_0(J)$ with respect to the action J_1' conjugate to $(h_1|\theta)$. It is easy to see[3] that $\tilde{\nu}_{11} \sim \mathcal{O}(|h_1^2|)$. Therefore, the most important resonances are those with higher A_{h_1} and lower $|h_1|$.

Once the term h_1 is selected, we change variables through a Lagrangian extended point transformation where we impose $\theta_1' = (h_1|\theta)$. Let it be, for example,

$$\begin{aligned} \theta_1' &= (h_1|\theta) & J_1' &= J_1/h_{1(1)} \\ \theta_\varrho' &= \theta_\varrho & J_\varrho' &= J_\varrho - (h_{1(\varrho)}/h_{1(1)})J_1 \end{aligned} \tag{4.93}$$

($\varrho = 2, \cdots, N$). Then, $\mathcal{F}_1$ becomes

$$\mathcal{F}_1 = H_0(J(J')) + \varepsilon A_{h_1}(J(J')) \cos\theta_1'. \tag{4.94}$$

This is the Hamiltonian of the Ideal Resonance Problem and we may construct its angle–action variables w_1, Λ_1 (see Sect. 4.5). Hence,

$$\theta_1' = \theta_1'(w_1, \Lambda_1; J_\varrho') \qquad J_1' = J_1'(w_1, \Lambda_1; J_\varrho'). \tag{4.95}$$

Since the given system has N degrees of freedom, we have to extend this transformation of one pair of variables to the whole set, which is done by imposing $J_\varrho' = \Lambda_\varrho$ and by using one of the algorithms of Sect. 2.4.4:

$$w_\varrho = \theta_\varrho' + \Xi_\varrho(w_1, \Lambda), \tag{4.96}$$

where, for instance,

$$\Xi_\varrho = \int_0^{w_1} \left(\frac{\partial\theta_1'}{\partial w_1}\frac{\partial J_1'}{\partial \Lambda_\varrho} - \frac{\partial J_1'}{\partial w_1}\frac{\partial\theta_1'}{\partial \Lambda_\varrho}\right) dw_1 \tag{4.97}$$

(Henrard-Lemaitre transformation).

Once we have completed the transformation, we go back to the given Hamiltonian H and perform the canonical transformation $(\theta', J') \Rightarrow (w, \Lambda)$. $\mathcal{F}_1$ will become a function of Λ only, and the remaining terms of (4.90), not included in $\mathcal{F}_1$, will be periodic functions of the angles w. They may be expanded in Fourier series so that, instead of H, we have a new Hamiltonian

$$H^* = H_0^*(\Lambda) + \varepsilon \sum_{h \in \mathbf{D}^*} A_h^*(\Lambda) \cos(h|w). \tag{4.98}$$

---

[3] From (4.93), we obtain $\tilde{\nu}_{11} = \sum_{j=1}^N \sum_{k=1}^N \nu_{jk} h_{1(j)} h_{1(k)}$.

120   4 Resonance

This completes one Delaunay–Morbidelli operation. We may, then, restart the procedure by choosing a new term in H^*:

$$\mathcal{F}_2 = H_0^*(\Lambda) + \varepsilon A_{h_2}^*(\Lambda) \cos(h_2|w). \tag{4.99}$$

We then introduce $w_1' = (h_2|w)$ and new angles w_ϱ' and momenta Λ' through a Lagrange point transformation; we construct new angle–action variables ψ_1, K_1 and complete the transformation to include the other degrees of freedom. We thus get a new H^{**} and continue as before.

We have, purposely, presented the Delaunay–Morbidelli operation without stressing that the angle–action variables are not globally valid and are not computed in the same way in circulations and librations. In fact, Morbidelli's successive elimination of harmonics is not meant to construct formal solutions (the chains of elliptic functions and integrals would make it impossible), but to map the geometry of the resonances in a given domain of the phase space. One important point is that, in H^*, the angles are w. The w_ϱ differ from the given θ_ϱ by the quantity Ξ_ϱ, which is small: the corresponding frequencies in $H_0(J)$ and $H_0^*(\Lambda)$ are of the same order. w_1 is the uniform angle associated with the libration (or circulation), and has the frequency of this motion. Thus, new resonances may appear in H^*, involving w_1 and some of the w_ϱ not appearing in the given H. The best known examples are the so-called secondary resonances in the Kirkwood gaps of the asteroid belt (see [77]). These gaps appear near initial conditions corresponding to asteroids with an orbital period commensurable with Jupiter's period. The motion of an asteroid inside the gap is a libration about a periodic orbit; the libration frequency may be approximately known by selecting the main term with the critical combination of the two longitudes, and using the Ideal Resonance Problem. The Hamiltonian H^* shows new critical terms in which the libration frequency is a multiple of the frequency of motion of the perihelion (one of the θ_ϱ). The overlap of these secondary resonances may be studied taking, in turn, each of these terms in H^* to compose the abridged Hamiltonian $\mathcal{F}_2$.

To circumvent the difficulties due to elliptic functions and integrals, it is possible to construct numerically all transformations mentioned in this section. We may use the direct techniques described in Sect. 2.2 to construct the angle–action variables. In such case, the result will not be written as formal functions, but as functions defined by a table or computer code allowing them to be known.

4.6.1 An Example

Let us consider an application of Morbidelli's elimination algorithm to the Hamiltonian function obtained at the end of Sect. 3.8. We discard terms of the order $\mathcal{O}(\varepsilon^2)$ and adopt the notation θ, J (without stars) for angles and actions. Also, for practical reasons, we interchange the subscripts 2 and 3 in the variables and adopt the particular value $A_0 = \alpha J_2$ for the secular

term of the perturbation. We also assume that α, B_1, L_0 and M_1 are positive constants. Hence,
$$H(\theta, J) = H_0 + \varepsilon H_2$$
where
$$H_0 = -\frac{1}{2J_1^2} + \nu_3(J_3 - 2J_1) \tag{4.100}$$
and
$$H_2 = \alpha J_2 + L_0\sqrt{-J_2}\cos\theta_2 + B_1\cos\theta_1 + M_1\sqrt{-J_2}\cos(\theta_1 + \theta_2). \tag{4.101}$$

The action J_3 is a constant (since the angle θ_3 is absent from the Hamiltonian) and the exact resonance value of J_1 is defined by
$$\nu_1^* = \nu_1(J_1^*) = \left.\frac{\partial H_0}{\partial J_1}\right|_{J_1=J_1^*} \stackrel{\text{def}}{=} 0; \tag{4.102}$$
that is
$$J_1^* = \frac{1}{\sqrt[3]{2\nu_3}}. \tag{4.103}$$
We also have
$$\nu_{11}^* = -6\nu_3\sqrt[3]{2\nu_3} = -3(2\nu_3)^{4/3}. \tag{4.104}$$
We recall that the example of Sect. 3.8 is founded on the asteroidal three-body problem and B_1 is, there, a quantity of the order of the orbital eccentricity of the disturbing planet.

In the neighborhood of $J_1 = J_1^*$, the Hamiltonian given by (4.101) has two resonant terms: $\varepsilon B_1 \cos\theta_1$ and $\varepsilon M_1\sqrt{-J_2}\cos(\theta_1 + \theta_2)$. Let us consider the Ideal Resonance Problems (IRPs) which they, separately, define:
$$\begin{aligned}\mathcal{F}_{1(a)} &= H_0(J) + \varepsilon B_1 \cos\theta_1 \\ \mathcal{F}_{1(b)} &= H_0(J) + \varepsilon M_1\sqrt{-J_2}\cos(\theta_1 + \theta_2).\end{aligned} \tag{4.105}$$

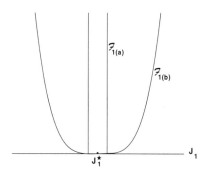

Fig. 4.2. Separatrices of the two IRPs of (4.105)

The widths (maximum libration amplitudes) of these resonances are, respectively,

$$\Delta J_{\text{sep}(a)} = 4\sqrt{\frac{-\varepsilon B_1}{\nu_{11}^*}}$$

$$\Delta J_{\text{sep}(b)} = 4\sqrt{\frac{-\varepsilon M_1}{\nu_{11}^*}} \sqrt[4]{-J_2}.$$

Figure 4.2 shows the locus of the separatrices of the two considered IRPs, in the plane $J_1, |J_2|$. If $|B_1| \ll |M_1|$, the strip corresponding to the resonance of $\mathcal{F}_{1(a)}$ is narrow (as shown in the figure) and the hierarchy of the two considered harmonics is well established. It is then possible to start the elimination of harmonics with the largest one, $\mathcal{F}_{1(b)}$.

Following the recipe given above, we perform, initially, the point transformation

$$\begin{aligned} \theta_2' &= \theta_1 & J_2' &= J_1 - J_2 - J_1^* \\ \theta_1' &= \theta_1 + \theta_2 & J_1' &= J_2; \end{aligned} \qquad (4.106)$$

$\mathcal{F}_1$ becomes

$$\mathcal{F}_1 = \mathcal{F}_{1(b)} = H_0(J(J')) + \varepsilon M_1 \sqrt{-J_1'} \cos \theta_1'. \qquad (4.107)$$

Let us consider the small-amplitude librations of this one-degree-of-freedom system about the libration center $J_1'^* = -J_2'$. They are given by (see 4.88 and 4.89):

$$\theta_1' = \sqrt{\frac{2\Lambda_1 \nu_{11}^*}{\dot{w}_1}} \sin w_1 \qquad (4.108)$$

$$J_1' = -J_2' - \sqrt{\frac{2\Lambda_1 \dot{w}_1}{\nu_{11}^*}} \cos w_1, \qquad (4.109)$$

where w_1, Λ_1 are the angle–action variables of the IRP defined by $\mathcal{F}_1$, ν_{11}^* is a known number and

$$\dot{w}_1 = \sqrt{-\varepsilon \nu_{11}^* M_1} \sqrt[4]{J_2'} - \frac{1}{8} \nu_{11}^* \Lambda_1. \qquad (4.110)$$

In order to have $\theta_1' = 0$ at the libration center, we assumed $M_1 > 0$ (we recall that $\nu_{11}^* < 0$ and $\Lambda_1 < 0$). The next step in Morbidelli's algorithm is to complete the canonical transformation $(\theta_1', \theta_2', J_1', J_2') \Rightarrow (w_1, w_2, \Lambda_1, \Lambda_2)$ through

$$\begin{aligned} \theta_2' &= w_2 - \Xi_2(w_1, \Lambda_1, \Lambda_2) \\ J_2' &= \Lambda_2, \end{aligned}$$

where

$$\Xi_2 = \int_0^{w_1} \left(\frac{\partial \theta_1'}{\partial w_1} \frac{\partial J_1'}{\partial \Lambda_2} - \frac{\partial J_1'}{\partial w_1} \frac{\partial \theta_1'}{\partial \Lambda_2} \right) dw_1.$$

We note that θ_1', J_1' depend on J_2', that is, on Λ_2, also through $\dot{w}_1$. The derivatives are

4.6 Morbidelli's Successive Elimination of Harmonics

$$\frac{\partial \theta_1'}{\partial w_1} = \sqrt{\frac{2\Lambda_1 \nu_{11}^*}{\dot{w}_1}} \cos w_1,$$

$$\frac{\partial J_1'}{\partial w_1} = \sqrt{\frac{2\Lambda_1 \dot{w}_1}{\nu_{11}^*}} \sin w_1,$$

$$\frac{\partial \theta_1'}{\partial \Lambda_2} = \frac{\nu_{11}^*}{8\Lambda_2^{3/4}} \sqrt{\frac{-2\varepsilon M_1 \Lambda_1}{\dot{w}_1^3}} \sin w_1,$$

$$\frac{\partial J_1'}{\partial \Lambda_2} = -1 - \frac{1}{8\Lambda_2^{3/4}} \sqrt{\frac{-2\varepsilon M_1 \Lambda_1}{\dot{w}_1}} \cos w_1.$$

Hence,

$$\Xi_2 = -\int_0^{w_1} \left(\frac{\partial \theta_1'}{\partial w_1} + \frac{|\Lambda_1|\sqrt{-\varepsilon M_1 \nu_{11}^*}}{4\Lambda_2^{3/4} \dot{w}_1} \right) dw_1,$$

or

$$\Xi_2 = -\theta_1' - \frac{\Lambda_1}{8\Lambda_2}, \tag{4.111}$$

where, for the sake of simplicity, we kept $\dot{w}_1$ restricted to its first approximation. The transformation is, now, complete and may be used to transform the given Hamiltonian.

With the new variables, $\mathcal{F}_1$ may depend only on the actions. The substitution of variables in $\mathcal{F}_1$ is cumbersome and the cancellation of periodic terms, in higher orders, is only partially achieved because of the many simplifications introduced. However, a shortcut exists. We know that, if we denote by $\widehat{\mathcal{F}}_1(\Lambda_1, \Lambda_2)$ the result of the transformation, by definition,

$$\dot{w}_1 = \frac{\partial \widehat{\mathcal{F}}_1}{\partial \Lambda_1}$$

or

$$\widehat{\mathcal{F}}_1 = \int \dot{w}_1 \, d\Lambda_1.$$

The problem with this shortcut is that the integration introduces an arbitrary additive function of Λ_2, for whose derivation, the direct transformation is necessary. Since this additive function cannot depend on Λ_1 and all involved functions are polynomials in $\sqrt{-\Lambda_1}$, we need just transform the parts of $\widehat{\mathcal{F}}_1$ independent of Λ_1 to obtain it. Hence

$$\widehat{\mathcal{F}}_1(\Lambda_1, \Lambda_2) = \sqrt{-\varepsilon \nu_{11}^* M_1} \sqrt[4]{\Lambda_2} \Lambda_1 - \frac{1}{16} \nu_{11}^* \Lambda_1^2 + \varepsilon M_1 \sqrt{\Lambda_2}, \tag{4.112}$$

where the two first terms resulted from the integration of $\dot{w}_1$ and the last one from a direct calculation. The constant terms (depending on J_3 and J_1^*) do not need to be taken into account since they do not contribute to the equations. The terms of H not considered in $\mathcal{F}_{1(b)}$ need to be written with

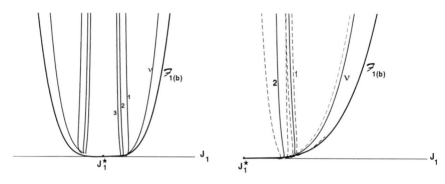

Fig. 4.3. Secular (ν) and secondary ($k = 1, 2, 3$) resonances

the new variables. The results are Fourier expansions in the angle $w_2 + kw_1$. The new $\widehat{H}_0$ is

$$\widehat{H}_0 = \widehat{\mathcal{F}}_1 - \varepsilon \alpha \Lambda_2. \quad (4.113)$$

To do a new Delaunay–Morbidelli operation, we have to select a new resonant periodic term to add to $\widehat{H}_0$. Let us, first, search the resonance locus of the main terms. To do this, we need the expressions for $\dot{w}_1$ and $\dot{w}_2$:

$$\dot{w}_1 = \sqrt{-\varepsilon \nu_{11}^* M_1} \sqrt[4]{\Lambda_2} - \frac{1}{8} \nu_{11}^* \Lambda_1 \quad (4.114)$$

$$\dot{w}_2 = \frac{\partial \widehat{H}_0}{\partial \Lambda_2} = \sqrt{-\varepsilon \nu_{11}^* M_1} \frac{\Lambda_1}{4\Lambda_2^{3/4}} + \frac{1}{2} \frac{\varepsilon M_1}{\sqrt{\Lambda_2}} - \varepsilon \alpha. \quad (4.115)$$

When numerical values are given to ε, ν_{11}^*, M_1 and α, the locus of the curves $\dot{w}_2 \pm k\dot{w}_1 = 0$ is easily found. It is convenient to show these curves in the plane $J_1, |J_2|$ instead of the plane Λ_1, Λ_2. The transformation $\Lambda_1, \Lambda_2 \Rightarrow J_1, J_2$, however, depends on w_1. It is, then, necessary to fix the value of w_1. We follow the same practice usual in resonant asteroid dynamics, and fix it at the boundaries of the librations of the action J_1' conjugate to the critical angle θ_1'. Thus, we assume $|\cos w_1| = 1$. As a consequence, to each point in the plane (Λ_1, Λ_2) we obtain two points in the plane (J_1, J_2), one on each side of the vertical line $J_1 = J_1^*$. Figure 4.3 (*left*) shows the lines falling inside the boundary of the libration domain. They are: the secular resonance $\dot{w}_2 = 0$ (indicated by ν following astronomers' classical notation); and the secondary resonances $\dot{w}_2 + k\dot{w}_1 = 0$ with $k > 0$ (the lines $k = 1, 2, 3$ are shown)[4].

To each of the resonances in Fig. 4.3 (*left*) there corresponds one libration zone defined by the separatrices of the Ideal Resonance Problem obtained when the corresponding perturbative term is selected and added to $\widehat{H}_0$. Let us introduce the new set of canonical variables

---

[4] For a Lie series study of secular and secondary resonances, see Sects. 9.4.5 and 9.4.6.

4.6 Morbidelli's Successive Elimination of Harmonics

$$\begin{aligned}\varphi_1 &= w_2 + kw_1 & K_1 &= \Lambda_2 \\ \varphi_2 &= w_1 & K_2 &= \Lambda_1 - k\Lambda_2\end{aligned} \quad (4.116)$$

and study the Ideal Resonance Problems

$$\mathcal{F}_{2(k)} = \widehat{H}_0(K(\Lambda)) + \mathcal{A}_k(K)\cos\varphi_1 \qquad (k=0,1,2,3), \qquad (4.117)$$

where the coefficients $\mathcal{A}_k(K)$ come from the expansion of those terms of $H(\theta, J)$ whose coefficients are B_1 and L_0.

Figure 4.3 (*right*) shows the separatrices of the IRPs corresponding to $k = 0, 1, 2$. The secular resonance $k = 0$ (ν) and the secondary resonance $k = 1$ are isolated in this figure, while the secondary resonances $k = 2$ and $k = 3$ (shown only in Fig. 4.3 *left*) are very close and overlap each other. (The secondary resonance $k = 3$ is inside the resonance zone of the secondary resonance $k = 2$.) If Fig. 4.3 (*right*) were the result of an exact numerical calculation, $\mathcal{F}_{2(0)}$ and $\mathcal{F}_{2(1)}$ could be considered as good candidates for elimination of further harmonics. On the contrary, because of the overlap of their libration domains, the isolated consideration of $\mathcal{F}_{2(2)}$ or $\mathcal{F}_{2(3)}$ would be unrealistic. However, Fig. 4.3 (*right*) is the result of analytical approximations valid only in a small neighborhood of $J_1 = J_1^*$, and we have to restrict our analysis to it. The motions in this neighborhood are far from the resonance lines of Fig. 4.3 (*left*) and we may use the original Delaunay operation of Sect. 4.1 to get rid of the harmonic remaining in the Hamiltonian. Maybe, in the case of the harmonic $k = 2$, given the broadness of its resonant zone, we should consider the expression of the circulations given by the Ideal Resonance Problem, since that given by the classical Delaunay operation assumes that the resonance is very far and do not influence the solution.

One important remark yet to be made concerns the numerical choice of the coefficients appearing in the Hamiltonian. To obtain Fig. 4.3 (*right*), we had to consider $L_0 \ll M_1$ and neglect the term $\varepsilon B_2 \cos\theta_1$. Otherwise, the libration zones of the $\mathcal{F}_{2(k)}$ would be so broad that they would overlap over almost the whole region shown in the figures. In that case, it would no longer be possible to select one domain in the plane for further studies with the technique discussed here. These limitations may not, however, be considered as a weak point. On the contrary, allowing us to map the overlap of resonances, Morbidelli's successive elimination of harmonics clearly shows the extreme limits where approximate regular solutions can exist.

The given example used the heavy analytical machinery of Garfinkel's Ideal Resonance Problem with the aim of allowing the reader to have a step-by-step view of the technique. But one should take advantage of the possibility of direct numerical construction of the transformations leading to particular angle–action variables, as discussed in Sect. 2.2, to have exact calculations and, as a consequence, an exact chart of resonances and libration domains, at every step.

5

Lie Mappings

5.1 Lie Transformations

There is a straightforward way of introducing Lie series mappings into the study of Hamiltonian systems. Proposition 5.3.1 shows that the mapping defined by the series (5.25)–(5.26) is canonical and may be used in the construction of canonical perturbation theories. Thus, the reader interested only in such applications can go straight to those equations and skip the first two sections whose aim is to introduce Lie derivatives, Lie mappings and the relationship between Lie mappings and Jacobian canonical transformations.

5.1.1 Infinitesimal Canonical Transformations

Let us consider a Jacobian generating function S, continuous and with continuous derivatives with respect to a parameter λ:

$$S = S(Q, q; \lambda) \qquad \lambda \in (a, b) \subset \mathbf{R} \qquad (5.1)$$

$Q \equiv (Q_1, Q_2, \cdots, Q_N)$, $q \equiv (q_1, q_2, \cdots, q_N)$. It spans a one-parameter family of canonical transformations:

$$p_i = \frac{\partial S(Q, q; \lambda)}{\partial q_i} \qquad P_i = -\frac{\partial S(Q, q; \lambda)}{\partial Q_i} \qquad (5.2)$$

fixed by the parameter λ. The transformation generated by $S(Q, q; \lambda)$ maps one point (Q, P) of the given phase space into one point (q, p) of the transformed phase space. In the same way, the transformation generated by $S(Q, q^*; \lambda^*)$ maps the same point (Q, P) into some other point (q^*, p^*). Since $S(Q, q; \lambda)$ is a continuous function of λ, to neighboring values of $\lambda \in (a, b)$ there correspond neighboring points in the phase space. Thus, the transformation determines, in the new phase space, an ordered set Γ which is the locus of the points into which (Q, P) is mapped when λ varies in (a, b) (Fig. 5.1).

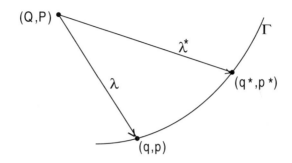

Fig. 5.1. The transformations generated by $S(Q, q; \lambda)$

Because of the group property of the canonical transformations, the transformation from (q^*, p^*) to (q, p) is itself canonical. When $\lambda - \lambda^*$ is an infinitesimal quantity, the transformation $(q^*, p^*) \Rightarrow (q, p)$ is a special type of canonical transformation in which each point of the phase space is transformed into a neighboring point. These transformations are called *infinitesimal canonical transformations* [97].

We use the Jacobian canonical condition of Sect. 1.3 to write

$$\begin{array}{l}(p \mid \delta q) - (P \mid \delta Q) = \delta S(Q, q; \lambda) \\ (p^* \mid \delta q^*) - (P \mid \delta Q) = \delta S(Q, q^*; \lambda^*),\end{array} \quad (5.3)$$

where the variations δ refer only to arbitrary changes in the canonical variables; the parameter is kept fixed during this operation. Forming the difference of the two equations above, we obtain the relation

$$(p^* \mid \delta q^*) - (p \mid \delta q) = \delta \left[S(Q, q^*; \lambda^*) - S(Q, q; \lambda) \right]. \quad (5.4)$$

We may then write the first-order Taylor expansion of S:

$$S(Q, q^*; \lambda^*) = S(Q, q; \lambda) + \left(\frac{\partial S(Q, q; \lambda)}{\partial q} \right) \bigg|\, (q^* - q) + \frac{\partial S(Q, q; \lambda)}{\partial \lambda}(\lambda^* - \lambda).$$

Substituting this expansion into (5.4) and taking (5.2) into account, we obtain, to first order,

$$(p^* \mid \delta q^*) - (p \mid \delta q) = \delta(p \mid q^* - q) + (\lambda^* - \lambda) \delta \left[\frac{\partial S(Q, q; \lambda)}{\partial \lambda} \right].$$

A straightforward calculation leads to the remarkable equation

$$(p - p^* \mid \delta q) - (q - q^* \mid \delta p) = (\lambda - \lambda^*) \delta \left[\frac{\partial S(Q, q; \lambda)}{\partial \lambda} \right], \quad (5.5)$$

where, in the first term, we have neglected the higher-order infinitesimal difference $(p - p^* \mid \delta q^* - \delta q)$. What makes this equation remarkable is that the differential form in the left-hand side of (5.5) involves the variations of the two old variables q, p, instead of mixing variations of new and old variables as with the differential forms of Sect. 1.3. The price to be paid for having this uniformity is that the transformation $(q, p) \Rightarrow (q^*, p^*)$ is only infinitesimally different from the identity. Thus, it is not so general as the Jacobian canonical transformations. In order to draw further conclusions, we introduce the function

$$S'_\lambda(Q, q; \lambda) = \frac{\partial S(Q, q; \lambda)}{\partial \lambda}. \tag{5.6}$$

The system (5.2) may be solved with respect to Q giving $Q = Q(q, p, \lambda)$. Thus, we may substitute it for Q in S'_λ and call $-W$ the resulting function:

$$W(q, p; \lambda) \stackrel{\text{def}}{=} -S'_\lambda(Q(q, p; \lambda), q; \lambda).$$

With the help of the function W, (5.5) is written in a simpler way,

$$(q - q^* \mid \delta p) - (p - p^* \mid \delta q) = (\lambda - \lambda^*)\, \delta W(q, p; \lambda), \tag{5.7}$$

and comparison of the two sides of this equation yields the transformation equations

$$q_i - q_i^* = (\lambda - \lambda^*) \frac{\partial W}{\partial p_i}, \qquad p_i - p_i^* = -(\lambda - \lambda^*) \frac{\partial W}{\partial q_i}. \tag{5.8}$$

These equations show that infinitesimal canonical transformations can be represented in explicit form. The relative variables $p - p^*$ and $q - q^*$ are explicitly expressed in terms of a single function $W(q, p; \lambda)$ characterizing the transformation.

Dividing both sides of (5.8) by $\lambda - \lambda^*$ and letting $\lambda - \lambda^*$ tend towards zero, gives

$$\frac{dq_i}{d\lambda} = \frac{\partial W}{\partial p_i}, \qquad \frac{dp_i}{d\lambda} = -\frac{\partial W}{\partial q_i}. \tag{5.9}$$

The general solution of these differential equations is a one-parameter group of canonical transformations. The function W determines this group uniquely. We shall call W the *Lie generating function* or *Lie generator* of the group.

Definition 5.1.1 (Lie mapping). *Lie mappings are all canonical transformations defined by solutions of the canonical equations (5.9).*

It is worthwhile mentioning that the reasoning with first-order infinitesimal quantities is necessary only to establish the equations. Once the limit operation is accomplished, the result is an exact equation whose solutions allow us to extend the domain of these transformations to finite intervals. They are usually presented in power series, as in Sect. 5.3, but the definition given above is more general.

Equation (5.9) serves also to show the duality existing between canonical equations and canonical transformations. Every property of one of them may be extended to the other. The exploration of this duality is beyond the scope of this work. However, the transposition of the results from canonical equations to canonical transformations and vice versa is easy.

5.2 Lie Derivatives

Definition 5.2.1 (Lie derivative [65]). *Let $\mathcal{O}$ be an open subset of the phase space and let $\mathcal{F}$ be the ring of all functions $f : \mathcal{O} \to \mathbf{R}$ of class C^∞. The Lie derivative generated by W is the application*

$$f \to D_W f = \{f, W\}, \tag{5.10}$$

where $\{f, W\}$ is the Poisson bracket of the functions f and $W \in \mathcal{F}$:

$$\{f, W\} = \sum_{i=1}^{N} \left(\frac{\partial f}{\partial q_i} \frac{\partial W}{\partial p_i} - \frac{\partial W}{\partial q_i} \frac{\partial f}{\partial p_i} \right). \tag{5.11}$$

The application (5.10) has the following characteristic properties:

$$\begin{aligned} D_W(f+g) &= D_W f + D_W g \\ D_W(fg) &= f D_W g + g D_W f \\ D_W c &= 0 \end{aligned} \tag{5.12}$$

for all $f \in \mathcal{F}, g \in \mathcal{F}$ and all constants c. These properties allow us to say that D_W defines a derivation of the ring $\mathcal{F}$.

The well-known *Jacobi identity*

$$\{\{f, g\}, h\} + \{\{g, h\}, f\} + \{\{h, f\}, g\} = 0$$

leads to another important property of the Lie derivatives:

$$D_W\{f, g\} = \{D_W f, g\} + \{f, D_W g\}. \tag{5.13}$$

This property of the Lie derivatives allows us to say that they form a *Lie algebra*.

The Jacobi identity also implies the existence of a property concerning the composition of the Lie derivatives generated by two functions W_1 and W_2:

$$D_{\{W_1, W_2\}} = D_{W_2} D_{W_1} - D_{W_1} D_{W_2}$$

Repeated application of the operator D_W gives

$$D_W^n f = D_W(D_W^{n-1} f).$$

By recurrence over n, the properties given by (5.12) and (5.13) can be extended to become

$$D_W^n(f+g) = D_W^n f + D_W^n g \qquad (5.14)$$

$$D_W^n(fg) = \sum_{k=0}^{n} \binom{n}{k} D_W^k f . D_W^{n-k} g \qquad (5.15)$$

$$D_W^n\{f,g\} = \sum_{k=0}^{n} \binom{n}{k} \{D_W^k f, D_W^{n-k} g\}. \qquad (5.16)$$

Lie derivatives have an important interpretation:

Proposition 5.2.1. *The Lie derivative of f generated by W is the derivative, with respect to λ, of the restriction of f to an integral curve of the canonical equations spanned by W.*

Proof. Let $\tilde{f} : \Gamma \to \mathbf{R}$ be a restriction of the function f defined by

$$\tilde{f} = f(q(\lambda), p(\lambda)), \qquad (5.17)$$

where $q(\lambda), p(\lambda)$ is a solution of the differential equations (5.9) for a given set of initial conditions. We then have

$$\frac{d\tilde{f}}{d\lambda} = \sum_{i=1}^{N} \left(\frac{\partial \tilde{f}}{\partial q_i} \frac{dq_i}{d\lambda} + \frac{\partial \tilde{f}}{\partial p_i} \frac{dp_i}{d\lambda} \right)$$

or, taking into account (5.9) and the definition of the Lie derivatives,

$$\frac{d\tilde{f}}{d\lambda} = \{f, W\} = D_W f. \qquad (5.18)$$

□

Exercise 5.2.1 (Homogeneity). Show that the Lie derivative D_W^n is homogeneous of degree n in W, that is, if $k \in \mathbf{R}$ is a constant,

$$D_{kW}^n f = k^n D_W f. \qquad (5.19)$$

5.3 Lie Series

We have defined the Lie mapping as a solution of the canonical system of differential equations (5.9). It follows, therefore, that the construction of a Lie mapping is the construction of solutions of (5.9). The usual integration by series involves the Cauchy existence theorem:

Theorem 5.3.1 (Cauchy[1]). *Given a system of ordinary differential equations in Weierstrass normal form:*

$$\frac{dy}{dx} = Y(y, x), \tag{5.20}$$

where $Y : \mathbf{R}^{n+1} \to \mathbf{R}^n$ *is analytical in some domain* $\mathbf{D} \subset \mathbf{R}^{n+1}$, *this system has a solution* $y(x)$ *analytical in a non-infinitesimal domain* $\mathbf{D}_1 \subset \mathbf{D}$ *that contains the given initial conditions* x_0, y_0.

Cauchy's existence theorem means that the solutions of (5.20) may be written as a convergent Taylor series:

$$y = y_0 + \sum_{k=1}^{\infty} \frac{1}{k!} \left(\frac{d^k y}{dx^k}\right)_{x=x_0} (x - x_0)^k, \tag{5.21}$$

where the values of the derivatives $d^k y/dx^k$ can be formally deduced from the given equations by repeated differentiation of (5.20). Hence,

$$y = y_0 + Y(y_0, x_0)(x - x_0) + \sum_{k=2}^{\infty} \frac{1}{k!} \left(\frac{d^{k-1} Y(y(x), x)}{dx^{k-1}}\right)_{x=x_0} (x - x_0)^k. \tag{5.22}$$

Now, let $\mathcal{O}$ be an open subset of the phase space and let the right-hand sides of (5.9) be analytic for $(q, p) \in \mathcal{O}$ and λ in a neighborhood $V(0)$ of the origin. The series

$$q_i = q_i^* + \sum_{k=1}^{\infty} \frac{\lambda^k}{k!} \left[\frac{d^{k-1}}{d\lambda^{k-1}}\left(\frac{\partial W}{\partial p_i}\right)\right]_{\lambda=0} \tag{5.23}$$

$$p_i = p_i^* - \sum_{k=1}^{\infty} \frac{\lambda^k}{k!} \left[\frac{d^{k-1}}{d\lambda^{k-1}}\left(\frac{\partial W}{\partial q_i}\right)\right]_{\lambda=0} \tag{5.24}$$

converge in some $\mathcal{O}_1 \subset \mathcal{O}$ and $V_1(0) \subset V(0)$ and represent the solution of (5.9) that assumes the values q^*, p^* when $\lambda = 0$.

The derivatives $\frac{d}{d\lambda}$ are just the Lie derivatives generated by the function W (see 5.18) and the above equations may be written as

$$q_i = \sum_{k=0}^{\infty} \frac{\lambda^k}{k!} \left[D_W^k q_i\right]_{\lambda=0} \tag{5.25}$$

$$p_i = \sum_{k=0}^{\infty} \frac{\lambda^k}{k!} \left[D_W^k p_i\right]_{\lambda=0}. \tag{5.26}$$

These relations define the Lie series.

---

[1] For a proof of Cauchy's theorem, see [91].

Definition 5.3.1 (Lie series). *For each function $f \in \mathcal{F}$, each point $(q, p) \in \mathcal{O}$ and a given Lie generator W of class C^∞ in $\mathcal{O}$, the application*

$$f \to E_W f = \sum_{k=0}^{\infty} \frac{\lambda^k}{k!} D_W^k f \tag{5.27}$$

is the Lie series of the function f, generated by W.

The following properties are immediate consequences of (5.14), (5.15) and (5.16):

$$\begin{aligned} E_W(f+g) &= E_W f + E_W g \\ E_W(fg) &= E_W f . E_W g \\ E_W c &= c. \end{aligned} \tag{5.28}$$

In addition, we have

Proposition 5.3.1 (Canonical condition).

$$E_W \{f, g\} = \{E_W f, E_W g\}. \tag{5.29}$$

Proof.

$$\{E_W f, E_W g\} = \left\{ \sum_{k=0}^{\infty} \frac{\lambda^k}{k!} D_W^k f, \sum_{k'=0}^{\infty} \frac{\lambda^{k'}}{k'!} D_W^{k'} g \right\} = \sum_{k=0}^{\infty} \sum_{k'=0}^{\infty} \frac{\lambda^{k+k'}}{k! k'!} \{D_W^k f, D_W^{k'} g\}$$

or, changing to $n = k + k'$ in the summations:

$$\{E_W f, E_W g\} = \sum_{n=0}^{\infty} \sum_{k=0}^{n} \frac{\lambda^n}{k!(n-k)!} \{D_W^k f, D_W^{n-k} g\}$$

$$= \sum_{n=0}^{\infty} \frac{\lambda^n}{n!} \sum_{k=0}^{n} \binom{n}{k} \{D_W^k f, D_W^{n-k} g\}.$$

Hence, because of (5.16),

$$\{E_W f, E_W g\} = \sum_{n=0}^{\infty} \frac{\lambda^n}{n!} D_W^n \{f, g\} = E_W \{f, g\}.$$

□

Proposition 5.3.1 states that Poisson brackets of two canonical variables are invariant to the transformation $(q, p) \Rightarrow (E_W q, E_W p)$ and therefore, this transformation is canonical Thus, the canonical nature of Lie mappings is a consequence of (5.29). When a direct introduction of Lie series is desirable, we may start the subject at this very point, just defining the Lie series and stating Proposition 5.3.1.

We may extend to the Lie series, the interpretation given to the Lie derivatives:

Proposition 5.3.2. *The Lie series of f, generated by $W(q,p)$, is the Taylor series expansion, around $\lambda = 0$, of the restriction of f to an integral curve of the canonical equations spanned by W.*

Proof. Let us consider, again, the restriction $\tilde{f} : \Gamma \to \mathbf{R}$ of the function f (see (5.17)). Then, (5.27) may be written

$$E_W f = \sum_{k=0}^{\infty} \frac{\lambda^k}{k!} \left(\frac{d^k \tilde{f}}{d\lambda^k} \right)_{\lambda=0}$$

□

The series representation of the Lie mapping generated by the function $W(q,p)$, given by (5.25)–(5.26), may be written in a very compact form as

$$z = E_{W^*} z^*, \qquad (5.30)$$

where $z \equiv (q,p)$, $z^* \equiv (q^*, p^*)$ and $W^* = W(z^*)$. Equation (5.30) gives the Taylor series expansion around $\lambda = 0$ of the restriction of the variables z to an integral curve of the canonical equations spanned by W. Proposition 5.3.2 shows that this result is not restricted to the variables and the transformation defined here is such that for any $f \in \mathcal{F}$ we have[2]

$$f(z) = E_{W^*} f(z^*). \qquad (5.31)$$

Canonical transformations given by Lie series have the practical advantage of avoiding cumbersome operations such as inversions and substitutions, which are always necessary when Jacobian transformations are used (as discussed in Sect. 3.12).

However, one must keep in mind that Lie series mappings are not universal in the sense that not every canonical mapping can be represented as a Lie series mapping. This representation is restricted to mappings in the neighborhood of the identity. This makes them very useful in perturbation theories. We recall that the classical theories discussed in Chap. 3 are always such that the canonical transformation is reduced to the identical transformation, when $\varepsilon = 0$ (the generating function is reduced to $(q \mid p^*)$).

5.4 Inversion of a Lie Mapping

Proposition 5.4.1. $E_W E_{-W}$ *is the identity operator.*

---

[2] This result is known as the *commutation theorem* [43]. Indeed, from (5.30) and (5.31), it follows that $f(E_{W^*} z^*) = E_{W^*} f(z^*)$.

Proof. Using the definition of Lie series, we obtain

$$E_W E_{-W} f(z) = \sum_{k'=0}^{\infty} \sum_{k=0}^{\infty} \frac{\lambda^{k'+k}}{k'!k!} D_W^{k'} D_{-W}^k f(z)$$

or, changing to $\ell = k + k'$,

$$E_W E_{-W} f(z) = \sum_{\ell=0}^{\infty} \sum_{k=0}^{\ell} \frac{\lambda^\ell}{\ell!} (-1)^k \binom{\ell}{k} D_W^\ell f(z).$$

The proof follows from the fact that $\sum_{k=0}^{\ell} (-1)^k \binom{\ell}{k} = 0$ for all $\ell \neq 0$, and then $E_W E_{-W} f(z) = f(z)$. □

Proposition 5.3.2 and Fig. 5.1 allow us to get a trivial insight into the above proposition. $f(z) = E_{W^*} f(z^*)$ (where $W^* = W(z^*)$) is the Taylor series of the restriction of f to the integral curve Γ about $\lambda = 0$ (that is, about $z^* \equiv (q^*, p^*)$) and gives the value of f at a generic λ (that is, at a generic $z \equiv (q, p)$). To get the inversion of the Lie series, we have just to invert the role played by z and z^* and the direction of motion along Γ (that is, the sign of the generating function). Then $f(z^*) = E_{-W} f(z)$ (where $W = W(z)$). It is worth recalling that, by construction, $W(z) = W(z^*)$ (W is the "Hamiltonian" of the equations (5.9) defining the Lie mapping.

5.5 Lie Series Expansions

In the applications to canonical perturbation theory, some authors (e.g. Deprit [23]) used, instead of the external parameter λ, the same parameter ε used to characterize the strength of the perturbation. This choice is not possible here because the Lie derivative may operate on functions depending on ε and several of the previous results, e.g. (5.18), no longer holds if f depends on the parameter. If it is desired to write Lie series expansions as power series in ε, it is enough to replace the Lie generator by εW and use the homogeneity property $D_{\varepsilon W}^n = \varepsilon^n D_W^n$ to obtain such a series (but keeping λ as the formal parameter of the transformation). Besides, since W is always introduced in perturbation theory as an arbitrary function to be determined so as to satisfy some given property, we may arbitrarily take whichever Lie generator suits us best. For instance, as Mersman [69], we adopt

$$\lambda = 1. \tag{5.32}$$

It is worth noting that the canonical condition given by (5.29) holds for all λ. It is not necessary to assume that λ is small. The decrease of the terms necessary to guarantee the convergence of the series may be obtained otherwise, for instance by choosing $||W||$ small. (Remember again that the operator D_W^n is homogeneous of degree n with respect to W.)

Let us, now, assume that f and W are the series

$$f = \sum_{k=0}^{\infty} f_k \qquad \text{and} \qquad W = \sum_{k=1}^{\infty} W_k. \tag{5.33}$$

We write these series in a more general form than just power series in some small parameter. The subscript k in f_k and W_k indicates that these terms are homogeneous functions of degree k in a given set of parameters, or, for short, of kth order. In the case where this set of parameters has just one element, say ε, we have the usual power series. But this is not the only possible case in Lie series perturbation theories and less restrictive possibilities will be widely explored in forthcoming chapters.

In what follows, the only restrictive assumption made is that the order of the terms (in the chosen set of parameters) is not affected by differentiation with respect to the variables. We also assume $W_0 = 0$ (or any constant); this is equivalent to saying that the Lie mapping reduces itself to the identity at order 0.

5.5.1 Lie Series Expansion of f

With the above assumptions, the Lie series expansion of f is

$$\begin{aligned}
E_W f = \;& f_0 \\
& + f_1 + D_{W_1} f_0 \\
& + f_2 + D_{W_1} f_1 + D_{W_2} f_0 + \tfrac{1}{2} D_{W_1} D_{W_1} f_0 \\
& + f_3 + D_{W_1} f_2 + D_{W_2} f_1 + D_{W_3} f_0 + \tfrac{1}{2} D_{W_1} D_{W_1} f_1 \\
& \quad + \tfrac{1}{2} D_{W_1} D_{W_2} f_0 + \tfrac{1}{2} D_{W_2} D_{W_1} f_0 + \tfrac{1}{6} D_{W_1} D_{W_1} D_{W_1} f_0 \\
& + \cdots,
\end{aligned} \tag{5.34}$$

where the terms were grouped in accordance with their orders (which increase by one unit from one row to the next).

When Poisson brackets are used instead of the D_W notation, we have

$$\begin{aligned}
E_W f = \;& f_0 \\
& + f_1 + \{f_0, W_1\} \\
& + f_2 + \{f_1, W_1\} + \{f_0, W_2\} + \tfrac{1}{2} \{\{f_0, W_1\}, W_1\} \\
& + f_3 + \{f_2, W_1\} + \{f_1, W_2\} + \{f_0, W_3\} \\
& \quad + \tfrac{1}{2} \{\{f_1, W_1\}, W_1\} + \tfrac{1}{2} \{\{f_0, W_2\}, W_1\} + \tfrac{1}{2} \{\{f_0, W_1\}, W_2\} \\
& \quad + \tfrac{1}{6} \{\{\{f_0, W_1\}, W_1\}, W_1\} \\
& + \cdots.
\end{aligned}$$

$$\tag{5.35}$$

In the Lie series perturbation theories of the forthcoming chapters, this expansion with explicit Poisson brackets will be preferred.

5.5.2 Deprit's Recursion Formula

The calculation of high-order Lie derivatives is a painstaking task. An important result, allowing extended calculations to be kept under control, is Deprit's recursion formula [23]. In order to derive it, we first consider the Lie derivative of f generated by W (as given by 5.33):

$$D_W f = \sum_{k=1}^{\infty} \sum_{k'=0}^{\infty} D_{W_k} f_{k'},$$

or, changing to $\ell = k + k'$:

$$D_W f = \sum_{\ell=1}^{\infty} \sum_{k=1}^{\ell} D_{W_k} f_{\ell-k}.$$

Let us, now, introduce the functions:

$$\Phi_\ell^1 = \sum_{k=1}^{\ell} D_{W_k} f_{\ell-k} \qquad (\ell \geq 1). \tag{5.36}$$

Hence,

$$D_W f = \sum_{\ell=1}^{\infty} \Phi_\ell^1.$$

We proceed similarly to obtain the second Lie derivative:

$$D_W^2 f = D_W \sum_{k'=1}^{\infty} \Phi_{k'}^1 = \sum_{k=1}^{\infty} \sum_{k'=1}^{\infty} D_{W_k} \Phi_{k'}^1,$$

or, changing again to $\ell = k + k'$:

$$D_W^2 f = \sum_{\ell=2}^{\infty} \sum_{k=1}^{\ell-1} D_{W_k} \Phi_{\ell-k}^1.$$

We then introduce the functions:

$$\Phi_\ell^2 = \sum_{k=1}^{\ell-1} D_{W_k} \Phi_{\ell-k}^1 \qquad (\ell \geq 2) \tag{5.37}$$

and obtain

$$D_W^2 f = \sum_{\ell=2}^{\infty} \Phi_\ell^2.$$

Continuing with the iterations, we obtain Deprit's recursion formula:

138   5 Lie Mappings

$$D_W^n f = \sum_{\ell=n}^{\infty} \Phi_\ell^n \quad (n \geq 1), \tag{5.38}$$

where

$$\Phi_\ell^n = \sum_{k=1}^{\ell-n+1} D_{W_k} \Phi_{\ell-k}^{n-1} \quad (\ell \geq n)$$

and

$$\Phi_\ell^0 = f_\ell \quad (\ell \geq 0).$$

From the above equations, we obtain:

$$\Phi_1^1 = D_{W_1} f_0$$
$$\Phi_2^1 = D_{W_1} f_1 + D_{W_2} f_0$$
$$\Phi_3^1 = D_{W_1} f_2 + D_{W_2} f_1 + D_{W_3} f_0$$
$$\Phi_4^1 = D_{W_1} f_3 + D_{W_2} f_2 + D_{W_3} f_1 + D_{W_4} f_0$$
$$\Phi_2^2 = D_{W_1} \Phi_1^1$$
$$\Phi_3^2 = D_{W_1} \Phi_2^1 + D_{W_2} \Phi_1^1$$
$$\Phi_4^2 = D_{W_1} \Phi_3^1 + D_{W_2} \Phi_2^1 + D_{W_3} \Phi_1^1$$
$$\Phi_3^3 = D_{W_1} \Phi_2^2$$
$$\Phi_4^3 = D_{W_1} \Phi_3^2 + D_{W_2} \Phi_2^2$$
$$\Phi_4^4 = D_{W_1} \Phi_3^3$$
$$\ldots \ldots$$

and the Lie series expansion of f becomes

$$\begin{aligned} E_W f = \ & f_0 \\ & + f_1 + \Phi_1^1 \\ & + f_2 + \Phi_2^1 + \tfrac{1}{2} \Phi_2^2 \\ & + f_3 + \Phi_3^1 + \tfrac{1}{2} \Phi_3^2 + \tfrac{1}{6} \Phi_3^3 \\ & + f_4 + \Phi_4^1 + \tfrac{1}{2} \Phi_4^2 + \tfrac{1}{6} \Phi_4^3 + \tfrac{1}{24} \Phi_4^4 \\ & + \cdots . \end{aligned} \tag{5.39}$$

6

Lie Series Perturbation Theory

6.1 Introduction

In 1966, the Poincaré and von Zeipel–Brouwer theories were rejuvenated by Hori [53] through the introduction of canonical transformations expressed by Lie series mappings instead of the classical Jacobian transformations. The use of mathematical operations in perturbation theories with the same properties as Lie derivatives and Lie series was already current among physicists [13] and, around 1960, in at least one instance (Sérsic [85]), it was suggested that Lie series could be used to represent the canonical transformations of Celestial Mechanics.

Hori theory takes full advantage of the invariance of Lie derivatives (i.e. Poisson brackets) to canonical changes of variables. The invariance of Lie derivatives allowed him to use unspecified canonical variables instead of angle–action variables and, as a consequence, to formulate a general perturbation theory. His theory is free of particularities associated with specific sets of canonical variables. Hori's general theory disclosed, in a natural way, the existence of a privileged dynamical system – the *auxiliary system* – hereafter called the *Hori kernel*. This system, defined from the homological partial differential equation, exists in every perturbation theory. Its topology, in some sense, freezes the phase space and constrains the solutions of the transformed system. It is the key to understanding the dynamics behind perturbation theories, and a necessary tool to solve more complex perturbation problems such as Bohlin's problem (see Chaps. 8 and 9).

In practical applications, we prefer to use, where possible, angle–action variables, in which case the theory follows closely what was done in the theories using Jacobian canonical transformations. The task of solving the homological partial differential equation becomes trivial and the development of the theory is much simpler (but hides the existence of a privileged dynamical system behind the perturbation equations). It is presented below and, then, compared to Poincaré theory by means of some examples.

6.2 Lie Series Theory with Angle–Action Variables

Let us start with the canonical system of equations:

$$\frac{d\theta_i}{dt} = \frac{\partial H}{\partial J_i} \qquad \frac{dJ_i}{dt} = -\frac{\partial H}{\partial \theta_i}, \tag{6.1}$$

where $H = H_0(J) + \varepsilon R(\theta, J)$ is a smooth time-independent Hamiltonian and $\theta \equiv (\theta_1, \theta_2, \cdots, \theta_N)$, $J \equiv (J_1, J_2, \cdots, J_N)$ are angle–action variables. Let us, then, consider the transformation $(\theta, J) \Rightarrow (\theta^*, J^*)$ defined by the generic equation

$$\phi(\theta, J) = E_{W^*}\phi(\theta^*, J^*) = \sum_{k=0}^{\infty} \frac{1}{k!} D_{W^*}^k \phi(\theta^*, J^*), \tag{6.2}$$

where $W^* = W(\theta^*, J^*)$[1]. For the reasons explained in Sect. 5.5, the parameter of the Lie mappings is fixed at $\lambda = 1$. Some authors prefer to adopt, here, the same parameter ε used to characterize the strength of the perturbation, and let it be free. However, such a choice leads to unnecessary discussions about transformations that depend explicitly on the parameter. These complications are avoided when we fix $\lambda = 1$ and no generality is lost because of the homogeneity properties of Lie derivatives.

The assumption that the Hamiltonian is time-independent is also made without loss of generality. We recall that, in the case of a time-dependent Hamiltonian, it is enough to extend the phase space and to introduce a new degree of freedom. The Hamiltonian of the resulting parametric equations does not depend on the new independent variable (see Sect. 1.6).

We follow, now, the same steps as in Chap. 3, but considering the canonical transformation defined by (6.2). The given Hamiltonian is transformed in the same way:

$$H(\theta, J) = E_{W^*} H(\theta^*, J^*) = \sum_{k=0}^{\infty} \frac{1}{k!} D_{W^*}^k H(\theta^*, J^*), \tag{6.3}$$

which will be written hereafter with the explicit Poisson brackets, instead of the D_W notation (see 5.35).

The conservation of the Hamiltonian in time-independent canonical transformations allows us to write

$$H^*(\theta^*, J^*) = H(\theta, J), \tag{6.4}$$

where H^* is the Hamiltonian of the transformed system. The combination of (6.4) and (6.3) gives

---

[1] We have written $(\theta, J) \Rightarrow (\theta^*, J^*)$ to retain the classical way of presenting the transformation, but, in fact, (6.2) gives the transformation $(\theta^*, J^*) \Rightarrow (\theta, J)$, which is the transformation that indeed matters in actual theories (see [73]).

6.2 Lie Series Theory with Angle–Action Variables

$$H^*(\theta^*, J^*) = E_{W^*} H(\theta^*, J^*). \tag{6.5}$$

Hori's perturbation equations are obtained by substituting into (6.5) the expansions

$$\begin{aligned} H &= H_0 + \varepsilon H_1 + \varepsilon^2 H_2 + \cdots + \varepsilon^n H_n + \cdots \\ H^* &= H_0^* + \varepsilon H_1^* + \varepsilon^2 H_2^* + \cdots + \varepsilon^n H_n^* + \cdots \\ W^* &= \quad\quad \varepsilon W_1^* + \varepsilon^2 W_2^* + \cdots + \varepsilon^n W_n^* + \cdots. \end{aligned} \tag{6.6}$$

The simplicity of these expansions is remarkable Now, there are no expansions due to the internal substitution of variables by the series defining the transformation as in classical theories and we just need the power series expansions of H and H^* in ε, in their explicit form.

We may compare the Lie series expansion of $H(\theta, J)$ with the expansion of $H^*(\theta^*, J^*)$, according to (6.6) and (5.35), and identify the terms in the same powers of ε. We obtain the perturbation equations

$$\begin{aligned} H_0^* &= H_0 \\ H_1^* &= H_1 + \{H_0, W_1^*\} \\ H_2^* &= H_2 + \{H_1, W_1^*\} + \tfrac{1}{2}\{\{H_0, W_1^*\}, W_1^*\} + \{H_0, W_2^*\} \\ H_3^* &= H_3 + \{H_2, W_1^*\} + \{H_1, W_2^*\} + \tfrac{1}{2}\{\{H_1, W_1^*\}, W_1^*\} \\ &\quad + \tfrac{1}{2}\{\{H_0, W_1^*\}, W_2^*\} + \tfrac{1}{2}\{\{H_0, W_2^*\}, W_1^*\} \\ &\quad + \tfrac{1}{6}\{\{\{H_0, W_1^*\}, W_1^*\}, W_1^*\} + \{H_0, W_3^*\} \\ &\cdots\cdots \\ H_n^* &= H_n + \{H_{n-1}, W_1^*\} + \tfrac{1}{2}\{\{H_{n-2}, W_1^*\}, W_1^*\} + \cdots + \{H_0, W_n^*\}. \end{aligned} \tag{6.7}$$

In these equations, the functions H_k are to be read as $H_k(\theta^*, J^*)$, that is, functions depending on (θ^*, J^*) in the same way as $H_k(\theta, J)$ depend on (θ, J). We note that, in the kth equation, W_k^* only appears through the additive term $\{H_0, W_k^*\}$ and H_k^* only appears in the left-hand side; all other W_j^* and H_j^*, in that equation, have subscripts smaller than k. The generic equation for $k \geq 1$ is the *homological* equation:

$$H_k^* = \Psi_k + \{H_0, W_k^*\}, \tag{6.8}$$

where $\Psi_k(\theta^*, J^*)$ is a known function when the previous k equations have already been solved[2].

---

[2] Some simplified expressions with fewer brackets to calculate are

$$\begin{aligned} \Psi_2 &= H_2 + \tfrac{1}{2}\{H_1^* + H_1, W_1^*\}, \\ \Psi_3 &= H_3 + \tfrac{1}{2}\{H_2^* + H_2, W_1^*\} + \tfrac{1}{2}\{H_1^* + H_1, W_2^*\} - \tfrac{1}{12}\{\{H_1^* - H_1, W_1^*\}, W_1^*\}, \\ \Psi_4 &= H_4 + \tfrac{1}{2}\{H_3^* + H_3, W_1^*\} + \tfrac{1}{2}\{H_2^* + H_2, W_2^*\} + \tfrac{1}{2}\{H_1^* + H_1, W_3^*\} \\ &\quad - \tfrac{1}{12}\{\{H_2^* - H_2, W_1^*\}, W_1^*\} - \tfrac{1}{12}\{\{H_1^* - H_1, W_2^*\}, W_1^*\} \\ &\quad - \tfrac{1}{12}\{\{H_1^* - H_1, W_1^*\}, W_2^*\}. \end{aligned}$$

Equation (6.8) is a partial differential equation in the unknown function W_k^*:

$$H_k^* = \Psi_k + \{H_0, W_k^*\} = \Psi_k + \sum_{i=1}^{N}\left(\frac{\partial H_0}{\partial \theta_i^*}\frac{\partial W_k^*}{\partial J_i^*} - \frac{\partial W_k^*}{\partial \theta_i^*}\frac{\partial H_0}{\partial J_i^*}\right). \qquad (6.9)$$

We have not yet made a definite assumption about the given Hamiltonian. If, as in von Zeipel–Brouwer theory, we assume that $H_0 = H_0(J_\mu)$ ($\mu = 1, \cdots, M \leq N$), the homological equation becomes

$$\sum_{\mu=1}^{M} \nu_\mu^* \frac{\partial W_k^*}{\partial \theta_\mu^*} = \Psi_k - H_k^*, \qquad (6.10)$$

where

$$\nu_\mu^* = \frac{\partial H_0}{\partial J_\mu^*}. \qquad (6.11)$$

6.2.1 Averaging

Equation (6.8) is the same homological equation of von Zeipel–Brouwer theory (see Sect. 3.4) and its solution may be obtained in exactly the same way. To solve the indetermination of (6.10), we adopt the averaging rule

$$H_k^* = <\Psi_k> = \left(\frac{1}{2\pi}\right)^M \int_0^{2\pi}\cdots\int_0^{2\pi} \Psi_k d\theta_1^* \cdots d\theta_M^*. \qquad (6.12)$$

The averaging involves only those angle variables whose associated frequency ν_μ^* is different from zero.

The multiperiodic functions Ψ_k are split into secular, long-period and short-period parts. These parts are indicated by the subscripts S, LP and SP, respectively, and are defined as in Sect. 3.4. Using this separation, (6.12) and (6.10) may be written

$$H_k^* = \Psi_{k(S)} + \Psi_{k(LP)}$$

and

$$\sum_{i=1}^{M} \nu_i^* \frac{\partial W_k^*}{\partial \theta_i^*} = \Psi_{k(SP)}. \qquad (6.13)$$

The solution of (6.13) introduces the divisors $\sum_{\mu=1}^{M} h_\mu \nu_\mu^*$. When one of them becomes close to zero, the theory fails. We assume, then, that a relation $\sum_{\mu=1}^{M} h_\mu \nu_\mu^* \approx 0$ does not hold for any of the h values present in the $\Psi_{k(SP)}$ [3].

---

[3] The terms with $h_1 = \cdots = h_M = 0$ were excluded from $\Psi_{k(SP)}$ and included in $\Psi_{k(S)}$ and $\Psi_{k(LP)}$.

The new Hamiltonian H^* does not depend on the *short-period* angles θ^*_μ ($\mu = 1, \cdots, M$). The system is thus reduced to $N - M$ degrees of freedom and the transformed Hamiltonian may be written

$$H^* = H_0^*(J^*) + \sum_{k \geq 1} H_k^*(J^*, \theta^*) \tag{6.14}$$

with $\theta^* \in \mathbf{T}^{N-M}$, $J^* \in \mathbf{R}^N$.

The results thus obtained are formally identical to those obtained with von Zeipel–Brouwer theory (Sect. 3.4). The only important difference is that Lie series theories are algebraically more straightforward than classical von Zeipel–Brouwer theory. Much of this simplification comes from the fact that Lie mappings are resolved with respect to the new (or the old) variables while Jacobian mappings have a mixed structure, with half of the equations resolved with respect to the new variables and half of them with respect to the old ones. The structure of Lie series theories is particularly suited to the use of algebraic manipulators, allowing programmable iteration schemes to be set.

6.2.2 High-Order Theories

In the case of high-order calculations, it is convenient to use Deprit's recursion formula and to substitute, in the comparison of both sides of (6.5), the Lie series expansion of H given by (5.39), instead of the one with explicit Poisson brackets. Hence, the perturbation equations become

$$\begin{aligned}
H_0^* &= H_0 \\
H_1^* &= H_1 + \Phi_1^1 \\
H_2^* &= H_2 + \Phi_2^1 + \tfrac{1}{2}\Phi_2^2 \\
H_3^* &= H_3 + \Phi_3^1 + \tfrac{1}{2}\Phi_3^2 + \tfrac{1}{6}\Phi_3^3 \\
&\cdots\cdots \\
H_n^* &= H_n + \sum_{k=1}^{n} \frac{1}{k!}\Phi_n^k
\end{aligned} \tag{6.15}$$

and the functions Ψ_k, in the homological equation $H_k^* = \Psi_k + \{H_0, W_k^*\}$, are

$$\begin{aligned}
\Psi_1 &= H_1 \\
\Psi_2 &= H_2 + \Upsilon_2^1 + \tfrac{1}{2}\Phi_2^2 \\
\Psi_3 &= H_3 + \Upsilon_3^1 + \tfrac{1}{2}\Phi_3^2 + \tfrac{1}{6}\Phi_3^3 \\
\Psi_4 &= H_4 + \Upsilon_4^1 + \tfrac{1}{2}\Phi_4^2 + \tfrac{1}{6}\Phi_4^3 + \tfrac{1}{24}\Phi_4^4 \\
&\cdots\cdots \\
\Psi_n &= H_n + \Upsilon_n^1 + \sum_{k=2}^{n} \frac{1}{k!}\Phi_n^k,
\end{aligned} \tag{6.16}$$

where we have introduced

$$\Upsilon_k^1 = \sum_{\ell=1}^{k-1} D_{W_\ell^*} H_{k-l} = \Phi_k^1 - D_{W_k^*} H_0. \tag{6.17}$$

The calculations up to an order n may be easily organized into cycles. Each cycle starts with the value of a generic W_k^* and ends when W_{k+1}^* is obtained and may be schematized as follows:

$$W_k^* \longrightarrow \left\{ \begin{array}{c} \Upsilon_{k+1}^1 \\ \Phi_k^1, \Phi_{k+1}^2, \Phi_{k+2}^3, \ldots \end{array} \right\} \longrightarrow \Psi_{k+1} \longrightarrow H_{k+1}^*, W_{k+1}^*.$$

Each calculation only depends on functions already calculated in previous cycles or in previous steps of the same cycle. This scheme is easily programmable and may be used up to very high orders of approximation.

Exercise 6.2.1. Use the above routines to calculate H_k^* and W_k^* (up to $k = 4$) in the case of the one-degree-of-freedom Hamiltonian

$$H = J_1 - \frac{1}{2} J_1^2 + \varepsilon \sqrt{2J_1} \cos \theta_1.$$

(For the results up to $k = 3$ see Sect. 6.6.1.)

6.3 Comparison to Poincaré Theory. Example I

The equivalence of Hori's Lie series perturbation theory to Poincaré and von Zeipel–Brouwer theories is a direct consequence of the fact that Lie mappings and Jacobian canonical transformations in the neighborhood of the identity (that is, infinitesimal) are equivalent, as shown in Sect. 5.1.1 (cf. [59], Chap. 7). As a consequence, the series solutions obtained *in angle–action variables* with theories using either one or another transformation are the same [54], [90]. However the paths to the solution are different in both cases and, depending on the problem being considered, may represent very different amounts of work.

Let us consider an example and solve it using both Hori's Lie series theory in angle–action variables and the Poincaré theory. Let us consider the system given by the Hamiltonian

$$H = H_0 + \varepsilon R(\theta, J) = -\frac{1}{2J_1^2} + J_2 + \varepsilon (J_1 J_2 + J_1 \cos 2\theta_1 + J_1 J_2 \cos 2\theta_2) \tag{6.18}$$

This is a non-degenerate (in Schwarzschild's sense) system with frequencies (for $\varepsilon = 0$)

$$\nu_1 = \frac{\partial H_0}{\partial J_1} = \frac{1}{J_1^3}, \qquad \nu_2 = \frac{\partial H_0}{\partial J_2} = 1;$$

and, also,

$$\nu_{11} = \frac{\partial^2 H_0}{\partial J_1^2} = -\frac{3}{J_1^4}, \qquad \nu_{22} = \nu_{12} = 0.$$

6.3 Comparison to Poincaré Theory. Example I

The homological equations are

LIE SERIES THEORY	POINCARÉ THEORY
$\sum \nu_i^* \frac{\partial W_k^*}{\partial \theta_i^*} = \Psi_k(\theta^*, J^*) - H_k^*$	$\sum \nu_i^* \frac{\partial S_k}{\partial \theta_i} = H_k^* - \Psi_k(\theta, J^*),$

where $\nu_1^* = \nu_1(J^*) = J_1^{*-3}$. We recall that $W_k^* = W_k(\theta^*, J^*)$ and $S_k = S_k(\theta, J^*)$. The functions Ψ_k are

$\Psi_1 = H_1 = R(\theta^*, J^*)$	$\Psi_1 = H_1 = R(\theta, J^*)$
$\Psi_2 = H_2 + \frac{1}{2}\{\{H_0, W_1^*\}, W_1^*\}$	$\Psi_2 = H_2 + \frac{1}{2}\sum_i \sum_j \nu_{ij}^* \frac{\partial S_1}{\partial \theta_i} \frac{\partial S_1}{\partial \theta_j}$
$\qquad +\{H_1, W_1^*\}$	$\qquad + \sum_i \frac{\partial H_1}{\partial J_i^*} \frac{\partial S_1}{\partial \theta_i},$
$= H_2 + \frac{1}{2}\{H_1 + H_1^*, W_1^*\},$	

where $R(\theta, J) = J_1 J_2 + J_1 \cos 2\theta_1 + J_1 J_2 \cos 2\theta_2$. Then, in both cases, we obtain the same H_1^*:

$$H_1^* = <\Psi_1> = J_1^* J_2^*$$

and the integration of the homological equation for $k = 1$ gives

$W_1^* = \frac{1}{2}J_1^{*4} \sin 2\theta_1^* + \frac{1}{2}J_1^* J_2^* \sin 2\theta_2^*;$	$S_1 = -\frac{1}{2}J_1^{*4} \sin 2\theta_1 - \frac{1}{2}J_1^* J_2^* \sin 2\theta_2.$

The following derivatives are necessary in the next step (an open box means that the corresponding derivative is not used and, then, not calculated):

$\frac{\partial W_1^*}{\partial \theta_1^*} = J_1^{*4} \cos 2\theta_1^*$	$\frac{\partial S_1}{\partial \theta_1} = -J_1^{*4} \cos 2\theta_1$
$\frac{\partial W_1^*}{\partial \theta_2^*} = J_1^* J_2^* \cos 2\theta_2^*$	$\frac{\partial S_1}{\partial \theta_2} = -J_1^* J_2^* \cos 2\theta_2$
$\frac{\partial W_1^*}{\partial J_1^*} = 2J_1^{*3} \sin 2\theta_1^* + \frac{1}{2}J_2^* \sin 2\theta_2^*$	☐
$\frac{\partial W_1^*}{\partial J_2^*} = \frac{1}{2}J_1^* \sin 2\theta_2^*$	☐
$\frac{\partial(H_1+H_1^*)}{\partial \theta_1^*} = -2J_1^* \sin 2\theta_1^*$	☐
$\frac{\partial(H_1+H_1^*)}{\partial \theta_2^*} = -2J_1^* J_2^* \sin 2\theta_2^*$	☐
$\frac{\partial(H_1+H_1^*)}{\partial J_1^*} = 2J_2^* + \cos 2\theta_1^* + J_2^* \cos 2\theta_2^*$	$\frac{\partial H_1}{\partial J_1^*} = J_2^* + \cos 2\theta_1 + J_2^* \cos 2\theta_2$
$\frac{\partial(H_1+H_1^*)}{\partial J_2^*} = 2J_1^* + J_1^* \cos 2\theta_2^*;$	$\frac{\partial H_1}{\partial J_2^*} = J_1^* + J_1^* \cos 2\theta_2.$

In the next set of calculations, we just show the left-hand sides of the equations. The non-written right-hand sides result from products of the above given trigonometric polynomials. We do not show them, but we indicate all of them; they are the lengthier parts in the calculations and determine the amount of work involved in the application of each of the theories.

$$\frac{\partial (H_1+H_1^*)}{\partial \theta_1^*}\frac{\partial W_1^*}{\partial J_1^*} = \circ \circ \circ$$

$$\frac{\partial (H_1+H_1^*)}{\partial \theta_2^*}\frac{\partial W_1^*}{\partial J_2^*} = \circ \circ \circ$$

$$-\frac{\partial (H_1+H_1^*)}{\partial J_1^*}\frac{\partial W_1^*}{\partial \theta_1^*} = \circ \circ \circ$$

$$-\frac{\partial (H_1+H_1^*)}{\partial J_2^*}\frac{\partial W_1^*}{\partial \theta_2^*} = \circ \circ \circ$$

$$\frac{\partial H_1}{\partial J_1^*}\frac{\partial S_1}{\partial \theta_1} = \circ \circ \circ$$

$$\frac{\partial H_1}{\partial J_2^*}\frac{\partial S_1}{\partial \theta_2} = \circ \circ \circ$$

$$\tfrac{1}{2}\nu_{11}^*\left(\frac{\partial S_1}{\partial \theta_1}\right)^2 = \circ \circ \circ.$$

Once these products are calculated, we may add them together to obtain Ψ_2. We obtain, in both cases, the same H_2^*:

$$H_2^* = <\Psi_2> = -\frac{5}{4}J_1^{*4} - \frac{1}{2}J_1^{*2}J_2^*$$

and the integration of the homological equation for $k=2$ yields

$$\begin{aligned}W_2^* = &-\tfrac{1}{2}J_1^{*7}J_2^*\sin 2\theta_1^*\\ &-\tfrac{1}{2}J_1^{*2}J_2^*\sin 2\theta_2^*\\ &+\tfrac{3}{16}J_1^{*7}\sin 4\theta_1^*\\ &+\\ &+\tfrac{1}{8}J_1^{*4}J_2^*\tfrac{1-J_1^{*3}}{1+J_1^{*3}}\sin(2\theta_1^*+2\theta_2^*)\\ &-\tfrac{1}{8}J_1^{*4}J_2^*\tfrac{1+J_1^{*3}}{1-J_1^{*3}}\sin(2\theta_1^*-2\theta_2^*);\end{aligned}$$

$$\begin{aligned}S_2 = &\tfrac{1}{2}J_1^{*7}J_2^*\sin 2\theta_1\\ &+\tfrac{1}{2}J_1^{*2}J_2^*\sin 2\theta_2\\ &+\tfrac{5}{16}J_1^{*7}\sin 4\theta_1\\ &+\tfrac{1}{8}J_1^{*2}J_2^*\sin 4\theta_2\\ &+\tfrac{1}{4}\tfrac{J_1^{*7}J_2^*}{1+J_1^{*3}}\sin(2\theta_1+2\theta_2)\\ &+\tfrac{1}{4}\tfrac{J_1^{*7}J_2^*}{1-J_1^{*3}}\sin(2\theta_1-2\theta_2).\end{aligned}$$

We have, thus, completed the transformation of the given Hamiltonian system up to $\mathcal{O}(\varepsilon^2)$. The amount of calculation is slightly larger with Hori's Lie series theory than with Poincaré theory. We may compute, now, the explicit values of the variables. We present here the calculation of J_1. The equations necessary to get J_1 up to the second-order are:

$$\begin{aligned}J_1 &= E_{W^*}J_1^*\\ &= J_1^* + \varepsilon\{J_1^*,W_1^*\} + \varepsilon^2\{J_1^*,W_2^*\}\\ &\quad + \tfrac{\varepsilon^2}{2}\{\{J_1^*,W_1^*\},W_1^*\},\end{aligned}$$

$$J_1 = J_1^* + \varepsilon\frac{\partial S_1}{\partial \theta_1} + \varepsilon^2\frac{\partial S_2}{\partial \theta_1}$$

$$\theta_1^* = \theta_1 + \varepsilon\frac{\partial S_1}{\partial J_1^*} + \varepsilon^2\frac{\partial S_2}{\partial J_1^*};$$

or

$$\begin{aligned}J_1 = &J_1^* - \varepsilon\frac{\partial W_1^*}{\partial \theta_1^*} - \varepsilon^2\frac{\partial W_2^*}{\partial \theta_1^*}\\ &-\tfrac{\varepsilon^2}{2}\sum_j\frac{\partial^2 W_1^*}{\partial \theta_1^*\partial \theta_j^*}\frac{\partial W_1^*}{\partial J_j^*}\\ &+\tfrac{\varepsilon^2}{2}\sum_j\frac{\partial^2 W_1^*}{\partial \theta_1^*\partial J_j^*}\frac{\partial W_1^*}{\partial \theta_j^*}\end{aligned}$$

$$\begin{aligned}J_1 = &J_1^* + \varepsilon\frac{\partial T_1}{\partial \theta_1^*} + \varepsilon^2\frac{\partial T_2}{\partial \theta_1^*}\\ &-\varepsilon^2\sum_j\frac{\partial^2 T_1}{\partial \theta_1^*\partial \theta_j^*}\frac{\partial T_1}{\partial J_j^*},\end{aligned}$$

(6.19)

where (in 6.19)
$$T_k \stackrel{\text{def}}{=} S_k(\theta^*, J^*).$$

The more visible practical advantage of Lie's over Jacobian mappings lies on the fact that Lie mappings do not mix old and new variables as Jacobian mappings do. Thus, the iteration leading to (6.19), which becomes extremely cumbersome in high-order theories, is not necessary to obtain the solutions in a Lie series theory. However, this advantage is reduced by the fact that the explicit calculation of the Lie series giving θ_1, J_1 involves the calculation of many brackets.

The not yet calculated derivatives necessary to obtain $J_1(\theta^*, J^*)$ are

$\dfrac{\partial^2 W_1^*}{\partial \theta_1^{*2}} = -2J_1^{*4} \sin 2\theta_1^*$ \qquad $\dfrac{\partial^2 T_1}{\partial \theta_1^{*2}} = 2J_1^{*4} \sin 2\theta_1^*$

$\dfrac{\partial^2 W_1^*}{\partial \theta_1^* \partial J_1^*} = 4J_1^{*3} \cos 2\theta_1^*$ \qquad $\dfrac{\partial T_1}{\partial J_1^*} = -2J_1^{*3} \sin 2\theta_1^* - \tfrac{1}{2} J_2^* \sin 2\theta_2^*$

$\dfrac{\partial W_2^*}{\partial \theta_1^*} = \circ \circ \circ$ \qquad $\dfrac{\partial T_2}{\partial \theta_1^*} = \circ \circ \circ$

(some null derivatives and the terms multiplying them are not written). The non-zero contributions come from the terms:

$\dfrac{\partial W_1^*}{\partial \theta_1^*} = \circ \circ \circ$ \qquad $\dfrac{\partial T_1}{\partial \theta_1^*} = \circ \circ \circ$

$-\dfrac{\partial^2 W_1^*}{\partial \theta_1^{*2}} \dfrac{\partial W_1^*}{\partial J_1^*} = \circ \circ \circ$ \qquad $\dfrac{\partial^2 T_1}{\partial \theta_1^*} \dfrac{\partial T_1}{\partial J_1^*} = \circ \circ \circ$

$-\dfrac{\partial^2 W_1^*}{\partial \theta_1^* \partial J_1^*} \dfrac{\partial W_1^*}{\partial \theta_1^*} = \circ \circ \circ$

$\dfrac{\partial W_2^*}{\partial \theta_1^*} = \circ \circ \circ,$ \qquad $\dfrac{\partial T_2}{\partial \theta_1^*} = \circ \circ \circ,$

where, as before, the right-hand sides were omitted.

In both cases, the result is

$$J_1 = J_1^* + 2\varepsilon^2 J_1^{*7} - \varepsilon J_1^{*4} \cos 2\theta_1^* + \varepsilon^2 J_1^{*7} J_2^* \cos 2\theta_1^* - \frac{3}{4}\varepsilon^2 J_1^{*7} \cos 4\theta_1^*$$
$$- \frac{1}{2}\varepsilon^2 \frac{J_1^{*4} J_2^*}{1 + J_1^{*3}} \cos(2\theta_1^* + 2\theta_2^*) + \frac{1}{2}\varepsilon^2 \frac{J_1^{*4} J_2^*}{1 - J_1^{*3}} \cos(2\theta_1^* - 2\theta_2^*). \quad (6.20)$$

In these equations, J_j^* are constants and θ_j^* are linear functions of t with time derivatives equal to $\partial H^*/\partial J_1^*$.

This example shows that the proper variable J_1^* is not the average of J_1. This noteworthy point will be discussed in Sect. 6.8.

6.4 Comparison to Poincaré Theory. Example II

Let us consider a second example, suggested to us by J. Henrard, and let us solve it using both theories. It is the case of the one-degree-of-freedom

Hamiltonian
$$H = H_0 + \varepsilon R(\theta_1, J_1) = J_1 + \varepsilon\sqrt{2J_1}\cos\theta_1 \qquad (6.21)$$
with the constant undisturbed frequency $\nu_1 = 1$, and $\nu_{11} = 0$.

The homological equations are

Lie series Theory	Poincaré Theory
$\nu_1^* \dfrac{\partial W_k^*}{\partial \theta_1^*} = \Psi_k(\theta_1^*, J_1^*) - H_k^*(J_1^*)$	$\nu_1^* \dfrac{\partial S_k}{\partial \theta_1} = H_k^*(J_1^*) - \Psi_k(\theta_1, J_1^*).$

We recall that $W_k^* = W_k(\theta^*, J^*)$ and $S_k = S_k(\theta, J^*)$. The functions Ψ_i are

$$\begin{aligned}
\Psi_1 &= H_1(\theta_1^*, J_1^*) = \sqrt{2J_1^*}\cos\theta_1^* \\
\Psi_2 &= \tfrac{1}{2}\{H_1^* + H_1, W_1^*\} \\
\Psi_3 &= \tfrac{1}{2}\{H_1^* + H_1, W_2^*\} \\
&\quad + \tfrac{1}{2}\{H_2^*, W_1^*\} \\
&\quad - \tfrac{1}{12}\{\{H_1^* - H_1, W_1^*\}, W_1^*\}
\end{aligned} \qquad \begin{aligned}
\Psi_1 &= H_1(\theta_1, J_1^*) = \sqrt{2J_1^*}\cos\theta_1 \\
\Psi_2 &= \tfrac{\partial H_1}{\partial J_1^*}\tfrac{\partial S_1}{\partial \theta_1} \\
\Psi_3 &= \tfrac{\partial H_1}{\partial J_1^*}\tfrac{\partial S_2}{\partial \theta_1} + \tfrac{1}{2}\tfrac{\partial^2 H_1}{\partial J_1^{*2}}\left(\tfrac{\partial S_1}{\partial \theta_1}\right)^2,
\end{aligned}$$

where we have already take into account that $H_2 = H_3 = 0$ and $\nu_{11}^* = 0$. In both cases we obtain
$$H_1^* = <\Psi_1> = 0$$
and the integration of the homological equation for $k = 1$ gives

$$W_1^* = \sqrt{2J_1^*}\sin\theta_1^*; \qquad S_1 = -\sqrt{2J_1^*}\sin\theta_1.$$

The derivatives necessary in the next step are

$$\begin{aligned}
\tfrac{\partial W_1^*}{\partial \theta_1^*} &= \sqrt{2J_1^*}\cos\theta_1^* \\
\tfrac{\partial W_1^*}{\partial J_1^*} &= \tfrac{1}{\sqrt{2J_1^*}}\sin\theta_1^* \\
\\
\tfrac{\partial H_1}{\partial \theta_1^*} &= -\sqrt{2J_1^*}\sin\theta_1^* \\
\tfrac{\partial H_1}{\partial J_1^*} &= \tfrac{1}{\sqrt{2J_1^*}}\cos\theta_1^*
\end{aligned} \qquad \begin{aligned}
\tfrac{\partial S_1}{\partial \theta_1} &= -\sqrt{2J_1^*}\cos\theta_1 \\
\\
\\
\\
\tfrac{\partial H_1}{\partial J_1^*} &= \tfrac{1}{\sqrt{2J_1^*}}\cos\theta_1
\end{aligned}$$

and their products are

$$\begin{aligned}
\tfrac{\partial H_1}{\partial \theta_1^*}\tfrac{\partial W_1^*}{\partial J_1^*} &= -\sin^2\theta_1^* \\
\tfrac{\partial H_1}{\partial J_1^*}\tfrac{\partial W_1^*}{\partial \theta_1^*} &= \cos^2\theta_1^*
\end{aligned} \qquad \tfrac{\partial H_1}{\partial J_1^*}\tfrac{\partial S_1}{\partial \theta_1} = -\cos^2\theta_1.$$

6.4 Comparison to Poincaré Theory. Example II

Hence

$$\Psi_2 = -\tfrac{1}{2} \qquad \qquad \Psi_2 = -\tfrac{1}{2} - \tfrac{1}{2}\cos 2\theta_1$$

and we obtain, in both cases, the same H_2^*:

$$H_2^* = <\Psi_2> = -\frac{1}{2}.$$

The integration of the homological equation, for $k = 2$, yields

$$W_2^* = 0 \qquad \qquad S_2 = \tfrac{1}{4}\sin 2\theta_1.$$

Let us continue and similarly calculate the next order terms. Now, the calculations in the two theories are very different and the parallel presentation of them no longer makes sense. Taking into account that $H_1^* = 0$, $H_2 = 0$ and $W_2^* = 0$, the expression of Ψ_3 in Lie series theory becomes

$$\Psi_3 = \frac{1}{2}\{H_2^*, W_1^*\} + \frac{1}{6}\{\{H_1, W_1^*\}, W_1^*\}.$$

The first bracket is equal to zero because H_2^* is a constant; the second term is also equal to zero because the bracket $\{H_1, W_1^*\}$ is also a constant. Then, $\Psi_3 = 0$. Similarly, we have $\Psi_k = 0$ and $H_k^* = 0$ for all $k \geq 3$, and the transformed Hamiltonian obtained by means of Lie series theory is exact.

In the application of the Poincaré theory, we obtain

$$\Psi_3 = \frac{\partial H_1}{\partial J_1^*}\frac{\partial S_2}{\partial \theta_1} + \frac{1}{2}\frac{\partial^2 H_1}{\partial J_1^{*2}}\left(\frac{\partial S_1}{\partial \theta_1}\right)^2$$

or

$$\Psi_3 = -\frac{1}{8\sqrt{2J_1^*}}(\cos\theta_1 - \cos 3\theta_1)$$

and, as in the Lie series theory, we obtain $H_3^* = <\Psi_3> = 0$. The integration of the homological equation gives

$$S_3 = \frac{1}{8\sqrt{2J_1^*}}\left(\sin\theta_1 - \frac{1}{3}\sin 3\theta_1\right).$$

The transformation is completed at the given order of approximation, and we do not have any hint of the next approximations. The only way to obtain S_k and H_k^* for higher values of k is through the actual calculations.

Let us compute, now, with both theories, the explicit value of J_1. The equations to third order are:

Lie Series Theory	Poincaré Theory
$\begin{aligned} J_1 &= E_{W^*} J_1^* \\ &= J_1^* + \varepsilon\{J_1^*, W_1^*\} + \varepsilon^2\{J_1^*, W_2^*\} \\ &\quad + \tfrac{\varepsilon^2}{2}\{\{J_1^*, W_1^*\}, W_1^*\} \\ &\quad + \varepsilon^3\{J_1^*, W_3^*\} \\ &\quad + \tfrac{\varepsilon^3}{2}\{\{J_1^*, W_2^*\}, W_1^*\} \\ &\quad + \tfrac{\varepsilon^3}{2}\{\{J_1^*, W_1^*\}, W_2^*\} \\ &\quad + \tfrac{\varepsilon^3}{6}\{\{\{J_1^*, W_1^*\}, W_1^*\}, W_1^*\} \end{aligned}$	$\begin{aligned} J_1 &= J_1^* + \varepsilon \tfrac{\partial S_1}{\partial \theta_1} + \varepsilon^2 \tfrac{\partial S_2}{\partial \theta_1} + \varepsilon^3 \tfrac{\partial S_3}{\partial \theta_1} \\ \theta_1^* &= \theta_1 + \varepsilon \tfrac{\partial S_1}{\partial J_1^*} + \varepsilon^2 \tfrac{\partial S_2}{\partial J_1^*} + \varepsilon^3 \tfrac{\partial S_3}{\partial J_1^*}. \end{aligned}$

Once more, in this example, the calculations in the two theories are very different and a parallel presentation makes no sense. In the Lie series theory, since $W_k^* = 0$ for all $k \geq 2$, the Lie series reduces to

$$J_1 = J_1^* + \varepsilon\{J_1^*, W_1^*\} + \frac{\varepsilon^2}{2}\{\{J_1^*, W_1^*\}, W_1^*\} + \frac{\varepsilon^3}{6}\{\{\{J_1^*, W_1^*\}, W_1^*\}, W_1^*\} + \cdots;$$

the first Lie derivative in this series is

$$\{J_1^*, W_1^*\} = -\frac{\partial W_1^*}{\partial \theta_1^*} = -\sqrt{2J_1^*}\cos\theta_1^*$$

and the second one is

$$\{\{J_1^*, W_1^*\}, W_1^*\} = -\frac{\partial}{\partial \theta_1^*}\left(\frac{\partial W_1^*}{\partial \theta_1^*}\right)\frac{\partial W_1^*}{\partial J_1^*} + \frac{\partial}{\partial J_1^*}\left(\frac{\partial W_1^*}{\partial \theta_1^*}\right)\frac{\partial W_1^*}{\partial \theta_1^*} = 1.$$

Since the second Lie derivative is equal to a constant, all the following ones will be equal to zero, and the calculation is completed. Hence, exactly,

$$J_1 = J_1^* - \varepsilon\sqrt{2J_1^*}\cos\theta_1^* + \frac{1}{2}\varepsilon^2. \tag{6.22}$$

In the case of Poincaré theory, the first step is to solve the implicit equations to obtain J_1 as a function of θ_1^*, J_1^*. This is a cumbersome task. The extended Lagrange formulas of Sect. 3.12 may be used to obtain

$$\begin{aligned} J_1 &= J_1^* + \varepsilon\frac{\partial T_1}{\partial \theta_1^*} + \varepsilon^2\left(\frac{\partial T_2}{\partial \theta_1^*} - \frac{\partial^2 T_1}{\partial \theta_1^{*2}}\frac{\partial T_1}{\partial J_1^*}\right) + \frac{1}{2}\varepsilon^3\frac{\partial^3 T_1}{\partial \theta_1^{*3}}\left(\frac{\partial T_1}{\partial J_1^*}\right)^2 \\ &\quad + \varepsilon^3\left(\frac{\partial^2 T_1}{\partial \theta_1^{*2}}\frac{\partial^2 T_1}{\partial \theta_1^* \partial J_1^*}\frac{\partial T_1}{\partial J_1^*} - \frac{\partial^2 T_1}{\partial \theta_1^{*2}}\frac{\partial T_2}{\partial J_1^*} - \frac{\partial^2 T_2}{\partial \theta_1^{*2}}\frac{\partial T_1}{\partial J_1^*} + \frac{\partial T_3}{\partial \theta_1^*}\right), \end{aligned}$$

where $T_k \stackrel{\text{def}}{=} S_k(\theta^*, J^*)$. To order $\mathcal{O}(\varepsilon^2)$ the above equation gives the same result as (6.22). All monomials yield terms

$$\text{coeff } \frac{1}{\sqrt{2J_1^*}}(\cos\theta_1^* - \cos 3\theta_1^*)$$

with coefficients $\frac{1}{8}, \frac{1}{4}, 0, -\frac{1}{2}, \frac{1}{8}$, respectively. As expected, the sum is zero; but, it can only be known after the actual calculations are done. The calculations at higher orders are more cumbersome, but (6.22) allows us to anticipate that they also will have a null contribution to the expression of J_1 as a function of θ_1^*, J_1^*.

Therefore, in the case of this example, the advantage of Lie series theory over Poincaré theory is enormous. Several reasons work together for this result, namely: (a) the given system may be easily integrated with elementary functions; (b) H_0 is trivial and leads to a constant ν_1; (c) the given Hamiltonian is a polynomial in the variables $\sqrt{2J_1}\cos\theta_1, \sqrt{2J_1}\sin\theta_1$. The last property, known as the d'Alembert property, is conserved by Lie derivatives, since the Poisson brackets of two such polynomials is also a polynomial in these variables (see Sect. 7.3). However, the d'Alembert property alone is far from being a guarantee for what has been shown. For instance, the Hamiltonian considered in Exercise 6.2.1, $H = J_1 - \frac{1}{2}J_1^2 + \varepsilon\sqrt{2J_1}\cos\theta_1$, satisfies the d'Alembert property and is integrable. However, the series cannot be so easily obtained as in the given example. The main advantage of Lie series theory, in that case, is limited to the recursion formulas allowing high-order solutions to be obtained.

The example considered in this section showed how a bad choice may introduce unnecessarily cumbersome calculations. Even the Lie series approach used here with angle–action variables is not a good choice in this case. Nonsingular variables allow this problem to be trivially solved (see Exercise 7.6.1). In fact, the transformation of this Hamiltonian becomes obvious when the right variables are used.

6.5 Hori's General Theory. Hori Kernel and Averaging

Let us consider, now, the canonical system of equations

$$\frac{dq_i}{dt} = \frac{\partial H}{\partial p_i} \qquad \frac{dp_i}{dt} = -\frac{\partial H}{\partial q_i}, \qquad (6.23)$$

where $H(q,p)$ is a time-independent Hamiltonian, $q \equiv (q_1, q_2, \cdots, q_N)$, $p \equiv (p_1, p_2, \cdots, p_N)$ are unspecified canonical variables and let us consider the transformation $(q,p) \Rightarrow (q^*, p^*)$ defined by a Lie series:

$$\phi(q,p) = E_{W^*}\phi(q^*, p^*) = \sum_{k=0}^{\infty} \frac{1}{k!} D_{W^*}^k \phi(q^*, p^*), \qquad (6.24)$$

where $W^* = W(q^*, p^*)$. As in Sect. 6.2, the conservation of the Hamiltonian leads to

$$H^*(q^*, p^*) = E_{W^*} H(q^*, p^*), \qquad (6.25)$$

152 6 Lie Series Perturbation Theory

where H^* is the Hamiltonian of the transformed system and the perturbation equations are obtained by substituting the expansions (6.6) into (6.25). The perturbation equations are the same as (6.7), but, now, the functions depend on (q^*, p^*) instead of (θ^*, J^*). The homological equation is

$$H_k^* = \Psi_k + \sum_{i=1}^{N} \left(\frac{\partial H_0}{\partial q_i^*} \frac{\partial W_k^*}{\partial p_i^*} - \frac{\partial W_k^*}{\partial q_i^*} \frac{\partial H_0}{\partial p_i^*} \right), \tag{6.26}$$

where $\Psi_k(q^*, p^*)$ is a known function if the previous k equations were already solved.

This homological equation can no longer be trivially solved as in Sect. 6.2.1. To solve this linear partial differential equation in the unknown function W_k^*, we use the Cauchy–Darboux theory of characteristics. We may apply the results of Theorem 6.5.1 below and go on straight to (6.30) and (6.31). However, since the homological equation is linear in the derivatives of W_k^* and includes W_k^* only through its derivatives, we may easily construct those equations. To do this, let us introduce a generic $2N$-dimensional variable $z^* \equiv (q^*, p^*)$. With it, the homological equation becomes

$$(-\mathsf{J} H'_{0z^*} \mid W_{kz^*}^{*\prime}) = \Psi_k - H_k^*, \tag{6.27}$$

where $H'_{0z^*} = \partial H_0 / \partial z^*$ and $W_{kz^*}^{*\prime} = \partial W_k^* / \partial z^*$ are the gradients of $H_0(z^*)$ and $W_k^*(z^*)$ in the $2N$-dimensional phase space, respectively, J is the symplectic matrix of rank $2N$:

$$\mathsf{J} = \begin{pmatrix} 0 & -\mathsf{E} \\ \mathsf{E} & 0 \end{pmatrix} \tag{6.28}$$

and E is the unit matrix of rank N. H'_{0z^*} and $H_k^* - \Psi_k$ are assumed to be continuous in the domain under consideration and do not vanish simultaneously.

Let $W_k^*(z^*)$ be a solution of (6.27) and let us consider the integral manifold $\mathcal{M}$ defined by $W_k^*(z^*) - W_k^* = 0$ in the $(2N + 1)$-dimensional space of the variables (z^*, W_k^*) (Fig. 6.1). We introduce a family of curves on $\mathcal{M}$ through the parametric equations

$$z^* = z^*(u) \qquad W_k^* = W_k^*(u).$$

These curves, called *characteristics* by Monge, define a vector field $\boldsymbol{T}$ tangent to the manifold. At every point P of $\mathcal{M}$, the vector $\boldsymbol{T}$ is proportional to $(\mathrm{d}z^*/\mathrm{d}u, \mathrm{d}W_k^*/\mathrm{d}u)$. We may also construct a vector $\boldsymbol{N}$, normal to $\mathcal{M}$ at P, by means of the gradient of the function $(W_k^*(z^*) - W_k^*)$. The gradient of this function in the $(2N + 1)$-dimensional space of the variables (z^*, W_k^*) is $(W_{kz^*}^{*\prime}, -1)$. Since the vectors $\boldsymbol{T}$ and $\boldsymbol{N}$ are orthogonal, $\boldsymbol{T}.\boldsymbol{N} = 0$ and, then,

$$\left(\frac{\mathrm{d}z^*}{\mathrm{d}u} \mid W_{kz^*}^{*\prime} \right) - \frac{\mathrm{d}W_k^*}{\mathrm{d}u} = 0. \tag{6.29}$$

Comparison of this equation, issued from a simple geometric construction, to (6.27) gives

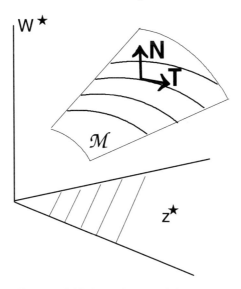

Fig. 6.1. The manifold $\mathcal{M}$ and a set of characteristic curves.

$$\frac{dW_k^*}{du} = \Psi_k - H_k^* \tag{6.30}$$

and

$$\frac{dz^*}{du} = -JH'_{0z^*} \tag{6.31}$$

or, in the variables q^*, p^*,

$$\frac{dq_i^*}{du} = \frac{\partial H_0}{\partial p_i^*} \qquad \frac{dp_i^*}{du} = -\frac{\partial H_0}{\partial q_i^*}. \tag{6.32}$$

Equations (6.30) and (6.31) (or 6.30 and 6.32) are the system of differential equations of the characteristic curves of the given partial differential equation.

Equations (6.31) (or 6.32) are the same for all values of k. They were called *auxiliary equations* by Hori and the system whose Hamiltonian is H_0 is often referred to as the *Hori kernel* of the perturbatrion theory thus constructed.

To solve the homological equation, for all k, (6.32) must be completely integrable, and we need to obtain the general solution:

$$\begin{aligned} q_i^* &= q_i^*(u + \gamma_1, \gamma_\ell, C_j) \\ p_i^* &= p_i^*(u + \gamma_1, \gamma_\ell, C_j) \end{aligned} \tag{6.33}$$

$(j = 1, \cdots, N; \ell = 2, \cdots, N)$; C_j and γ_j are arbitrary constants of integration.

The solution of the homological equation is completed by the integration of (6.30). Its left-hand side contains the unknown function $W_k(q^*, p^*)$ and the right-hand side contains the undetermined H_k^*. This last function is chosen to be such that

$$H_k^* = <\Psi_k> \tag{6.34}$$

and, thus, $<\mathrm{d}W_k^*/\mathrm{d}u> = 0$. We note that, when the solutions given by (6.33) are bounded, $\Psi_k[q^*(u), p^*(u)]$ is an almost periodic function. Bohr's mean-value theorem for almost periodic functions [9] may then be used to average Ψ_k:

$$<\Psi_k> = \lim_{T\to\infty} \frac{1}{T}\int_0^T \Psi_k \mathrm{d}u. \tag{6.35}$$

$W_k^*(u)$ is, then, immediately determined from (6.30) through

$$W_k^*(u) = \int (\Psi_k - H_k^*)\mathrm{d}u. \tag{6.36}$$

In these operations, we use the solutions (6.33) to write Ψ_k as a function of u and then we perform the integration; the arbitrary constant is set to be such that $<W_k^*(u)> = 0$. Finally, to know H_k^* and W_k^* as functions of q^*, p^*, we replace $u + \gamma_1$ and the integration constants γ_ℓ, C_j by the inverses of (6.33):

$$\begin{aligned} u + \gamma_1 &= g_1(q^*, p^*) \\ \gamma_\ell &= g_\ell(q^*, p^*) \\ C_j &= g_{N+j}(q^*, p^*) \qquad (j = 1, \cdots, N; \ell = 2, \cdots, N). \end{aligned} \tag{6.37}$$

Because of this inversion, the actual application of Hori's theory with unspecified canonical variables to general problems is, generally, cumbersome. When possible, it is always convenient to use angle–action variables or variables close to them.

It is worth noting that the resulting transformed Hamiltonian $H^* = \sum H_k^*$ is a function of the variables q^*, p^*. Eventually, when the angle–action variables of the Hori kernel are introduced, at least one of the angles becomes ignorable and the reduction of the system becomes evident. However, while the variables q^*, p^* are used, the reduction of the system comes from the existence of a new formal first integral as shown in Sect. 6.7. One example with variables that are not angle–action variables is presented in Sect. 7.8.

6.5.1 Cauchy–Darboux Theory of Characteristics

Definition 6.5.1 (Characteristic curves). *Consider the partial differential equation*

$$F(z, W_z, W) = 0, \tag{6.38}$$

where $z \equiv (z_1, z_2, \cdots, z_n)$, $W_z \equiv (W_{z_1}, W_{z_2}, \cdots, W_{z_n})$, $W : \mathbf{R}^n \to \mathbf{R} \in C^2$ *and* $F : \mathbf{R}^{2n+1} \to \mathbf{R} \in C^2$ *in the neighborhood of one point where* $\sum_1^n \left(\frac{\partial F}{\partial W_{z_i}}\right)^2 \neq 0$.

The characteristics of the given partial differential equations are the solutions $z(u), \pi(u), W(u)$ of the system of $2n+1$ ordinary differential equations:

$$\begin{aligned}
\frac{\mathrm{d}z_i}{\mathrm{d}u} &= \frac{\partial F(z,\pi,W)}{\partial \pi_i} \\
\frac{\mathrm{d}\pi_i}{\mathrm{d}u} &= -\frac{\partial F(z,\pi,W)}{\partial z_i} - \pi_i \frac{\partial F(z,\pi,W)}{\partial W} \\
\frac{\mathrm{d}W}{\mathrm{d}u} &= \sum_{i=1}^{n} \pi_i \frac{\partial F(z,\pi,W)}{\partial \pi_i},
\end{aligned} \qquad (6.39)$$

where $\pi \equiv (\pi_1, \pi_2, \cdots, \pi_n)$.

Theorem 6.5.1. *All solutions $W(z) : \mathbf{R}^n \to \mathbf{R} \in C^2$ of the partial differential equation $F(z, W_z, W) = 0$ can be obtained from the characteristic curves. These functions are, in general, uniquely determined by prescribing their values at the points of an $(n-1)$-dimensional manifold.*

For simplicity we have adopted notations similar to those used in Hori's general theory. The construction of the function $W(z)$, in the general case, follows the same steps as in the previous section: elimination of integration constants between the solutions of (6.39). The proof of the theorem is classical in the theory of first-order partial differential equations (see [18], Chap. 3).

Hori used the notation t^* for the parameter u of the equations. The interpretation of the Cauchy–Darboux parameter as a *pseudo time* [53] hid its actual meaning and, worst, allowed some noxious misunderstandings to become widespread.

6.6 Topology and Small Divisors

Hori's general theory is conceptually important because it allows us to understand a basic operation involved in perturbation theories, which usually remains hidden by the very particular form of the equations in angle–action variables. It shows the existence of a privileged dynamical system – the Hori kernel. The Hori kernel is the projection on the phase space (q^*, p^*) of the characteristic curves of the homological equation. It is the same *for all k*. For different values of k, the characteristic curves differ only in the $(2N+1)^{\text{th}}$ coordinate W_k^*.

To understand the role played by the Hori kernel, let us consider a Hamiltonian system with a non-degenerate H_0. In angle–action variables, the corresponding Hori kernel equations are:

$$\frac{\mathrm{d}\theta_i^*}{\mathrm{d}u} = \frac{\partial H_0}{\partial J_i^*} = \nu_i^* \neq 0 \qquad \frac{\mathrm{d}J_i^*}{\mathrm{d}u} = -\frac{\partial H_0}{\partial \theta_i^*} = 0 \qquad (6.40)$$

$(i = 1, 2, \cdots, N)$. The solutions of this system lie over N-tori defined by the equations $J_i^*(q^*, p^*) = \text{const}$. The transformed Hamiltonian H^* is also a function of J_i^* (and only of them) and, thus, their solutions lie over the same N-tori, only with different frequencies.

A consequence of this fact is that a perturbation theory is not suitable to disclose the actual topology of the given perturbed Hamiltonian system. It is only good for the calculation of solutions with the same topology as its Hori kernel.

6.6.1 Topological Constraint. The Rise of Small Divisors

The overall geometry of the transformed system is the same as that of the Hori kernel and it is so, regardless of the perturbation represented by the given H_k and of the bifurcations that they may have introduced in the flow of the given Hamiltonian [30]. Moreover, the Lie series mapping is a diffeomorphism and cannot introduce any topological change. In general, the bifurcations of the given perturbed Hamiltonian lead to small divisors whose unbound increase in number, from one order to the next, is responsible by the non-convergence of the results in an open set when $n \to \infty$.

Let us illustrate the rise of a small divisor at the place where a bifurcation should occur with a simple example. Let us consider the Hamiltonian

$$H = J_1 - \frac{1}{2}J_1^2 + \varepsilon\sqrt{2J_1}\cos\theta_1. \tag{6.41}$$

This Hamiltonian is one well-known particular case of the Ideal Resonance Problem thoroughly studied by Andoyer (see Appendix C). Its portrait in the (θ_1, J_1) plane is shown in the left side of Fig. 6.2. The application of Lie series theory to this Hamiltonian gives:

$$W^* = \frac{\varepsilon}{\nu_1^*}\sqrt{2J_1^*}\sin\theta_1^* + \frac{\varepsilon^2 J_1^*}{4\nu_1^{*3}}\sin 2\theta_1^*$$
$$+ \frac{\varepsilon^3}{48}\sqrt{2J_1^*}\left(\frac{22}{\nu_1^{*4}}\sin\theta_1^* + \frac{3J_1^*}{\nu_1^{*5}}(5\sin\theta_1^* + \sin 3\theta_1^*)\right) + \mathcal{O}(\varepsilon^4),$$

where

$$\nu_1^* = 1 - J_1^*, \tag{6.42}$$

and

$$H^* = J_1^* - \frac{1}{2}J_1^{*2} - \frac{\varepsilon^2}{2\nu_1^{*2}} + \mathcal{O}(\varepsilon^4) \tag{6.43}$$

with the proper frequency

$$g_1 = \frac{\partial H^*}{\partial J_1^*} = \nu_1^* - \frac{\varepsilon^2}{\nu_1^{*3}} + \mathcal{O}(\varepsilon^4).$$

The portrait of H^* in the (θ_1^*, J_1^*) plane is shown in the right side of Fig. 6.2. It mimics the phase portrait of $H_0(J)$. At the place where a bifurcation

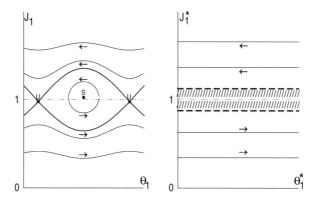

Fig. 6.2. Phase portraits of H (left) and H^* (right)

occurs in the complete Hamiltonian ($J_1 \simeq 1$), the transformed Hamiltonian H^* undergoes a complex change of sign of its proper frequency g_1 ($g_1 = 0$ on both bold dashed lines). This change cannot be correctly studied with the given equations, since the series giving W^* and H^* have the divisor ν_1^* and become singular at $J_1^* = 1$.

The above discussion shows that when the quantitative study of a feature is aimed at, it is necessary to engineer the Hori kernel and introduce that feature in its topology.

Exercise 6.6.1. Consider a one-degree-of-freedom system and introduce the canonical variables ϕ_1, E_1, where ϕ_1 is a uniform angle with unit frequency and E_1 is the Hori kernel energy. Show that $H = H(E_1)$ and that the set of curves $H = $ const and $E_1 = $ const coincide. Extend the reasoning to two degrees of freedom. *Hint:* Introduce the canonical variables ϕ_1, ϕ_2, E_1, J_2 and consider the set of curves $H = $ const and $E_1 = $ const in the manifold $J_2 = $ const.

6.7 Hori's Formal First Integral

Theorem 6.7.1 (Hori [53]). *The function $F(q,p) = H_0(q^*(q,p), p^*(q,p))$ is, at the order of approximation of the canonical transformation, a first integral of the given perturbed system.*

The proof of this theorem is very simple and just a simple chain of calculations. We have to show that the time derivative of F is equal to zero *at the order of approximation of the canonical transformation*. Indeed,

$$\frac{dF}{dt} = \sum_{i=1}^{N} \left(\frac{\partial H_0}{\partial q_i^*} \frac{dq_i^*}{dt} + \frac{\partial H_0}{\partial p_i^*} \frac{dp_i^*}{dt} \right).$$

We may then use: (1) the equations of the transformed dynamical system to replace the time derivatives of q_i^*, p_i^* by partial derivatives of H^*; (2) the equations of the Hori kernel to replace the partial derivatives of H_0 by derivatives of q_i^*, p_i^* with respect to u. Hence,

$$\frac{\mathrm{d}F}{\mathrm{d}t} = \sum_{i=1}^{N}\left(-\frac{\mathrm{d}p_i^*}{\mathrm{d}u}\frac{\partial H^*}{\partial p_i^*} - \frac{\mathrm{d}q_i^*}{\mathrm{d}u}\frac{\partial H^*}{\partial q_i^*}\right)$$

or

$$\frac{\mathrm{d}F}{\mathrm{d}t} = -\frac{\mathrm{d}H^*}{\mathrm{d}u},$$

which is equal to zero since H^* is defined by definite integrations over u.

□

This integral is only formal, not a true one, since the remainder $\mathcal{R}_n(q^*, p^*, \varepsilon)$ of the calculation of H^* was not considered in the above demonstration, and it is not independent of u. The order of approximation of the formal first-integral F is ε^n.

6.8 "Average" Hamiltonians

The word "average" and its variations became popular in the past century, implicitly carrying the idea that methods founded on "averaging" operations lead to "average" Hamiltonians governing the secular variation of the given system. However, in more than one instance, second-order solutions such as that given by (6.20) (at the end of Sect. 6.3), were found showing that their average is not equal to the solution of the "averaged" equations – in (6.20), we have $<J_1> \neq J_1^*$. The non-periodic terms appearing in the solution were often a source of disappointment.

In Lie series perturbation theory, the solutions have the general form

$$\phi = E_W \phi^* = \phi^* + D_W \phi^* + \frac{1}{2!}D_W^2 \phi^* + \frac{1}{3!}D_W^3 \phi^* + \cdots, \qquad (6.44)$$

where ϕ denotes a generic variable and $W(\theta^*, J^*)$ the Lie generating function resulting from the theory. By construction, the generating function is a zero-average periodic function of the angles $\theta_1, \theta_2, \cdots, \theta_N$. A glance at the above equation is enough to see that, notwithstanding the zero average of W, the terms of order 2, and higher, involve products of derivatives of W between themselves, and, in these operations, non-periodic terms are eventually generated.

A consequence of these non-periodic terms is that H^* is not an average. The actual solutions of the given Hamiltonian system oscillate about the solutions of the Hamiltonian system defined by H^*, but with a non-zero average.

For $k = 1$, (6.44) is reduced to $\phi = \phi^* + \{\phi^*, W\}$ including only one bracket: $\{\phi^*, W\}$, which is equal to a derivative of W and, therefore, a zero-average function. This means that, to first-order, the transformed H^* behaves as an average. This fact certainly played a role in the introduction of the word "average" and its variations to designate perturbation theories of this kind.

In the above discussed case, we may suspect that the non-periodic terms come from the definition of the canonical transformations through a Lie generating function, but it is possible to see that for any canonical transformation $(\theta, J) \Rightarrow (\theta^*, J^*)$ defined explicitly by

$$\begin{aligned}\theta_i &= \theta_i^* + Q_1^i(\theta^*, J^*) + Q_2^i(\theta^*, J^*) + \cdots \\ J_i &= J_i^* + P_1^i(\theta^*, J^*) + P_2^i(\theta^*, J^*) + \cdots,\end{aligned} \qquad (6.45)$$

it is not possible to have simultaneously H_k^* independent of θ^* and $<P_k^i> = <Q_k^i> = 0$ (for all i and $k > 1$) [34].

6.8.1 On Secular Theories and Proper Elements

Given the large number of degrees of freedom of the equations of planetary motion, it is usual, since the work of Laplace and Lagrange, to reduce the equations of motion to first-order averaged ones. The classical "secular theory" of Laplace and Lagrange is the analysis of the solutions of the Hamiltonian resulting from the elimination of short-period terms by means of first-order perturbations theory (see Sect. 3.7). In the case of asteroids, canonical perturbation theories may be used to define "proper actions", which are choice parameters for the identification of asteroid families. For practical reasons, they are often replaced by average values of elements calculated numerically. Even if, strictly speaking, averages differ for proper actions, it is evident that in non-degenerate systems, averages are functions of the proper actions and may show the same time invariance as the proper actions themselves.

7

Non-Singular Canonical Variables

7.1 Singularities of the Actions

The actions J_i defined by the phase integrals $J_i = \frac{1}{2\pi} \oint p_i \mathrm{d}q_i$ may become singular. The simplest example is given by the actions of a Hamiltonian depending on the squares of the momenta. In this case, p_i is proportional to $\dot{q}_i$ and, as a consequence, the integral $\int p_i \mathrm{d}q_i$ is proportional to $\int \dot{q}_i^2 \mathrm{d}t$ and, thus, sign definite. In other words, the integration path is always circulated in the same direction and the sign of J_i may not be reversed (Fig. 7.1). Consequently, the equations of motion in this variable are singular at $J_i = 0$.

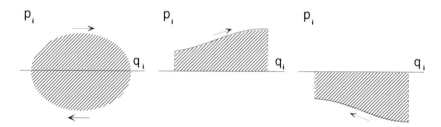

Fig. 7.1. Integration paths

Examples are abundant in Celestial Mechanics. The Delaunay actions

$$\begin{aligned} L &= \sqrt{\mu a} \\ G &= L\sqrt{1-e^2} \\ H &= G\cos i \end{aligned} \qquad (7.1)$$

are singular for $a = 0$, $e = 0$ (or $e = 1$), and $\sin i = 0$.

The singularities at $a = 0$ and $e = 1$ correspond to critical physical situations. At $a = 0$, the orbit degenerates into one point. We recall that the attracting force becomes infinite for $r = 0$ and specific regularizing techniques are needed to deal with motions in which r becomes close to zero (see [6], [95]). For $e = 1$, non-rectilinear motions cease being bounded. The other two singularities ($e = 0$, and $i = 0$), often found in practical applications, are, however, just geometric and do not correspond to singular physical situations. They can be overcome with the pure geometrical tools described in this chapter.

7.2 Poincaré Non-Singular Variables

When the actions J_i are small, the corresponding angle–action variables may lead to algebraic difficulties as, for instance, division by zero. (The equations of motion, generally, include these actions as denominators.) If the singularity is just the geometrical singularity of the angle–action variables, we know, since Lagrange, that a transformation of variables from the polar-like angle–action variables θ_i, J_i to the associated non-singular rectangular-like coordinates $J_i \cos \theta_i, J_i \sin \theta_i$ is enough to avoid numerical difficulties. However, for the needs of the Hamiltonian theories discussed in this book, this is not sufficient, since Lagrange's variables are not canonical. Nevertheless, similar canonical variables were proposed by Poincaré by taking $\sqrt{2|J_i|}$ instead of J_i.

When $J_i < 0$, the Poincaré non-singular canonical variables associated with θ_i, J_i are

$$x_i = \sqrt{-2J_i} \cos \theta_i$$
$$y_i = \sqrt{-2J_i} \sin \theta_i. \tag{7.2}$$

In this case, the inverse transformation is given by

$$J_i = -\frac{1}{2}(x_i^2 + y_i^2) \tag{7.3}$$

$$\theta_i = \arctan \frac{y_i}{x_i}. \tag{7.4}$$

The Poisson bracket of the new variables with respect to the old ones is

$$\{x_i, y_i\} = \{\theta_i, J_i\} = +1$$

and, therefore, the pair of canonical equations

$$\dot{\theta}_i = \frac{\partial H}{\partial J_i} \qquad \dot{J}_i = -\frac{\partial H}{\partial \theta_i} \tag{7.5}$$

becomes

$$\dot{x}_i = \frac{\partial \widehat{H}}{\partial y_i} \qquad \dot{y}_i = -\frac{\partial \widehat{H}}{\partial x_i}, \tag{7.6}$$

where $\widehat{H} = H(\theta_i(x_i, y_i), J_i(x_i, y_i))$.

We started with the case $J_i < 0$ for two important reasons. The first one is that, in this case, we have $\{x_i, y_i\} = +1$. The second reason is that

7.2 Poincaré Non-Singular Variables

this is often the case in Celestial Mechanics. Indeed, near the singularities $e = 0, i = 0$, the origin of the angles ℓ and w, canonically conjugate to L and G, becomes indeterminate (see Sect. 2.5.2). It is them common usage to replace the ordinary Delaunay elements by the set

$$\begin{aligned} \lambda &= \ell + w + \Omega & L & \\ \varpi &= w + \Omega & G - L &= L(\sqrt{1 - e^2} - 1) \\ \Omega & & H - G &= G(\cos i - 1). \end{aligned} \quad (7.7)$$

The new angles are longitudes and take their origin at the same point, e.g. the point O of Fig. 2.5, or, in Astronomy, the equinox[1]. The actions now associated with ϖ and Ω are both negative ($G - L < 0$, $H - G < 0$).

When $J_i > 0$, instead of (7.2), we have

$$\begin{aligned} x_i &= \sqrt{2J_i} \cos \theta_i \\ y_i &= \sqrt{2J_i} \sin \theta_i \end{aligned} \quad (7.8)$$

and the inverse transformation is

$$J_i = \frac{1}{2}(x_i^2 + y_i^2) \quad (7.9)$$

$$\theta_i = \arctan \frac{y_i}{x_i}. \quad (7.10)$$

In this case, $\{x_i, y_i\} = -1$ and, thus, to keep equations written in the same order as through this whole book, we have to change to $\{y_i, x_i\} = +1$. The corresponding canonical equations are, now,

$$\dot{y}_i = \frac{\partial \widehat{H}}{\partial x_i} \qquad \dot{x}_i = -\frac{\partial \widehat{H}}{\partial y_i}. \quad (7.11)$$

(Compare the signs of (7.6) and (7.11).)

It is easy to see how to modify the given definitions to deal with cases where the singularity of the actions occurs for non-zero values.

Exercise 7.2.1. Show that, in Poincaré variables, the action variables are

$$\Lambda = \frac{-s}{2\pi} \oint y \mathrm{d}x = \frac{+s}{2\pi} \oint x \mathrm{d}y, \quad (7.12)$$

where $s = \pm 1$ is the sign of J (s does not change over the path). Hint: Using the given definitions show that $J \mathrm{d}w = \frac{1}{2} s(x \mathrm{d}y - y \mathrm{d}x)$.

---

[1] These variables are sometimes called equinoctial.

7.3 The d'Alembert Property

When a function regular in a domain about the origin is written with polar coordinates, pure geometrical singularities may appear at $r = 0$ because of these coordinates. This singularity disappears when rectangular coordinates x, y are used instead of the polar ones. This situation is current in Celestial Mechanics and occurs with the pairs of polar-like variables e, ϖ (eccentricity, longitude of the periapsis) and i, Ω (inclination, longitude of the ascending node). The non-singular variables used instead of them, since Lagrange, are the associated rectangular pairs $e \cos \varpi, e \sin \varpi$ and $i \cos \Omega, i \sin \Omega$. (More usual definitions have $\sin i$, $\sin(i/2)$ or $\tan i$ instead of i.)

Let $g(x, y) : \mathbf{R}^2 \to \mathbf{R}$ be a regular function in a domain $\mathcal{O}$ about the origin and let $f(\alpha, r)$ be the expression of this function in polar coordinates:

$$f(\alpha, r) = g(x, y).$$

As the function $g(x, y)$ is regular in $\mathcal{O}$, it may be expanded in a power series in x, y,

$$g(x, y) = \sum_{i,j \geq 0} a_{ij} x^i y^j,$$

convergent in $\mathcal{O}$. Hence

$$f(\alpha, r) = \sum_{n \geq 0} \sum_{j=0}^{[n/2]} r^n \left\{ C_{jn} \cos[(n - 2j)\alpha] + S_{jn} \sin[(n - 2j)\alpha] \right\}, \quad (7.13)$$

where $[n/2]$ means the integer part of $n/2$ and C_{jn}, S_{jn} are numerical coefficients.

The features shown by this expression of $f(\alpha, r)$ are part of a set of rules found in the expansion of the disturbing potential in planetary theory known as *d'Alembert properties* (or d'Alembert characteristics). In the case of the above expansion, they may be expressed as follows: for each n, the coefficients of the multiples of α in the trigonometric part have the same parity as n and are at most equal to n.

However, the simple polar-to-rectangular transformation is not canonical and we have rather to consider the transformation defined by (7.2) (or 7.8). The d'Alembert property appears, then, in a slightly modified form: If we have

$$f(\theta, J) = g(x, y),$$

where, now, x, y are Poincaré non-singular variables and θ, J the corresponding angle–action variables, the power series in x, y becomes

$$f(\theta, J) = \sum_{n \geq 0} \sum_{j=0}^{[n/2]} |J|^{\frac{n}{2}} \left\{ \tilde{C}_{jn} \cos[(n - 2j)\theta] + \tilde{S}_{jn} \sin[(n - 2j)\theta] \right\}. \quad (7.14)$$

The d'Alembert property still holds and the only difference with respect to (7.13) is the $n/2$ exponent of $|J|$.

The situation described above is trivial; but it may become more complex when other variable transformations are added. In order to avoid the accidental transformation of a geometrical singularity into a singularity whose origin is not easily recognized, one transformation must comply with some simple rules:

Theorem 7.3.1 (Henrard [47]). *A transformation from an angle–action pair of variables to another preserves the d'Alembert property of a function if it is a Lie series mapping whose generating function has the d'Alembert property.*

Indeed, Lie series mappings are defined by the equation

$$f(\theta, J) = E_{W^*} f(\theta^*, J^*),$$

where $W^* = W(\theta^*, J^*)$. They involve only the computation of Poisson brackets, which are invariant to canonical transformations. Since the necessary and sufficient condition for a Poisson bracket of two functions to be regular is that these functions are regular, if f and W^* have the d'Alembert property then $D_{W^*} f$ and $E_{W^*} f$ also have the d'Alembert property.

7.4 Regular Integrable Hamiltonians

The perturbation techniques of Celestial Mechanics always consider that the undisturbed Hamiltonian is completely integrable. (See the statement of Delaunay's problem in Sect. 3.1.) The series expansion of a non-singular integrable Hamiltonian, about the origin, in terms of its angle–action variables (θ_i, J_i) is

$$H_0 = \sum_{i=1}^{N} \nu_i^\circ J_i + \frac{1}{2} \sum_{i=1}^{N} \sum_{j=1}^{N} \nu_{ij}^\circ J_i J_j + \cdots, \qquad (7.15)$$

where

$$\nu_i^\circ = \nu_i(0) = \left(\frac{\partial H_0}{\partial J_i}\right)_{J=0} \qquad (7.16)$$

and

$$\nu_{ij}^\circ = \nu_{ij}(0) = \left(\frac{\partial^2 H_0}{\partial J_i \partial J_j}\right)_{J=0}. \qquad (7.17)$$

The angles θ_i may not appear in the Hamiltonian by the very definition of the angle–action variables. On the other hand, half-integer powers of J_i cannot appear because of the regularity hypothesis. Indeed, non-singular functions must satisfy the d'Alembert property, which means that any half-integer power of J_i should necessarily appear multiplied by a trigonometric function of θ_i, at variance with the previous statement.

7 Non-Singular Canonical Variables

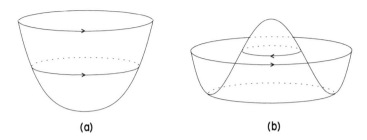

Fig. 7.2. Regular integrable Hamiltonians near the origin

In non-singular variables, H_0 is

$$H_0 = \frac{1}{2}\sum_{i=1}^{N} \nu_i^\circ (x_i^2 + y_i^2) + \frac{1}{8}\sum_{i=1}^{N}\sum_{j=1}^{N} \nu_{ij}^\circ (x_i^2 + y_i^2)(x_j^2 + y_j^2) + \cdots \quad (7.18)$$

(where, for the sake of simplicity, we assumed $J_i > 0$).

Let us consider, for a moment, the case $N = 1$. In this case, H_0 is the Hamiltonian of a differential rotator:

$$H_0 = \frac{1}{2}\nu_1^\circ (x_1^2 + y_1^2) + \frac{1}{8}\nu_{11}^\circ (x_1^2 + y_1^2)^2 + \cdots \quad (7.19)$$

Figure 7.2 shows the function H_0 in the neighborhood of the origin in the two possible cases:

(a.) ν_1° and ν_{11}° have the same sign ($\nu_1^\circ > 0$, $\nu_{11}^\circ > 0$);
(b.) ν_1° and ν_{11}° have opposite signs ($\nu_1^\circ < 0$, $\nu_{11}^\circ > 0$).

(If $\nu_{11}^\circ < 0$, the figures would be equal, but turned upside down.) When $\nu_1^\circ = 0$, the figure is similar to Fig. 7.2(a), but the curvature at the vertex is equal to zero since, in this case, the origin is a zero of fourth order.

The motions on these surfaces are circular and have constant velocities. Their frequencies are

$$\nu_1 = \frac{\partial H_0}{\partial J_1} = \nu_1^\circ + \nu_{11}^\circ J_1 + \cdots . \quad (7.20)$$

Thus, in the neighborhood of the origin of the (x_1, y_1) plane, in (a) the motions are direct ($\nu_1^\circ > 0$ and $\nu_{11}^\circ > 0$). In (b), the motions near the origin are retrograde up to the distance where the minimum of H_0 is reached, and direct beyond this minimum (up to the distance where another extremum of the function H_0, if it exists, is reached)[2].

In the most frequent case, $J_1 < 0$, we have

---

[2] In the case $J_1 > 0$, the motion in the (x_1, y_1) plane is retrograde (resp. direct) when the motion of θ_1 is direct (resp. retrograde). See Fig. 7.3.

$$H_0 = -\frac{1}{2}\nu_1^\circ(x_1^2 + y_1^2) + \frac{1}{8}\nu_{11}^\circ(x_1^2 + y_1^2)^2 + \cdots \tag{7.21}$$

and the situations invert with respect to the previous case. Now we have

(a'.) ν_1° and ν_{11}° have opposite signs ($\nu_1^\circ < 0$, $\nu_{11}^\circ > 0$);
(b'.) ν_1° and ν_{11}° have the same sign ($\nu_1^\circ > 0$, $\nu_{11}^\circ > 0$).

(again, we assumed $\nu_{11}^\circ > 0$). The directions of the motions are reversed with respect to those of the case $J_1 > 0$. The motions in (a') are always retrograde (on the (x_1, y_1) plane) while, in (b'), they are direct near the origin and retrograde outside the minimum of H_0.

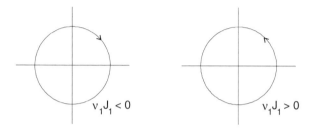

Fig. 7.3. Directions of motion in the (x_1, y_1) plane

At this point, let it be pointed out that the transformation to Poincaré non-singular variables defined by (7.2) (or 7.8) has regularizing properties that are more powerful than those of the simple geometric transformation $x_1 = |J_1|\cos\theta_1$, $y_1 = |J_1|\sin\theta_1$ (which is sufficient only to eliminate the geometrical singularity). Indeed, in the general case, we have $\nu_1^\circ \neq 0$ and this means that the first derivative of the function H_0 has a finite limit at the origin. If the ordinary polar-to-rectangular geometric transformation were used instead of the Poincaré transformation, the first derivative of H_0 would have no limit at the origin; Figs. 7.2(a) and (b) would have a cone-like structure near the vertex (except for $\nu_1^\circ = 0$). This additional regularizing property of the transformations defined in Sect. 7.2 arises from the fact that they involve $\sqrt{|J_1|}$ instead of J_1.

7.5 Lie Series Expansions About the Origin

Following the definitions given in Sect. 5.3, for each $f \in \mathcal{F}$, each point x, y in $\mathcal{O}$ and a given Lie generator W of class C^∞ in $\mathcal{O}$, the application

$$f \to E_W f = \sum_{k=0}^\infty \frac{\lambda^k}{k!} D_W^k f \tag{7.22}$$

is the Lie series expansion of the function f, generated by W. The Lie derivatives $D_W^k f$ are defined recursively from

$$f \to D_W f = \{f, W\}. \tag{7.23}$$

In the theories of Poincaré, von Zeipel–Brouwer and Hori, all expansions were done in series of powers of ε (or $\sqrt{\varepsilon}$ in Delaunay theory). In the neighborhood of the origin, we assume that the variables x, y are small quantities of order $\mathcal{O}(\varepsilon^d)$ (generally $d \leq 1$) and we have to adopt new rules for the comparison of terms. We will no longer use the powers of ε (or $\sqrt{\varepsilon}$), but the degree of homogeneity of the function with respect to the elements of the set $\mathcal{S} \equiv (x, y, \varepsilon^d)$.

To write the Lie series expansion of the function f in the neighborhood of the origin, let it be assumed that f is a homogeneous function of the elements of $\mathcal{S}$ and also that

$$W = \sum_{k \geq 1} W_k(x, y, \varepsilon), \tag{7.24}$$

where the $W_k(x, y, \varepsilon)$ have degree k in the elements of $\mathcal{S}$. Then

$$\begin{aligned} E_W f = {} & f + \{f, W_1\} + \{f, W_2\} + \{f, W_3\} + \cdots \\ & + \tfrac{1}{2}\{\{f, W_1\}, W_1\} + \tfrac{1}{2}\{\{f, W_1\}, W_2\} + \tfrac{1}{2}\{\{f, W_2\}, W_1\} + \cdots \\ & + \tfrac{1}{6}\{\{\{f, W_1\}, W_1\}, W_1\} + \cdots, \end{aligned} \tag{7.25}$$

where, as in previous theories, we assumed $\lambda = 1$. The law of formation of the terms for the kth row is very simple:

$$D_W^k f = \sum \{\{\ldots \{\{f, W_{\ell_1}\}, W_{\ell_2}\}, \ldots\}, W_{\ell_k}\},$$

where the sum extends over all combinations $(\ell_1, \ell_2, \ldots, \ell_k) \in \mathbf{Z}^k$.

We have to take into account, now, that the Lie derivative D_W modifies the order of the terms by subtracting two units, because of the differentiations with respect to x and y in each term of the Poisson bracket defining the Lie derivative. Therefore, if L is the degree of f, the degree of each term in $D_W^k f$ is

$$L + \sum_{i=1}^{k} \ell_i - 2k.$$

This means that there are, in the derivatives, terms with degree less than L (and even negative). To avoid this inconvenience, we assume $W_1 = 0$. Another difficulty resulting from the order losses in the derivatives is that the collection of the terms of the same degree of homogeneity as f,

$$f + \{f, W_2^*\} + \frac{1}{2}\{\{f, W_2^*\}, W_2^*\} + \frac{1}{6}\{\{\{f, W_2^*\}, W_2^*\}, W_2^*\} + \cdots,$$

has an unlimited number of terms. One practical requirement in the construction of the perturbation equations of a Lie series theory is that the number of terms in the Lie series $E_{W*}f$, of a given order (or degree of homogeneity), is finite. We then assume $W_2^* = 0$ and the Lie series expansion of f becomes

$$\begin{aligned}E_W f = f &+ \{f, W_3\} \\ &+ \{f, W_4\} + \tfrac{1}{2}\{\{f, W_3\}, W_3\} \\ &+ \{f, W_5\} + \tfrac{1}{2}\{\{f, W_3\}, W_4\} + \tfrac{1}{2}\{\{f, W_4\}, W_3\} \\ &\quad + \tfrac{1}{6}\{\{\{f, W_3\}, W_3\}, W_3\} \\ &+ \cdots,\end{aligned} \quad (7.26)$$

where, now, the terms have been ordered following their degree of homogeneity: L and $L+1$ in the first row, $L+2$ in the second row, $L+3$ in the third row, etc.

This series is very similar to those given in previous chapters; the only difference lies in the subscripts of W, which, in the expansion about the origin, are two units larger than in ordinary expansions.

7.6 Lie Series Perturbation Theory in Non-Singular Variables

Let us use Hori theory to study the solutions of a perturbed regular Hamiltonian, in the neighborhood of the origin. This can be done because Hori theory is valid for any set of canonical variables and thus may be used with non-singular canonical variables. We recall that the classical theories of Chap. 3 apply only to problems stated in angle–action variables and may not be straightforwardly used here.

Let us consider the Hamiltonian system given by

$$H = H_0(J) + \sum_{k \geq 1} \varepsilon^k H_k(\theta, J). \quad (7.27)$$

We assume that the undisturbed Hamiltonian $H_0(J)$ is regular in a domain around the origin and that $\nu_i \neq 0$, for all i, in this domain (there is no resonance at the origin). Hence

$$H_0 = \sum_{k \geq 1} X_{2k}(J) = \sum_{i=1}^{N} \nu_i^\circ J_i + \frac{1}{2}\sum_{i=1}^{N}\sum_{j=1}^{N} \nu_{ij}^\circ J_i J_j + \cdots \quad (7.28)$$

(see Sect. 7.4), or, with non-singular variables (assuming by default the case $J_i < 0$),

$$H_0 = -\frac{1}{2}\sum_{i=1}^{N} \nu_i^\circ (x_i^2 + y_i^2) + \frac{1}{8}\sum_{i=1}^{N}\sum_{j=1}^{N} \nu_{ij}^\circ (x_i^2 + y_i^2)(x_j^2 + y_j^2) + \cdots. \quad (7.29)$$

The disturbing terms will be written

$$H_k = \sum_{k' \geq 1} \mathcal{V}_k^{k'}(x, y) \tag{7.30}$$

with $\mathcal{V}_k^{k'}$ denoting the part of H_k of degree k' with respect to x, y. As already stated, the orders of magnitude will no longer be the powers of ε (or $\sqrt{\varepsilon}$), but the degrees of homogeneity in the elements of a given set $\mathcal{S}$. In this theory, we assume
$$x_i = \mathcal{O}(\varepsilon) \qquad y_i = \mathcal{O}(\varepsilon)$$
and
$$\mathcal{S} \equiv (x, y, \varepsilon).$$

The functions expand as indicated in the previous section with $d = 1$.

Let us introduce, now, the canonical transformation $\phi_n : (x, y) \Rightarrow (x^*, y^*)$ defined by
$$f(x, y) = E_{W^*} f(x^*, y^*), \tag{7.31}$$
where $E_{W^*} f$ is the Lie series expansion of $f(x, y)$ about the origin. Following the same development as in Chap. 6, we introduce
$$W^* = \sum_{k=3}^{n} W_k^*(x^*, y^*, \varepsilon), \tag{7.32}$$
where the quantities W_k^* are homogeneous functions of degree k in the elements of $\mathcal{S}$. Since the given canonical transformation is conservative, we have
$$H(x, y) = H^*(x^*, y^*) + \mathcal{R}_n(x^*, y^*),$$
that is,
$$H^*(x^*, y^*) + \mathcal{R}_n(x^*, y^*) = E_{W^*} H(x^*, y^*). \tag{7.33}$$
We then introduce, in these equations, the expansions already given for W^* and H as well as
$$H^* = \sum_{k \geq 2} H_k^*(x^*, y^*, \varepsilon), \tag{7.34}$$
where the quantities H_k^* are homogeneous functions of degree k in the elements of $\mathcal{S}$.

Comparing the parts of the same degree in both sides of (7.33), we obtain the equations of the Hori perturbation theory in the case under consideration:

$$\begin{aligned}
H_2^* &= X_2 + \varepsilon \mathcal{V}_1^1 \\
H_3^* &= \varepsilon \mathcal{V}_1^2 + \varepsilon^2 \mathcal{V}_2^1 + \{X_2 + \varepsilon \mathcal{V}_1^1, W_3^*\} \\
H_4^* &= X_4 + \varepsilon \mathcal{V}_1^3 + \varepsilon^2 \mathcal{V}_2^2 + \varepsilon^3 \mathcal{V}_3^1 + \{X_2 + \varepsilon \mathcal{V}_1^1, W_4^*\} \\
&\quad + \{\varepsilon \mathcal{V}_1^2 + \varepsilon^2 \mathcal{V}_2^1, W_3^*\} + \tfrac{1}{2}\{\{X_2 + \varepsilon \mathcal{V}_1^1, W_3^*\}, W_3^*\} \\
\cdots & \\
H_n^* &= \mathcal{V}_0^n + \sum_{k=1}^{n-1} \varepsilon^k \mathcal{V}_k^{n-k} + \sum_{k=0}^{n-2} \varepsilon^k \{\mathcal{V}_k^{n-k-1}, W_3^*\} + \cdots + \{X_2 + \varepsilon \mathcal{V}_1^1, W_n^*\}.
\end{aligned}$$
$$\tag{7.35}$$

7.6 Lie Series Perturbation Theory in Non-Singular Variables

In the last equation of this system, we introduced $\mathcal{V}_0^n$ with the definition $\mathcal{V}_0^n = X_n$ for n even and $\mathcal{V}_0^n = 0$ for n odd. The rules of construction of the right-hand sides are very simple. It is enough to replace the generic function f of (7.26) by the expansions of H and its parts and take into account the degree of homogeneity of every part. These rules are simple and, when computer algebraic manipulators are used, they allow simple iterative schemes to be introduced.

If the equations are used in turn to simplify those of higher orders, we have

$$H_2^* = X_2 + \varepsilon \mathcal{V}_1^1$$
$$H_3^* = \varepsilon \mathcal{V}_1^2 + \varepsilon^2 \mathcal{V}_2^1 + \{H_2^*, W_3^*\}$$
$$H_4^* = \sum_{k=0}^{3} \varepsilon^k \mathcal{V}_k^{4-k} + \frac{1}{2}\{H_3^* + \varepsilon \mathcal{V}_1^2 + \varepsilon^2 \mathcal{V}_2^1, W_3^*\} + \{H_2^*, W_4^*\}$$
$$\cdots \qquad (7.36)$$
$$H_n^* = \sum_{k=0}^{n-1} \varepsilon^k \mathcal{V}_k^{n-k} + \frac{1}{2}\{H_{n-1}^* + \sum_{k=0}^{n-2} \varepsilon^k \mathcal{V}_k^{n-k-1}, W_3^*\} + \cdots$$
$$+ \frac{1}{2}\{H_3^* + \varepsilon \mathcal{V}_1^2 + \varepsilon^2 \mathcal{V}_2^1, W_{n-1}^*\} + \{H_2^*, W_n^*\}.$$

Let it be recalled that all functions X_k and $\mathcal{V}_k^{k'}$ in the preceding systems are understood as $X_k(x^*, y^*)$ and $\mathcal{V}_k^{k'}(x^*, y^*)$, that is, $X_k(x,y)|_{x=x^*, y=y^*}$ and $\mathcal{V}_k^{k'}(x,y)|_{x=x^*, y=y^*}$.

Equation (7.36) may be synthesized in the homological partial differential equation

$$\{H_2^*, W_k^*\} = H_k^* - \Psi_k \qquad (7.37)$$

and the Hori kernel associated to it is

$$\frac{dx_i^*}{du} = \frac{\partial H_2^*}{\partial y_i^*} \qquad \frac{dy_i^*}{du} = -\frac{\partial H_2^*}{\partial x_i^*}, \qquad (7.38)$$

where the signs in the equations were chosen in accordance with the assumption $J_i < 0$ (that is, $\{x_i, y_i\} = +1$). The first feature to be considered in (7.38) is that these equations are separable into N second-order systems, since H_2^* is composed of N parts, each depending on only one pair of variables x_i^*, y_i^*. In addition, each separated system is easily integrated. Indeed, we have

$$\frac{dx_i^*}{du} = -\nu_i^\circ y_i^* + \varepsilon b_i' \qquad \frac{dy_i^*}{du} = \nu_i^\circ x_i^* - \varepsilon b_i, \qquad (7.39)$$

where we have assumed

$$\mathcal{V}_1^1(x_i^*, y_i^*) = \sum_{i=1}^{N}(b_i x_i^* + b_i' y_i^*). \qquad (7.40)$$

The general solutions of (7.39) are

$$x_i^* = C_i \cos(\nu_i^\circ u + \gamma_i) + \frac{\varepsilon b_i}{\nu_i^\circ}$$

$$y_i^* = C_i \sin(\nu_i^\circ u + \gamma_i) + \frac{\varepsilon b_i'}{\nu_i^\circ}, \qquad (7.41)$$

where the integration constants γ_i are chosen such that $C_i > 0$. The function $\Psi_k(x^*, y^*)$ is then a quasiperiodic function of u and one of the averaging operations of Chap. 6 may be used to obtain the corresponding H_k^*.

7.6.1 Solutions Close to the Origin (Case $J_1 < 0$)

The study of the solutions close to the origin may be simplified if, beforehand, we perform the canonical transformation

$$\widehat{x}_i = x_i - \frac{\varepsilon b_i}{\nu_i^\circ} \qquad \widehat{y}_i = y_i - \frac{\varepsilon b_i'}{\nu_i^\circ}. \qquad (7.42)$$

With these variables, the given problem is transformed into a modified one where

$$\widehat{X}_2 = -\frac{1}{2} \sum_{i=0}^{N} \nu_i^\circ (\widehat{x}_i^2 + \widehat{y}_i^2); \qquad \widehat{V}_1^1 = 0.$$

(The hat is used to indicate the functions transformed by means of (7.42)). The first of the perturbation equations is, now, simply

$$\widehat{H}_2^* = \widehat{X}_2$$

and the solutions of the modified Hori kernel are

$$\widehat{x}_i^* = C_i \cos(\nu_i^\circ u + \gamma_i)$$
$$\widehat{y}_i^* = C_i \sin(\nu_i^\circ u + \gamma_i). \qquad (7.43)$$

When there is no commensurability among the frequencies ν_i°, we may use the mean-value theorem of quasiperiodic functions to obtain the averages:

$$\widehat{H}_k^* = <\widehat{\Psi}(\widehat{x}^*, \widehat{y}^*)>, \qquad (7.44)$$

where $< \cdots >$ stands for the average over all angles $\nu_i^\circ u + \gamma_i$ from 0 to 2π. For all $k \geq 3$, $\widehat{H}_k^*$ will be a function of

$$\widehat{x}_i^{*2} + \widehat{y}_i^{*2} = C_i^2 \qquad (7.45)$$

only. The transformed Hamiltonian system is easily integrated. The new equations are

$$\frac{\mathrm{d}\widehat{x}_i^*}{\mathrm{d}t} = \frac{\partial \widehat{H}^*}{\partial \widehat{y}_i^*} \qquad \frac{\mathrm{d}\widehat{y}_i^*}{\mathrm{d}t} = -\frac{\partial \widehat{H}^*}{\partial \widehat{x}_i^*}. \tag{7.46}$$

Equation (7.45) shows that the solutions are circles and a simple calculation shows that the motions on these circles are uniform with frequencies

$$\nu_i^* = -\frac{1}{C_i}\frac{\partial \widehat{H}^*}{\partial C_i}. \tag{7.47}$$

Exercise 7.6.1. Consider the Hamiltonian that served as an example for a practical comparison of Poincaré and Lie series theories in Sect. 6.4. Show that the study of that Hamiltonian with non-singular variables is trivial ($H^* = H$ and $W^* = 0$).

7.6.2 Angle–Action Variables of H_2^* (Case $J_1 < 0$)

Notwithstanding the simplicity of the above calculations, one may easily verify that the integration constants (γ_i, C_i) are not canonical. This means that, in the calculation of the Poisson brackets of the next-order perturbation equation, these constants may not be used and one has to use the inverse of the general solutions to write all concerned functions again as functions of (x_i^*, y_i^*). This task can be avoided by using canonical integration constants (as given by the solution of the corresponding Hamilton–Jacobi equation), or, simply, by introducing the angle–action variables of H_2^*. In the case under study, they are trivially obtained:

$$w_i = |\nu_i^\circ u + \gamma_i|$$
$$\Lambda_i = \pm\frac{1}{2}C_i^2,$$

where the sign of Λ_i is equal to the sign of ν_i°. We recall that $s = -1$ ($J_1 < 0$) and that (7.12) gives, in this case, $\Lambda = \frac{1}{2\pi}\oint y\mathrm{d}x$, whose sign is opposite to the sign of $J_1\nu_1^\circ$ (see Fig. 7.3). Hence

$$\begin{aligned} x_i^* &= \sqrt{2|\Lambda_i|}\cos w_i + \varepsilon\frac{b_i}{\nu_i^\circ} \\ y_i^* &= \pm\sqrt{2|\Lambda_i|}\sin w_i + \varepsilon\frac{b_i'}{\nu_i^\circ}, \end{aligned} \tag{7.48}$$

where the sign in the last equation is opposite to the sign of ν_i°.

7.7 The Non-Resonance Condition

The condition for the use of the averaging rule fixed by (7.44) is a *non-resonance condition* analogous to that of Sect. 3.5: $(h \mid \nu^\circ) \neq 0$ for all integer vectors $h \in \mathbf{Z}^N$ appearing in the arguments of the given Hamiltonian or formed through the successive multiplication of trigonometric polynomials during the calculation of the Ψ_k ($k \leq n$). When a resonance is approached, we have to proceed as in von Zeipel–Brouwer theory: The von Zeipel averaging rule is written in the same way as before, but $< \cdots >$ stands for the average over the non-resonant short-period angles only.

To explain the procedures to follow in this case, we introduce the angle–action variables $(\widehat{\theta}_i^*, \widehat{J}_i^*)$ associated with $(\widehat{x}_i^*, \widehat{y}_i^*)$ by means of the Poincaré relations

$$\widehat{x}_i^* = \sqrt{2|\widehat{J}_i^*|} \cos \widehat{\theta}_i^* \qquad (7.49)$$
$$\widehat{y}_i^* = \sqrt{2|\widehat{J}_i^*|} \sin \widehat{\theta}_i^*.$$

The average of a function $\widehat{\Psi}_k^*$ is now given by

$$< \widehat{\Psi}_k^*(\widehat{\theta}^*, \widehat{J}^*) > = \widehat{\Psi}_{k(S)}^*(\widehat{J}^*) + \widehat{\Psi}_{k(K)}^*(\bar{h} \mid \widehat{\theta}^*, \widehat{J}^*),$$

where the subscripts S, K mean *secular* and *critical*, respectively. The critical terms are those depending on the angles $(\bar{h} \mid \widehat{\theta}^*)$, $\bar{h} \in \mathbf{Z}^N$, such that $(\bar{h} \mid \nu^\circ) = 0$.

If we assume that there are $L = N - M$ commensurability relations

$$(h_\varrho \mid \nu^\circ) = 0 \qquad (\varrho = M+1, \cdots, N), \qquad (7.50)$$

and construct the Lagrange point transformation

$$\begin{aligned}\widehat{\phi}_\varrho &= (h_\varrho \mid \widehat{\theta}^*) & (\varrho = M+1, \cdots, N) \\ \widehat{\phi}_\mu &= (h_\mu \mid \widehat{\theta}^*) & (\mu = 1, \cdots, M = N - L),\end{aligned} \qquad (7.51)$$

the canonical transformation is completed through the introduction of new actions $\widehat{I}_i$ such that

$$\sum_{i=1}^N \widehat{J}_i^* \, \delta \widehat{\theta}_i^* = \sum_{i=1}^N \widehat{I}_i \, \delta \widehat{\phi}_i.$$

The transformed Hamiltonian is now

$$\widehat{H}^* = \widehat{H}_2^*(\widehat{I}) + \sum_{k=3}^n \widehat{H}_k^*(\widehat{\phi}_\varrho, \widehat{I}, \varepsilon)$$

and it is independent of the angles $\widehat{\phi}_\mu$. Therefore, the $\widehat{I}_\mu$ are constants and

$$\mathcal{H}(\widehat{\phi}_\varrho, \widehat{I}_\varrho, \varepsilon) = \widehat{H}^*(\widehat{\phi}_\varrho, \widehat{I}, \varepsilon)$$

is the Hamiltonian of a canonical system with $L = N - M$ degrees of freedom; the commensurability relations given by (7.50) are, in this new system, simply

$$\widehat{\nu}_\varrho^\circ = 0, \qquad (7.52)$$

exactly as was assumed in the presentation of the von Zeipel–Brouwer theory. Thus, at variance with what was seen in Sect. 7.6, in the non-resonant case, the averaging does not reduce the system to a completely integrable one, but only to a reduced system with $L = N - M$ degrees of freedom.

7.8 Example

A complete non-singular Hamiltonian, as used in the theory of Sect. 7.6, is not often found in Celestial Mechanics because the main Keplerian term of the Hamiltonian depends only on the semi-major axis and is not affected by pure geometrical singularities. Non-singular Hamiltonians generally do appear after the averaging over the mean longitudes, as in the linear secular theory considered in Sect. 3.7. Some other important examples are studied in Chap. 10. The most common problems in Celestial Mechanics mix angle–action and non-singular Poincaré variables. Thus, the example given below is not a mere application of the previous theory, but rather an application of the principles used to construct it together with those of the previous chapter. It is the continuation of the example treated in Sect. 3.8.

The Hamiltonian given by (3.95), up to order $\mathcal{O}(\varepsilon)$, may be written as

$$H = -\frac{1}{2I_1^2} - 2I_1 + \varepsilon \left(a + bI_3 + L\sqrt{-2I_3}\cos\phi_3 + B\cos\phi_1 \right. \\ \left. + M\sqrt{-2I_3}\cos(\phi_1 + \phi_3) \right) \tag{7.53}$$

with several modifications: (a) the stars and coefficient subscripts were dropped; (b) the constant term $\nu_2 I_2$ was dropped (it does not contribute to the differential equations, since ϕ_2 is ignorable); (c) some factors $\sqrt{2}$ were introduced to get simpler coefficients in the forthcoming calculations; (d) the value $\nu_2 = 1$ was adopted; and (e) A_0^* was assumed to be linear in I_3: $A_0^*(I_1, I_3) = a + bI_3$.

First, we expand $H_0 = H\,|_{\varepsilon=0}$ about a reference value I_1°:

$$H_0 = \mathrm{const} + \nu_1^\circ \Xi + \frac{1}{2}\nu_{11}^\circ \Xi^2 + \frac{1}{6}\nu_{111}^\circ \Xi^3 + \cdots, \tag{7.54}$$

where

$$\Xi = I_1 - I_1^\circ \tag{7.55}$$

and

$$\nu_1^\circ = \frac{1}{I_1^{\circ 3}} - 2, \qquad \nu_{11}^\circ = \frac{-3}{I_1^{\circ 4}}, \qquad \nu_{111}^\circ = \frac{12}{I_1^{\circ 5}}, \cdots. \tag{7.56}$$

The coefficients $a(I_1)$, $b(I_1)$, $B(I_1)$, $L(I_1)$ and $M(I_1)$ are expanded in the same way. Then,

$$H = H_0 + \varepsilon \left(b_0 I_3 + L_0 \sqrt{-2I_3}\cos\phi_3 + B_0\cos\phi_1 + M_0\sqrt{-2I_3}\cos(\phi_1+\phi_3) \right) \\ + \varepsilon\Xi \left(a_1 + b_1 I_3 + L_1\sqrt{-2I_3}\cos\phi_3 + B_1\cos\phi_1 \right. \\ \left. + M_1\sqrt{-2I_3}\cos(\phi_1+\phi_3) \right) + \cdots, \tag{7.57}$$

where, now, all coefficients are calculated at the point $I_1 = I_1^\circ$ and, therefore, are constants. The subscript 1 in the coefficients denotes that they are first derivatives with respect to I_1. (The term εa_0 was dropped; it is constant and does not contribute to the equations.)

The next step is to assess the order of magnitude of the results. This assessment is critical because it determines the Hori kernel of the perturbation theory and, thus, constrains the solution. This is done by comparing the leading terms of H_0 with those of the perturbation. To see this, it is worth recalling that this example is suggested by the planar motion of an asteroid disturbed by Jupiter, L, B are proportional to Jupiter's eccentricity and I_3 is proportional to the squared asteroid eccentricity (see Sect. 3.6). Thus, if the eccentricities of Jupiter and the asteroid are assumed to be comparable, the leading terms of the perturbation have coefficients εB_0 and εM_0 (M_0 is a finite quantity)[3]. Then, we assume $\Xi = \mathcal{O}(\varepsilon \sqrt{-I_3})$ and $B_0 = \mathcal{O}(\sqrt{-I_3})$. We have, also, to assume a relationship between the orders of ε and $\sqrt{-I_3}$ and we assume $\sqrt{-I_3} = \mathcal{O}(\varepsilon)$. This choice is not the only one possible and, depending on the problem under study, may not even be a good one. However, it is the simplest when the actual calculations are concerned.

The given Hamiltonian may be expanded in a series ordered according to the degree of homogeneity of the elements of the set

$$\mathcal{S} \equiv (\sqrt{\Xi}, \sqrt{-I_3}, \varepsilon).$$

With the notation of Sect. 7.6, we have

$$H_0 = X_2(\Xi) + X_4(\Xi) + \cdots, \tag{7.58}$$

where

$$X_2 = \nu_1^\circ \Xi, \qquad X_4 = \frac{1}{2}\nu_{11}^\circ \Xi^2, \qquad \ldots, \tag{7.59}$$

and

$$H_1 = \sum_{k \geq 1} \mathcal{V}_1^k(\phi_1, \phi_3, \Xi, I_3), \tag{7.60}$$

where

$$\begin{aligned}
\mathcal{V}_1^1 &= B_0 \cos \phi_1 + M_0 \sqrt{-2I_3} \cos(\phi_1 + \phi_3) \\
\mathcal{V}_1^2 &= b_0 I_3 + L_0 \sqrt{-2I_3} \cos \phi_3 + a_1 \Xi \\
\mathcal{V}_1^3 &= B_1 \Xi \cos \phi_1 + M_1 \Xi \sqrt{-2I_3} \cos(\phi_1 + \phi_3) \\
\mathcal{V}_1^4 &= b_1 \Xi I_3 + L_1 \Xi \sqrt{-2I_3} \cos \phi_3 + \frac{1}{2} a_2 \Xi^2 \\
&\cdots.
\end{aligned} \tag{7.61}$$

The perturbation equations are (7.36). It is worth noting that those equations were obtained under the hypothesis that every Poisson bracket loses two units in its degree of homogeneity because of the differentiations. This is the

---

[3] It is useful to have in mind the orders of the various quantities present in this equation: a, b, M and their derivatives are finite quantities; B, L and their derivatives are of order $\mathcal{O}(\varepsilon)$; Ξ, I_3 are of order $\mathcal{O}(\varepsilon^2)$.

case also in this example for both pairs of variables (ϕ_1, Ξ) and (ϕ_3, I_3), regardless of the fact that we have not yet introduced the non-singular variables.

The homological equation is

$$\{H_2^*, W_k^*\} = H_k^* - \Psi_k(\phi_1^*, \phi_3^*, \Xi^*, I_3^*), \tag{7.62}$$

where

$$\begin{aligned} H_2^* &= X_2(\Xi_1^*) + \varepsilon \mathcal{V}_1^1(\phi_1^*, \phi_3^*, \Xi^*, I_3^*) \\ &= \nu_1^\circ \Xi^* + \varepsilon \left(B_0 \cos \phi_1^* + M_0 \sqrt{-2I_3^*} \cos(\phi_1^* + \phi_3^*) \right). \end{aligned} \tag{7.63}$$

The two terms of H_2^* have arguments including the angle ϕ_1^*. When Jupiter is on a circular orbit, only $M\sqrt{-2I_3^*}\cos(\phi_1^* + \phi_3^*)$ remains. It is, then, chosen as the main one and we define a new set of canonical variables (θ^*, J^*) through

$$\begin{aligned} \theta_1^* &= \phi_1^* + \phi_3^* & J_1^* &= I_3^* \\ \theta_2^* &= \phi_1^* & J_2^* &= \Xi^* - I_3. \end{aligned} \tag{7.64}$$

We are interested in solutions with $|I_3|$ small. We then replace θ_1^*, J_1^* by the non-singular variables

$$\begin{aligned} x_1^* &= \sqrt{-2J_1^*} \cos \theta_1^* \\ y_1^* &= \sqrt{-2J_1^*} \sin \theta_1^*. \end{aligned} \tag{7.65}$$

With the new variables, H_2^* becomes

$$H_2^* = \nu_1^\circ \Xi^* + \varepsilon \left(B_0 \cos \theta_2^* + M_0 x_1^* \right), \tag{7.66}$$

where

$$\Xi^* = J_2^* - \frac{1}{2}(x_1^{*2} + y_1^{*2}). \tag{7.67}$$

The Hori kernel is

$$\begin{aligned} \frac{dx_1^*}{du} &= \frac{\partial H_2^*}{\partial y_1^*} = -\nu_1^\circ y_1^* & \frac{dy_1^*}{du} &= -\frac{\partial H_2^*}{\partial x_1^*} = \nu_1^\circ x_1^* - \varepsilon M_0 \\ \frac{d\theta_2^*}{du} &= \frac{\partial H_2^*}{\partial J_2^*} = \nu_1^\circ & \frac{dJ_2^*}{du} &= -\frac{\partial H_2^*}{\partial \theta_2^*} = \varepsilon B_0 \sin \theta_2^* \end{aligned} \tag{7.68}$$

whose general solutions are

$$\begin{aligned} x_1^* &= C \cos \gamma + \frac{\varepsilon M_0}{\nu_1^\circ} & y_1^* &= C \sin \gamma \\ \theta_2^* &= \nu_1^\circ u + \theta_2^\circ & J_2^* &= -\frac{\varepsilon B_0}{\nu_1^\circ} \cos \theta_2^* + J_2^\circ, \end{aligned} \tag{7.69}$$

where

$$\gamma = \nu_1^\circ u + \theta_1^\circ.$$

(θ_1° and θ_2° are two independent integration constants.) Before continuing, it is worthwhile noting that the integration constants C and J_2° are, respectively, of orders $\mathcal{O}(\varepsilon)$ and $\mathcal{O}(\varepsilon^2)$, so that $\Xi^* = \mathcal{O}(\varepsilon^2)$ as assumed.

First Perturbation Equation

The first perturbation equation is

$$\{H_2^*, W_3^*\} = H_3^* - \Psi_3, \tag{7.70}$$

where (with the new variables)

$$\Psi_3 = \varepsilon V_1^2 = \varepsilon\,[b_0 J_1^* + L_0(x_1^* \cos\theta_2 + y_1^* \sin\theta_2) + a_1 \Xi^*] \tag{7.71}$$

(since $V_2^1 = 0$). Once the solution of the Hori kernel is substituted into Ψ_3, we get

$$\begin{aligned} H_3^* &= <\Psi_3> \tag{7.72}\\ &= -\frac{1}{2}\varepsilon(b_0 + a_1)C^2 - \frac{1}{2}\varepsilon^3(b_0 + a_1)\frac{M_0^2}{\nu_1^{\circ 2}} + \varepsilon a_1 J_2^\circ + \varepsilon L_0 C \cos(\theta_1^\circ - \theta_2^\circ) \end{aligned}$$

and

$$\begin{aligned} W_3^* &= \int (\Psi_3 - H_3^*)\,\mathrm{d}u \tag{7.73}\\ &= \frac{\varepsilon^2}{\nu_1^{\circ 2}}(M_0 L_0 - a_1 B_0)\sin(\nu_1^\circ u + \theta_2^\circ) - \frac{\varepsilon^2}{\nu_1^{\circ 2}}(b_0 + a_1)C M_0 \sin(\nu_1^\circ u + \theta_1^\circ). \end{aligned}$$

The only necessary condition is that ν_1° is not a small quantity (non-resonance condition).

The integration constants $\theta_1^\circ, \theta_2^\circ, C, J_2^\circ$ are not canonical and the above functions may be transformed into $H_3^*(x_1^*, y_1^*, \theta_2^*, J_2^*)$ and $W_3^*(x_1^*, y_1^*, \theta_2^*, J_2^*)$ before the next step, since Ψ_4 includes the calculation of $\{H_3^*, W_3^*\}$. In this example, this task may be accomplished trivially, but in more complex examples, this may not be the case.

Angle–Action Variables of H_2^*

The frequent use of the inverse of the general solutions of the Hori kernel may be avoided if those solutions are written in terms of the parameters (α, β) of the corresponding Hamilton–Jacobi equation or, equivalently, the angle–action variables of H_2^*. Poisson brackets are invariant to canonical transformations and we may calculate them using the parameters (α, β) or the angle–action variables of H_2^*. In the given example, H_2^* is separated into

$$H_2^* = K_1(x_1^*, y_1^*) + K_2(\theta_2^*, J_2^*), \tag{7.74}$$

where

$$\begin{aligned} K_1 &= -\frac{1}{2}\nu_1^\circ(x_1^{*2} + y_1^{*2}) + \varepsilon M_0 x_1^*\\ K_2 &= \nu_1^\circ J_2^* + \varepsilon B_0 \cos\theta_2^* \end{aligned} \tag{7.75}$$

and both sets of canonical parameters may be easily constructed. We will use, in this example, the angle–action variables:

- *Angle–action variables of K_1*: K_1 is a harmonic oscillator with frequency ν_1°. Then, keeping the same constants as before,

$$w_1 = |\nu_1^\circ|u + \theta_1^\circ \tag{7.76}$$

and

$$\Lambda_1 = \frac{1}{2\pi}\oint y_1^* \mathrm{d}x_1^* = \mp\frac{C^2}{2} = \mp\frac{1}{2}\left[y_1^{*2} + \left(x_1^* - \frac{\varepsilon M_0}{\nu_1^\circ}\right)^2\right]. \tag{7.77}$$

The sign of Λ_1 may be chosen as opposite to the sign of ν_1°.

- *Angle–action variables of K_2*: All solutions of K_2 are isochronous circulations with frequency ν_1°. Then

$$w_2 = |\nu_1^\circ|u + \theta_2^\circ \tag{7.78}$$

and

$$\Lambda_2 = \frac{1}{2\pi}\oint J_2^* \mathrm{d}\theta_2^* = \pm J_2^\circ = \pm\left(J_2^* + \frac{\varepsilon B_0}{\nu_1^\circ}\cos\theta_2^*\right). \tag{7.79}$$

The sign in front of J_2° may be chosen as equal to the sign of ν_1°.

With the angle–action variables thus introduced, the energy is

$$E_2 = |\nu_1^\circ|(\Lambda_1 + \Lambda_2). \tag{7.80}$$

This system is degenerate and one more change of variables, in the direction contrary to that given by (7.64), is useful:

$$\begin{aligned} \tilde{w}_1 &= w_2 & \tilde{\Lambda}_1 &= \Lambda_1 + \Lambda_2 \\ \tilde{w}_3 &= w_1 - w_2 & \tilde{\Lambda}_3 &= \Lambda_1 \end{aligned} \tag{7.81}$$

(see Sect. 2.7.1), where we restored the subscript 3 to make evident the correspondence with the variables of the given problem. The energy becomes

$$\tilde{E}_2 = |\nu_1^\circ|\tilde{\Lambda}_1. \tag{7.82}$$

If, to avoid unnecessary complicated notation with double signs, we assume $\nu_1^\circ > 0$, we may write the general solution of (7.68) as

$$\begin{aligned} x_1^* &= \sqrt{-2\Lambda_1}\cos w_1 + \frac{\varepsilon M_0}{\nu_1^\circ} & y_1^* &= \sqrt{-2\Lambda_1}\sin w_1 \\ \theta_2^* &= w_2 & J_2^* &= \Lambda_2 - \frac{\varepsilon B_0}{\nu_1^\circ}\cos w_2, \end{aligned} \tag{7.83}$$

instead of (7.69). Once these solutions are substituted into Ψ_3, instead of (7.72) and (7.73), we get

$$H_3^* = <\Psi_3> \tag{7.84}$$
$$= \varepsilon(b_0+a_1)\tilde{\Lambda}_3 - \frac{1}{2}\varepsilon^3(b_0+a_1)\frac{M_0^2}{\nu_1^{\circ 2}} + \varepsilon a_1(\tilde{\Lambda}_1-\tilde{\Lambda}_3) + \varepsilon L_0\sqrt{-2\tilde{\Lambda}_3}\cos\tilde{w}_3$$

and

$$W_3^* = \int(\Psi_3 - H_3^*)\,du \tag{7.85}$$
$$= \frac{\varepsilon^2}{\nu_1^{\circ 2}}(M_0 L_0 - a_1 B_0)\sin\tilde{w}_1 - \frac{\varepsilon^2}{\nu_1^{\circ 2}}(b_0+a_1)M_0\sqrt{-2\tilde{\Lambda}_3}\sin(\tilde{w}_1+\tilde{w}_3).$$

□

We will not continue the calculations, since they are, now, simple applications of the given routines. We just recall that the Poisson brackets in Ψ_k ($k \geq 4$) are more easily computed through

$$\{f,g\} = \sum_{i=1}^{2}\left(\frac{\partial f}{\partial \tilde{w}_i}\frac{\partial g}{\partial \tilde{\Lambda}_i} - \frac{\partial g}{\partial \tilde{w}_i}\frac{\partial f}{\partial \tilde{\Lambda}_i}\right). \tag{7.86}$$

8

Lie Series Theory for Resonant Systems

8.1 Bohlin's Problem (The Single-Resonance Problem)

The integration of the homological equation of the general perturbation theories of Chaps. 3 and 6 is only possible when the short-period frequencies ν_μ^* ($\mu = 1, \cdots, M$) obey the non-resonance condition

$$(h|\nu^*) = \sum_{\mu=1}^{M} h_\mu \nu_\mu^* \neq 0 \tag{8.1}$$

for all $h \equiv (h_1, \cdots, h_M) \in \mathbf{Z}^M$ appearing in the right-hand-side trigonometric polynomials. The strong restriction introduced by this condition is the very reason for which, in general, those theories cannot be extended to an arbitrarily high order. As discussed in Sect. 3.3.1, the set $D_k \subset \mathbf{Z}^M$ of values of h grows with the order of approximation k and values of $(h|\nu^*)$ smaller than any given limit may be formed as the set D_k grows. However, in the applications, we are often interested in a phase space domain where $(h|\nu^*) = 0$ for some h present in the given perturbation $\varepsilon R(\theta, J; \varepsilon)$. We have, then, to extend canonical perturbation theories to such cases and learn how to construct formal solutions valid in the neighborhood of resonances.

The general Hamiltonian in perturbation theory is

$$H = H_0(J_\mu) + \varepsilon R(\theta, J; \varepsilon) \tag{8.2}$$

with $\theta \equiv (\theta_1, \cdots, \theta_N)$, $J \equiv (J_1, \cdots, J_N)$ and $\mu = 1, \cdots, M \leq N$. R is a smooth function in $\mathbf{T}^N \times \mathcal{O} \times I$ ($\mathcal{O} \subset \mathbf{R}^N$ is an open set and $I \subset \mathbf{R}$) represented by a trigonometric polynomial in θ. We, generally, write,

$$R = R_{(S)} + R_{(LP)} + R_{(SP)},$$

where the subscripts (S), (LP), (SP) mean secular, long period and short period, following the definitions given at the end of Sect. 3.4. At variance with the assumed non-resonance condition of general theories, we assume, now, that, for some $\bar{h} \in \mathbf{Z}^M$ and some $J^* \in \mathcal{O}$, we have, simultaneously (a.) $\sum_{\mu=1}^{M} \bar{h}_\mu \nu_\mu^* = 0$;

(b.) either the trigonometric polynomial $R_{(SP)}$ includes one term $A_{\bar{h}} \cos(\lambda \bar{h}|\theta)$ ($\lambda \in \mathbf{N}$) or one such term will be formed in the calculation of some Ψ_k ($k \leq n$).

This means that one resonance has to be considered in the solution of the given problem.

For simplicity, we assume that all non-critical short-period angles were eliminated beforehand from H with the help of one of the previously discussed general theories using von Zeipel's averaging rule. We also assume that a Lagrangian point transformation such as (3.51) transformed the Hamiltonian of the resulting system into

$$H = H_0(J_1) + \varepsilon R(\theta, J; \varepsilon). \tag{8.3}$$

Our problem is then stated as the search for formal solutions of the Hamiltonian (8.3) in a neighborhood of the value $J_1 = J_1^*$, where

$$\nu_1^* = \left.\frac{\mathrm{d}H_0}{\mathrm{d}J_1}\right|_{J_1=J_1^*} = 0. \tag{8.4}$$

The problem of finding a formal canonical transformation able to eliminate the angle corresponding to the critical frequency $(\bar{h} \mid \nu^*)$, from the Hamiltonian, was first proposed by Bohlin (see Appendix A) and is referred hereafter as *Bohlin's problem*. As discussed in Appendix A, it is shown that all attempts at solving Bohlin's problem in the presence of the degenerate actions J_ϱ ($\varrho = 2, \cdots, N$), with classical theories, lead to an unsolved singularity (Poincaré singularity). The solution of Bohlin's problem in the presence of degenerate actions using Lie series theory is the subject of this chapter and the next.

8.2 Outline of the Solution

There is no general recipe to solve the single-resonance problem. We know that Hori theory, as given in Chap. 6, cannot be used to construct a formal solution of the stated single-resonance problem because $H_0(J_1)$ is not a topologically adequate Hori kernel in the neighborhood of a resonance. In the simple one-degree-of-freedom case, the Poincaré–Birkhoff theorem states that the perturbation R may change the topology of the phase plane by introducing a finite set of new equilibrium points. The new stable equilibrium points are centers of libration lobes separated from the general flow by asymptotic motions emanating from unstable equilibrium points. A simple example of such a flow bifurcation is shown in Fig. 6.2 (left). General theories fail in the neighborhood of a resonance because they consider the undisturbed Hamiltonian as the Hori kernel, regardless of the topological differences between the flows of disturbed and undisturbed Hamiltonians. This diagnosis of the origin of the small divisors appearing in general theories, at resonances, is the only clue that we have to attempt a solution of Bohlin's problem: The only way to study resonant problems is to get rid of $H_0(J_1)$ as the Hori kernel and to choose a new one whose flow reproduces the main topological features of the given flow in the neighborhood of the resonance. At this point, it is not superfluous to emphasize that this is <u>not</u> enough to get rid of <u>all</u> problems. It is only good to get formal solutions with the assigned topological features. The real nature of perturbed Hamiltonian flows is much more complex. For instance, when the order of the solutions inside a libration zone is pushed too far, small divisors due to other resonances are unavoidable. In practice, these other resonances generally appear as commensurabilities amongst the low-frequency terms of $R_{(LP)}$ (secular resonances) or among the low frequencies and the proper frequency of libration around one stable equilibrium point (secondary resonances) (see Fig. 4.3). These resonances are, generally, the only ones taken into account; but one may be aware that the process of elimination of "non-critical" short-period angles considered as previously done is valid in a domain $\mathcal{O}$ whose measure decreases as the formal accuracy of the theory increases. These terms, considered a priori as non-critical, are very dangerous, as they do not appear in the reduced Hamiltonian (8.3) and may remain unnoticed.

The general principles to be followed when dealing with the single-resonance problem are simple:

1. to select from the given Hamiltonian the terms that determine the main integrable topological features of the flow in the considered domain;
2. to solve the Hamilton–Jacobi equation of the integrable Hamiltonian H_{lead} formed with the selected terms or, equivalently, to construct its angle–action variables;
3. to introduce the canonical variables thus obtained in the given Hamiltonian and to construct a Lie series theory whose Hori kernel is H_{lead}.

Great difficulties are involved in steps 1 and 2. H_{lead} must not only be integrable, but we must be able to write its solutions as well as to construct its angle–action variables (or to completely solve its Hamilton–Jacobi equation). These requirements limit our choice to separable Hamiltonians whose reduction to the phase plane of the critical variables is a well-known one-degree-of-freedom Hamiltonian (such as the simple pendulum, the Ideal Resonance Problem, the Andoyer Hamiltonians, and a few others). In the study of spin-orbit resonance, a quadratic Hamiltonian as studied in Sect. 2.9, with three degrees of freedom, has been shown to be a suitable Hori kernel [62].

This procedure involves the two canonical transformations represented by single-line arrows in the scheme below (using the angle–action variables w, Λ of H_{lead} as new variables):

$$(\theta, J) \longrightarrow \longrightarrow (w, \Lambda)$$

$$\Downarrow \quad \searrow \quad \downarrow$$

$$\Downarrow \quad \searrow \quad \downarrow$$

$$(\theta^*, J^*) \Longrightarrow \Longrightarrow (w^*, \Lambda^*)$$

Resonant problems are almost always very peculiar and it is difficult to conceive of a very general H_{lead} for which a complete theory can be written. The actual Lie series theory for a resonant problem closely depends on the selected H_{lead} and, in fact, we have to tailor a theory for each problem. In accordance with the general principle stated in step 1, H_{lead} may be the term determining the main integrable features of the flow. To do this, orders of magnitude have to be assigned to variables and parameters so that the main terms are all present in H_{lead} on the same footing, while the remaining non-trivial terms of H are of higher orders. For this reason, we have to adopt new rules for the comparison of terms in the series expansions. In classical theories (Poincaré, von Zeipel–Brouwer, Bohlin, Delaunay) the perturbation equations are obtained through the identification of both sides of the energy conservation equation according to the powers of the small parameter (ε or $\sqrt{\varepsilon}$). The same construction was adopted in the general Lie series perturbation theory of Chap. 6. But for the Lie series expansions about the origin (Sect. 7.5), the structure of the Poisson brackets used in Lie series allowed us a different choice, which is very useful to avoid the mixing of orders that impaired the application of the classical theories to resonant systems (see Sect. A.2). We use, now, the fact that the excursions of the variable J_1 (action conjugate to the critical angle θ_1) in the neighborhood of the resonant value J_1^* are small, replace J_1 by the new variable

$$\xi = J_1 - J_1^* \tag{8.5}$$

and assume that $\xi = \mathcal{O}(\varepsilon^d)$, $d \leq 1$. (For instance, in Delaunay theory, $d = \frac{1}{2}$.) In the following, the orders of magnitude are no longer taken as the powers of ε (or $\sqrt{\varepsilon}$), but as the degree of homogeneity of the functions with respect to the elements of the set $\mathcal{S} \equiv (\xi, \varepsilon^d)$.

One last word concerning step 2 is necessary. The aim of canonical perturbation theories is, always, to get rid of degrees of freedom associated with fast angles. This is obtained, generally, by systematically averaging (over the fast angles) the right-hand side terms of the homological equation. However, the term H_{lead} remains itself unaltered in this process and, thus, it may have, *ab initio*, the reduced form sought for the new Hamiltonian. In the general case, the two sets of variables mentioned in step 2 satisfy this condition, but, given the practical difficulties to get them, it is useful to have in mind that, for a given problem, other possibilities may exist. For instance, when H_{lead} appears separated into two terms:

$$H_{\text{lead}} = H_{\text{lead}(1)}(\theta_1, J_1) + H_{\text{lead}(\varrho)}(\theta_2, \theta_3, \cdots, \theta_N, J_2, J_3, \cdots, J_N). \quad (8.6)$$

we do not need to deal with $H_{\text{lead}(\varrho)}$; the fast variable is included only in the one-degree-of-freedom $H_{\text{lead}(1)}$, making it easier to perform the operation indicated in step 2. It may appear, at a first sight, that the above splitting is very restrictive. However, the problem that arises after the introduction of the angle–action variables of $H_{\text{lead}(1)}$, is more general than the problem stated in the introduction of von Zeipel–Brouwer theory (see Sect. 3.4). There, the Hamiltonian H_0 depends only on the variables J_μ while, here, the zero-order Hamiltonian appears split into two separate parts: one like H_0 and another one depending, in an arbitrary way, of the remaining variables $\theta_\varrho, J_\varrho$.

At this point, it is worth warning against the apparent possibility of a more general separated case:

$$H_{\text{lead}(1)}(\theta_1, J_1, J_2, \cdots, J_N) + H_{\text{lead}(\varrho)}(\theta_2, \theta_3, \cdots, \theta_N, J_2, J_3, \cdots, J_N)$$

together with the set of angle–action variables of $H_{\text{lead}(1)}(\theta_1, J)$ given by the algorithms of Sect. 2.4.4. The variables obtained with those algorithm are the angle–action variables of $H_{\text{lead}(1)}(\theta_1, J)$, but not of H_{lead}. The transformation functions $\Xi(w_1, \Lambda)$ introduce the angle w_1 into $H_{\text{lead}(\varrho)}$ and the sought elimination of the fast angle (w_1) may be frustrated.

8.3 Functions Expansions

The Lie series of a function is a Taylor series and, therefore, unique. However, if we intend to group the terms according to the degrees of homogeneity in the elements of $\mathcal{S}$, we have to take into account that the Lie derivative of a function yields terms with different degrees of homogeneity. Indeed, given two functions $\psi_1(\theta, \xi, J_\varrho)$ and $\psi_2(\theta, \xi, J_\varrho)$ ($\varrho = 2, \ldots, N$), homogeneous in the elements of $\mathcal{S} \equiv (\xi, \varepsilon^d)$, their Poisson bracket is

$$\{\psi_1, \psi_2\} = \left(\frac{\partial \psi_1}{\partial \theta_1}\frac{\partial \psi_2}{\partial \xi} - \frac{\partial \psi_2}{\partial \theta_1}\frac{\partial \psi_1}{\partial \xi}\right) + \sum_{\varrho=2}^{N}\left(\frac{\partial \psi_1}{\partial \theta_\varrho}\frac{\partial \psi_2}{\partial J_\varrho} - \frac{\partial \psi_2}{\partial \theta_\varrho}\frac{\partial \psi_1}{\partial J_\varrho}\right). \quad (8.7)$$

The second part of this Poisson bracket is an ordinary operation and the degree of homogeneity of the result is equal to the sum of the degrees of homogeneity of ψ_1 and ψ_2. However, in the first part of the right-hand side, the operation $\frac{\partial}{\partial \xi}$ subtracts one unit from the degree of homogeneity; as a result, the degree of homogeneity of the first parenthesis, in the elements of $\mathcal{S} \equiv (\xi, \varepsilon^d)$, is one unit less than in the rest of the terms.

For the sake of the forthcoming developments, we introduce the notation

$$\{\psi_1, \psi_2\}_1 = \frac{\partial \psi_1}{\partial \theta_1}\frac{\partial \psi_2}{\partial \xi} - \frac{\partial \psi_2}{\partial \theta_1}\frac{\partial \psi_1}{\partial \xi}, \quad (8.8)$$

$$\{\psi_1, \psi_2\}_\varrho = \sum_{\varrho=2}^{N} \left(\frac{\partial \psi_1}{\partial \theta_\varrho} \frac{\partial \psi_2}{\partial J_\varrho} - \frac{\partial \psi_2}{\partial \theta_\varrho} \frac{\partial \psi_1}{\partial J_\varrho} \right), \tag{8.9}$$

and write the Poisson bracket as

$$\{\psi_1, \psi_2\} = \{\psi_1, \psi_2\}_1 + \{\psi_1, \psi_2\}_\varrho. \tag{8.10}$$

From (8.10), in the case of two embedded Poisson brackets involving three functions, we obtain

$$\begin{aligned}\{\{\psi_1, \psi_2\}, \psi_3\} &= \{\{\psi_1, \psi_2\}_1, \psi_3\}_1 + \{\{\psi_1, \psi_2\}_1, \psi_3\}_\varrho \\ &\quad + \{\{\psi_1, \psi_2\}_\varrho, \psi_3\}_1 + \{\{\psi_1, \psi_2\}_\varrho, \psi_3\}_\varrho,\end{aligned} \tag{8.11}$$

where the last term has the full degree of homogeneity (the sum of the degrees of homogeneity of the three functions). The degree of homogeneity of the two brackets showing both subscripts 1 and ϱ and the degree of homogeneity of the bracket showing the subscript 1 twice are, respectively, 1 and 2 units less than that of the last term.

Now, let $f(\theta, \xi, J_\varrho)$ be a homogeneous function of degree L in the elements of $\mathcal{S} \equiv (\xi, \varepsilon^d)$ and let us consider the canonical transformation

$$\phi_n : (\theta, \xi, J_\varrho) \Rightarrow (w^*, \Lambda^*).$$

This transformation does not lie in the neighborhood of the identity, and it cannot be written as a Lie series. Let us, then, introduce an auxiliary set of variables $(\theta^*, \xi^*, J_\varrho^*)$ related to (w^*, Λ^*) through the same equations linking (θ, ξ, J_ϱ) to (w, Λ). The transformation

$$\phi_{\text{aux}} : (\theta, \xi, J_\varrho) \Rightarrow (\theta^*, \xi^*, J_\varrho^*)$$

is nearly identical and may be given by a Lie series. Let ϕ_{aux} be determined by the Lie generator

$$W^* = \sum_{k=1}^{n} W_k^*(\theta^*, \xi^*, J_\varrho^*), \tag{8.12}$$

where the functions $W_k^*(\theta^*, \xi^*, J_\varrho^*)$ are homogeneous with degree k in the elements of $\mathcal{S}$. In accordance with the expansion given in Sect. 5.5.1, we have

$$\begin{aligned} f(\theta, \xi, J_\varrho) &= E_{W^*} f(\theta^*, \xi^*, J_\varrho^*) \\ &= f(\theta^*, \xi^*, J_\varrho^*) + \{f, W_1^*\} + \{f, W_2^*\} + \{f, W_3^*\} + \cdots \\ &\quad + \frac{1}{2}\{\{f, W_1^*\}, W_1^*\} + \frac{1}{2}\{\{f, W_2^*\}, W_1^*\} + \frac{1}{2}\{\{f, W_1^*\}, W_2^*\} \\ &\quad + \cdots + \frac{1}{6}\{\{\{f, W_1^*\}, W_1^*\}, W_1^*\} + \cdots. \end{aligned} \tag{8.13}$$

We may use the invariance of Poisson brackets to calculate them in terms of the averaged angle–action variables w^*, Λ^*. Because of such invariance, only $f(\theta^*, \xi^*, J_\varrho^*)$ needs to be explicitly changed. Hence

$$f(\theta, \xi, J_\varrho) = f(\theta^*(w^*, \Lambda^*), \xi^*(w^*, \Lambda^*), J_\varrho^*(w^*, \Lambda^*)) + \&c,$$

where &c. indicates the same brackets as in the previous equation with the only difference that, now, we consider all functions in the brackets as functions of the variables (w^*, Λ^*). It is worth noting that the practical procedure adopted in the construction of the canonical transformation ϕ_n does not follow the single-line arrows of the scheme given in Sect. 8.2, but rather the alternative path shown by double-line arrows.

Before grouping the terms of this modified Lie series according to their degrees of homogeneity with respect to the elements of $\mathcal{S}$, we note that, because of the assumptions made on the order of ξ, at the end of Sect. 8.2, the brackets $\{f, W_k^*\}$ yield terms with different degrees of homogeneity. The decomposition given by (8.10) may be used here. One difficulty resulting from the fact that the brackets with subscript 1 have a degree one unit less than the sum of the degrees of the two functions involved in the bracket is that the collection of terms of the same degree as f:

$$f + \{f, W_1^*\}_1 + \frac{1}{2}\{\{f, W_1^*\}_1, W_1^*\}_1 + \frac{1}{6}\{\{\{f, W_1^*\}_1, W_1^*\}_1, W_1^*\}_1 + \cdots.$$

has an unlimited number of terms. (It is easy to see that if $W_1^* \neq 0$, there are infinite terms in each homogeneous subseries of $E_{W^*}f$.) We then assume

$$W_1^* = 0$$

and the Lie series expansion of f becomes

$$\begin{aligned}
E_{W^*}f = \ & f(\theta^*(w^*, \Lambda^*), \xi^*(w^*, \Lambda^*), J_\varrho^*(w^*, \Lambda^*)) \\
& + \{f, W_2^*\}_1 \\
& + \{f, W_3^*\}_1 + \{f, W_2^*\}_\varrho + \frac{1}{2}\{\{f, W_2^*\}_1, W_2^*\}_1 \\
& + \{f, W_4^*\}_1 + \{f, W_3^*\}_\varrho + \frac{1}{2}\{\{f, W_2^*\}_1, W_3^*\}_1 + \frac{1}{2}\{\{f, W_3^*\}_1, W_2^*\}_1 \\
& + \frac{1}{2}\{\{f, W_2^*\}_\varrho, W_2^*\}_1 + \frac{1}{2}\{\{f, W_2^*\}_1, W_2^*\}_\varrho \\
& + \frac{1}{6}\{\{\{f, W_2^*\}_1, W_2^*\}_1, W_2^*\}_1 \\
& + \{f, W_5^*\}_1 + \{f, W_4^*\}_\varrho + \cdots,
\end{aligned} \qquad (8.14)$$

where we have put terms of degree L in the first row, terms of degree $L+1$ in the second row, and so on.

In the above expansion, the partial brackets in the Lie series have to be calculated, necessarily, with the variables $(\theta^*, \xi^*, J_\varrho^*)$ (the partial brackets are not invariant to canonical transformations). However, in the case of a Hamiltonian split as (8.6), the canonical transformation $(\theta^*, \xi^*, J_\varrho^*) \Rightarrow (w^*, \Lambda^*)$ can be separated into two parts: $(\theta_1^*, \xi^*) \Rightarrow (w_1^*, \Lambda_1^*)$ and $(\theta_\varrho^*, J_\varrho^*) \Rightarrow (w_\varrho^*, \Lambda_\varrho^*)$. In this case, the partial brackets $\{f, W_k^*\}_1$ and $\{f, W_k^*\}_\varrho$ are true Poisson brackets and, thus, invariant to the above partial canonical transformations and the brackets $\{f, W_k^*\}$ may be decomposed as indicated in (8.10) for every set of variables.

Exercise 8.3.1. The assumption $W_1^* = 0$, in the previous section, is stronger than necessary to kill the brackets $\{\cdot, W_1^*\}_1$. It would be enough to assume that W_1^* is independent of θ_1^*, ξ^*. How should (8.14) be written if this weaker assumption were adopted?

8.4 Perturbation Equations

To construct the perturbation equations of resonant systems, all functions will be expanded in power series with terms grouped following their degree of homogeneity in the elements of $\mathcal{S} \equiv (\xi, \varepsilon^d)$. The expansions of the functions H_0 and R are

$$H_0 = H_0(J_1^*) + X_2(\xi) + X_3(\xi) + \cdots \tag{8.15}$$

and

$$R(\theta, J; \varepsilon) = R^{(0)}(\theta, J_1^*, J_\varrho) + R^{(1)}(\theta, \xi, J_\varrho) + R^{(2)}(\theta, \xi, J_\varrho) + \cdots, \tag{8.16}$$

where $X_{k'}$ and $R^{(k')}$ are homogeneous functions of degree k' in the elements of $\mathcal{S} \equiv (\xi, \varepsilon^d)$. The term $X_1(\xi)$ is absent from the series for H_0, since, by Taylor's theorem,

$$X_1(\xi) = \frac{\mathrm{d}H_0(J_1^*)}{\mathrm{d}J_1^*}\xi = \nu_1(J_1^*)\xi$$

and, by hypothesis, $\nu_1(J_1^*) = \nu_1^* = 0$.

These expansions may be introduced into (8.3), giving

$$H = H_0(J_1^*) + \sum_{k \geq 2} F_k(\theta, \xi, J_\varrho; \varepsilon), \tag{8.17}$$

where we have grouped in F_k all homogeneous terms of degree k in the elements of $\mathcal{S} \equiv (\xi, \varepsilon^d)$. In particular, the leading non-trivial term of the given Hamiltonian is

$$F_2 = X_2(\xi) + \varepsilon R^{(0)}(\theta, J_1^*, J_\varrho)$$

and the application of the theory depends on the possibility of obtaining its angle–action variables or, alternatively, the splitting of F_2 into $F_{2(1)}(\theta_1, \xi) + F_{2(\varrho)}(\theta_\varrho, J_\varrho)$.

The canonical transformation ϕ_n is, now, given by (8.14) with the Lie generator

$$W^* = \sum_{k=2}^{n} W_k^*(\theta^*, \xi^*, J_\varrho^*) \tag{8.18}$$

8.4 Perturbation Equations

and the transformed Hamiltonian H^* is assumed to be expanded in the form

$$H^* = \sum_{k=1}^{n} H_k^*, \qquad (8.19)$$

where, again, the subscripts indicate the degree of homogeneity in the elements of $\mathcal{S}$.

The substitution of these expansions into the conservation equation (6.5) and comparison of the terms of the same degree of homogeneity in the elements of $\mathcal{S}$ yield the perturbation equations. These equations are similar to their analogs in the general Hori theory (6.7). Some differences are:

(a.) $F_0 \stackrel{\text{def}}{=} H_0(J^*)$ is a constant and, thus, all Poisson brackets including F_0 vanish;
(b.) $F_1 = X_1(\xi) \equiv 0$ and, thus, all Poisson brackets including F_1 vanish;
(c.) $W_1^* = 0$;
(d.) Poisson brackets are split into two parts, one of which, with subscript 1, has a degree of homogeneity in ξ one unit less than the other.

The perturbation equations for the resonant systems are [33]

$$H_0^* = H_0(J_1^*)$$
$$H_1^* = 0$$
$$H_2^* = F_2$$
$$H_3^* = F_3 + \{F_2, W_2^*\}_1$$
$$H_4^* = F_4 + \{F_3, W_2^*\}_1 + \{F_2, W_2^*\}_\varrho + \frac{1}{2}\{\{F_2, W_2^*\}_1, W_2^*\}_1 + \{F_2, W_3^*\}_1$$
$$\cdots \qquad \cdots$$
$$H_k^* = F_k + \{F_{k-1}, W_2^*\}_1 + \{F_{k-2}, W_2^*\}_\varrho + \frac{1}{2}\{\{F_{k-2}, W_2^*\}_1, W_2^*\}_1$$
$$+ \frac{1}{2}\{\{F_{k-3}, W_2^*\}_1, W_2^*\}_\varrho + \frac{1}{2}\{\{F_{k-3}, W_2^*\}_\varrho, W_2^*\}_1$$
$$+ \frac{1}{2}\{\{F_{k-4}, W_2^*\}_\varrho, W_2^*\}_\varrho + \cdots + \{F_2, W_{k-1}^*\}_1, \qquad (8.20)$$

where, in all functions, the variables are $(\theta^*, \xi^*, J_\varrho^*)$. When the transformation $(\theta^*, \xi^*, J_\varrho^*) \Rightarrow (w^*, \Lambda^*)$ may be separated into two parts as discussed at the end of Sect. 8.3, the variables may be (w^*, Λ^*) (or even the mixed set $w_1, \theta_\varrho^*, \Lambda_1, J_\varrho^*$).

These equations are synthesized in the homological partial differential equation

$$\{H_2^*, W_{k-1}^*\}_1 = H_k^* - \Psi_k \qquad (k \geq 3) \qquad (8.21)$$

or, in explicit form,

$$\frac{\partial H_2^*}{\partial \theta_1^*} \frac{\partial W_{k-1}^*}{\partial \xi^*} - \frac{\partial W_{k-1}^*}{\partial \theta_1^*} \frac{\partial H_2^*}{\partial \xi^*} = H_k^* - \Psi_k. \qquad (8.22)$$

When the variables w_1^*, Λ_1^* are used instead of θ_1^*, ξ^*, we have a homological equation like that of von Zeipel–Brouwer theory:

$$\frac{\partial H_2^*}{\partial \Lambda_1^*}\frac{\partial W_{k-1}^*}{\partial w_1^*} = \Psi_k - H_k^*. \tag{8.23}$$

One remarkable consequence of the homological equation thus obtained is that we have to consider up to the order k in the perturbation (Ψ_k includes F_k) to obtain the $(k-1)^{\text{th}}$ component (W_{k-1}^*) of the Lie generator of the canonical transformation.

8.5 Averaging

The averaging is to be done exactly as in the von Zeipel–Brouwer theory. We include in H_k^* and then exclude, from the homological equation, all terms of Ψ_k that may give rise to small divisors. We separate Ψ_k into two parts: $\Psi_{k(1)}$ and $\Psi_{k(\varrho)}$, where

- $\Psi_{k(1)}$ is the sum of all periodic terms of Ψ_k dependent on the fast angle w_1^*,
- $\Psi_{k(\varrho)}$ is the sum of the remaining terms, secular or dependent only on the slow angles θ_ϱ^* (or w_ϱ^*),

and use the averaging rule

$$H_k^* = \Psi_{k(\varrho)}. \tag{8.24}$$

This is equivalent to $H_k^* = <\Psi_k>$, where $<\cdots>$ means average over w_1^*. At least in theory, we may do the averaging and proceed with the construction of formal solutions to higher orders. Actually, the integration of H_{lead} is very difficult and, when achieved, involves elliptic integrals. Because of these integrals, calculations beyond the equation for W_2^* are almost impossible. It is then necessary to use alternative ways to construct the angle–action variables, as discussed in Sect. 2.2.

8.6 An Example

Let us reconsider the example studied in Sect. 7.8. Some of calculations are long, but, given its realistic formulation gathering terms of different kinds, they serve to prove the feasibility of the procedure outlined in Sect. 8.2

The Hamiltonian is

$$H(\phi, I) = H_0 + \varepsilon R, \tag{8.25}$$

where

$$H_0 = -\frac{1}{2I_1^2} - 2I_1 \tag{8.26}$$

and
$$R = a(I_1) + b(I_1) I_3 + B(I_1) \cos \phi_1 + L(I_1) \sqrt{-2I_3} \cos \phi_3$$
$$+ M(I_1) \sqrt{-2I_3} \cos (\phi_1 + \phi_3). \tag{8.27}$$

In the example of Sect. 7.8, the reference value I_1° was just a value such that $\xi = I_1 - I_1^\circ$ could be considered as a small quantity. In addition, $\nu_1^\circ = \nu_1(I_1^\circ)$ was assumed to be finite (non-resonance condition). In this example, on the contrary, we assume that the sought solution lies in the neighborhood of the exact resonance, the case in which the integration done in Sect. 7.8 is no longer valid. The exact resonance is defined by

$$\nu_1^* = \left(\frac{dH_0}{dI_1}\right)_{I_1=I_1^*} = \left(\frac{1}{I_1^*}\right)^3 - 2 = 0, \tag{8.28}$$

that is,
$$I_1^* = \frac{1}{\sqrt[3]{2}}. \tag{8.29}$$

The function H_0 may be expanded as a Taylor series in powers of $\xi = I_1 - I_1^*$. Hence,
$$H_0 = H_0(I^*) + X_2 + X_3 + X_4 + \cdots, \tag{8.30}$$
where
$$\begin{aligned}
X_2 &= \tfrac{1}{2}\nu_{11}^* \xi^2 &= -3\sqrt[3]{2}\xi^2 \\
X_3 &= \tfrac{1}{6}\nu_{111}^* \xi^3 &= +4\sqrt[3]{4}\xi^3 \\
X_4 &= \tfrac{1}{24}\nu_{1111}^* \xi^4 &= -10\xi^4.
\end{aligned} \tag{8.31}$$

In the same way, we may expand the disturbing potential εR into
$$\varepsilon R = \varepsilon \left(R^{(0)} + R^{(1)} + R^{(2)} + \cdots\right), \tag{8.32}$$
where
$$R^{(0)} = Y_0 \qquad R^{(1)} = Y_1 \xi \qquad R^{(2)} = \frac{1}{2} Y_2 \xi^2, \tag{8.33}$$
etc., where, for simplicity, we wrote
$$Y_k = a_k + b_k I_3 + B_k \cos \phi_1 + L_k \sqrt{-2I_3} \cos \phi_3 + M_k \sqrt{-2I_3} \cos (\phi_1 + \phi_3).$$

The subscripts 1 and 2 in the coefficients denote that they are, respectively, first and second derivatives with respect to I_1. All coefficients are calculated at the exact resonance value I_1^* and are, thus, constants.

The leading terms of H are the first non-constant term of H_0, that is, X_2, and the leading terms of εR. As pointed out in Sect. 7.8, the given Hamiltonian is suggested by the planar motion of an asteroid disturbed by Jupiter. The coefficients L, B are proportional to Jupiter's eccentricity and are small. I_3 is proportional to the squared asteroid eccentricity. We do not put any restriction on the value of I_3, which is considered, in this example, as a finite quantity. We assume that the coefficients a, b, M are finite. Then

$$H_{\text{lead}} = \frac{1}{2}\nu_{11}^*\xi^2 + \varepsilon\left(a_0 + b_0 I_3 + M_0\sqrt{-2I_3}\cos(\phi_1 + \phi_3)\right). \tag{8.34}$$

In order to have all terms of H_{lead} on the same footing, we assume $\xi = \mathcal{O}(\sqrt{\varepsilon})$, that is, $d = \frac{1}{2}$ or $\mathcal{S} \equiv (\xi, \sqrt{\varepsilon})$. We also assume $B, L = \mathcal{O}(\sqrt{\varepsilon})$. The remaining functions F_k $(k > 2)$ are

$$\begin{aligned}
F_3 &= X_3 + \varepsilon\left(B_0\cos\phi_1 + L_0\sqrt{-2I_3}\cos\phi_3\right) \\
&\quad + \varepsilon\xi\left(a_1 + b_1 I_3 + M_1\sqrt{-2I_3}\cos(\phi_1 + \phi_3)\right) \\
F_4 &= X_4 + \varepsilon\xi\left(B_1\cos\phi_1 + L_1\sqrt{-2I_3}\cos\phi_3\right) \\
&\quad + \tfrac{1}{2}\varepsilon\xi^2\left(a_2 + b_2 I_3 + M_2\sqrt{-2I_3}\cos(\phi_1 + \phi_3)\right)
\end{aligned} \tag{8.35}$$

etc. The Hamiltonian H_{lead} is completely integrable and step 2 of the solution outlined in Sect. 8.2 may be accomplished.

Let us introduce, now, the canonical transformation

$$\begin{aligned}
\theta_1 &= \phi_1 + \phi_3 & J_1 &= I_3 \\
\theta_2 &= \phi_1 & J_2 &= \xi - I_3 & (\xi = J_1 + J_2).
\end{aligned} \tag{8.36}$$

Then,

$$H_{\text{lead}} = \frac{1}{2}\nu_{11}^*(J_1 + J_2)^2 + \varepsilon\left(b_0 J_1 + M_0\sqrt{-2J_1}\cos\theta_1\right), \tag{8.37}$$

where we have discarded the constant term εa_0 since it does not contribute to the canonical equations. In this Hamiltonian, the angle θ_2 is ignorable and, thus, $J_2 = \text{const}$. The resulting one-degree-of-freedom Hamiltonian is an Andoyer Hamiltonian whose integration is presented in Sect. C.5. However, since J_1 is a finite quantity and $J_1 + J_2 = \xi$ is of order $\mathcal{O}(\sqrt{\varepsilon})$, it may be approximated by a simple pendulum. We change H_{lead} into

$$H_{\text{lead}} = \frac{1}{2}\nu_{11}^*(J_1 + J_2)^2 + \varepsilon\left(-b_0 J_2 + M_0\sqrt{2J_2}\cos\theta_1\right) \tag{8.38}$$

and add to F_k the differences

$$\begin{aligned}
\delta F_3 &= \varepsilon\xi\left(b_0 - \frac{M_0}{\sqrt{2J_2}}\cos\theta_1\right), \\
\delta F_4 &= -\tfrac{1}{2}\varepsilon\xi^2 M_0 (2J_2)^{-3/2}\cos\theta_1.
\end{aligned} \tag{8.39}$$

Figure 8.1 shows the libration domains corresponding to the simple pendulum (solid line) and to the Andoyer Hamiltonian (dashed line) given, respectively, by (8.38) and (8.37), in the case $b_0 = 0$. The axes were chosen to reproduce closely what is seen in the (semi-major axis, eccentricity) plane, in the resonant asteroid problem. In both axes, the unit is $4(\varepsilon M_0/\nu_{11}^*)^{2/3}$.

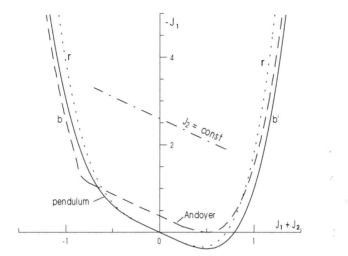

Fig. 8.1. Libration domains of two possible Hori kernels and one typical libration path. r is the locus of a secular resonance

One may note that the limits of the libration domain of the two Hamiltonians are very different for $|J_1|$ small (showing, even, a violation of the restriction $J_1 = I_3 < 0$ by the pendulum approximation); but, as $|J_1|$ grows, this difference almost reduces itself to a small shift to the right of the limits (b and b') of the pendulum libration zone with respect to those of the Andoyer Hamiltonian. It is worth emphasizing that the x-axis in Fig. 8.1 represents the action $\xi = J_1 + J_2$. The librations occur over the lines $J_2 = \text{const}$ and are segments with length proportional to the libration amplitude and center at $\xi = 0$. At the point where this segment crosses the $\xi = 0$ axis, the velocity of the pendulum is equal to zero: it corresponds to the borders of the libration of θ_1. In the same way, the extremities of the segment are points where $\dot\xi = 0$, that is, the points of maximum speed: they correspond to the transit through the origin $\theta_1 = 0$. The limiting curves b and b' are defined by the points of maximum speed of the motion over the pendulum separatrices.

In this case, not only is H_{lead} integrable, but its angle–action variables can be easily obtained. We limit ourselves to the case of small oscillations about the stable equilibrium and introduce the pendulum angle–action variables w_1, Λ_1. We assume $M_0 > 0$ (if $M_0 < 0$, since $\nu_{11}^* < 0$, the center of libration would be at $\theta_1 = \pi$ and we should add a phase π to θ_1). At the first approximation (see Sect. B.3.1):

$$\theta_1 = \sqrt{\frac{2\nu_{11}^* \Lambda_1}{\omega_1}} \sin w_1$$
$$J_1 = -J_2 - \sqrt{\frac{2\omega_1 \Lambda_1}{\nu_{11}^*}} \cos w_1, \quad (8.40)$$

where
$$\omega_1 = \sqrt{-\varepsilon \nu_{11}^* M_0} \sqrt[4]{2J_2} \tag{8.41}$$

($\Lambda_1 < 0$). In the construction of the above solutions, J_2 is kept constant.

At the end of Sect. 8.2, we have excluded the possibility of using the algorithms of Sect. 2.4.4 in the case of some non-separable Hori kernels. This case is, however, not included in that warning, since the whole H_{lead} does not depend on θ_2. Therefore, we may write

$$\begin{aligned} \theta_2 &= w_2 - \Xi_2(w_1, \Lambda) \\ J_2 &= \Lambda_2, \end{aligned} \tag{8.42}$$

where
$$\Xi_2(w_1, \Lambda) = \frac{\partial S(\theta_1, J)}{\partial \Lambda_2}$$

and
$$S = \Lambda_2(\theta_2 - \theta_1) + \frac{\omega_1}{\nu_{11}} \int_0^{\theta_1} \sqrt{\frac{2\nu_{11}\Lambda_1}{\omega_1} - \theta_1^2} \, d\theta_1$$

or
$$S = \Lambda_2(\theta_2 - \theta_1) + \frac{\omega_1 \theta_1}{2\nu_{11}} \sqrt{\frac{2\nu_{11}\Lambda_1}{\omega_1} - \theta_1^2} - \Lambda_1 \arcsin\left(\sqrt{\frac{\omega_1}{2\nu_{11}\Lambda_1}} \, \theta_1\right).$$

After differentiation with respect to Λ_2 and substitution of (8.40) to get rid of θ_1, we obtain

$$\theta_2 = w_2 + \sqrt{\frac{2\nu_{11}^*\Lambda_1}{\omega_1}} \sin w_1 - \frac{\Lambda_1}{8\Lambda_2} \sin 2w_1. \tag{8.43}$$

The next step is the introduction of the angle–action variables in the given Hamiltonian. The transformation of H_{lead} is easy, giving,

$$H_{\text{lead}} = \omega_1 \Lambda_1 - \varepsilon b_0 \Lambda_2 + \varepsilon M_0 \sqrt{2\Lambda_2}. \tag{8.44}$$

(It is important to stress the fact that (8.44) is not exact, since terms of order $\mathcal{O}(\theta_1^4)$, and higher, were not considered in the calculation; we are just using an approximate solution.)

In the sequence, we introduce the quantity

$$\mathcal{Q} = \sqrt{\frac{-\Lambda_1}{\omega_1}},$$

proportional to the libration amplitude of the angle θ_1. It is worth emphasizing that both Λ_1 and ω_1 are of order $\mathcal{O}(\sqrt{\varepsilon})$, but $\mathcal{Q}$ is limited, since the libration amplitude was assumed small. (For the sake of comparison with the solution near the singularity $\sqrt{-J_1} = 0$, in Sect. 9.4, we note that $\gamma = \sqrt{-2\nu_{11}^*} \, \mathcal{Q}$.)

The transformation of the next-order terms gives

$$F_3 + \delta F_3 = -\sqrt[3]{4}\left(\frac{2\Lambda_1\omega_1}{\nu_{11}^*}\right)^{3/2}(3\cos w_1 + \cos 3w_1)$$
$$-\varepsilon\left(b_0 + a_1 - b_1\Lambda_2 - \frac{M_0}{\sqrt{2\Lambda_2}} + M_1\sqrt{2\Lambda_2}\right)\sqrt{\frac{2\Lambda_1\omega_1}{\nu_{11}^*}}\cos w_1$$
$$+\frac{1}{2}\varepsilon(M_1\Lambda_2 - \frac{M_0}{2})\mathcal{Q}^2\sqrt{\frac{\Lambda_1\omega_1\nu_{11}^*}{\Lambda_2}}(\cos w_1 - \cos 3w_1)$$
$$+\frac{1}{2}\varepsilon B_0 \mathcal{Q}\sqrt{-2\nu_{11}^*}\left[\cos(w_1 + w_2) - \cos(w_1 - w_2)\right]$$
$$-\frac{1}{4}\varepsilon B_0 \mathcal{Q}^2 \nu_{11}^* \left[\cos(2w_1 - w_2) + \cos(2w_1 + w_2)\right]$$
$$+\varepsilon\left(B_0 + L_0\sqrt{2\Lambda_2} + \frac{1}{2}B_0\mathcal{Q}^2\nu_{11}^*\right)\cos w_2.$$

This expression was computed up to $\mathcal{O}(\varepsilon^{3/2}\mathcal{Q}^2)$. Terms in $\varepsilon B_0 \Lambda_1$, $\varepsilon L_0 \Lambda_1$, of order $\mathcal{O}(\varepsilon^2)$, were not written and should be included in the next order equations. The terms coming from the quantity $\Lambda_1/8\Lambda_2$ of (8.43) also do not give a contribution at this order.

First Perturbation Equation

We shall take into account that Λ_1 is of order $\mathcal{O}(\sqrt{\varepsilon})$ and, thus, the partial bracket $\{\ ,\ \}_1$ of two functions loses one unit of homogeneity in the elements of $\mathcal{S}$; the order of the remaining partial brackets is not affected. Therefore, the perturbation equations are those given by (8.20). The first perturbation equation is
$$\{F_2, W_2^*\}_1 = H_3^* - [F_3(w^*, \Lambda^*) + \delta F_3(w^*, \Lambda^*)] \tag{8.45}$$
with the Hori kernel
$$F_2(w^*, \Lambda^*) = \omega_1^* \Lambda_1^* - \varepsilon b_0 \Lambda_2^* + \varepsilon M_0 \sqrt{2\Lambda_2^*}, \tag{8.46}$$
where
$$\omega_1^* = \sqrt{-\varepsilon \nu_{11}^* M_0} \sqrt[4]{2\Lambda_2^*}. \tag{8.47}$$

The averaging is done over the fast angle w_1^*:
$$H_3^* = \frac{1}{2\pi}\int_0^{2\pi}(F_3 + \delta F_3)\,dw_1^*,$$
or
$$H_3^* = \varepsilon\left(B_0 + L_0\sqrt{2\Lambda_2^*} + \frac{1}{2}B_0\mathcal{Q}^{*2}\nu_{11}^*\right)\cos w_2^*, \tag{8.48}$$
where
$$\mathcal{Q}^* = \sqrt{\frac{-\Lambda_1^*}{\omega_1^*}}. \tag{8.49}$$

The perturbation equation may be written as
$$-\omega_1^* \frac{\partial W_2^*}{\partial w_1^*} = H_3^* - [F_3(w^*, \Lambda^*) + \delta F_3(w^*, \Lambda^*)] \tag{8.50}$$
whose integration gives:

$$W_2^* = -\sqrt[3]{4}\left(\frac{2\Lambda_1^*}{\nu_{11}^*}\right)^{3/2}\sqrt{\omega_1^*}\left(3\sin w_1^* + \tfrac{1}{3}\sin 3w_1^*\right)$$
$$-\varepsilon\left(b_0 + a_1 - b_1\Lambda_2^* - \frac{M_0}{\sqrt{2\Lambda_2^*}} + M_1\sqrt{2\Lambda_2^*}\right)\sqrt{\frac{-2}{\nu_{11}^*}}\,Q^*\sin w_1^*$$
$$+\varepsilon B_0 Q^*\sqrt{\frac{-\nu_{11}^*}{2\omega_1^*}}\left[\sin(w_1^* + w_2^*) - \sin(w_1^* - w_2^*)\right]$$
$$-\tfrac{1}{8}\varepsilon B_0 Q^{*2}\frac{\nu_{11}^*}{\omega_1^*}\left[\sin(2w_1^* - w_2^*) + \sin(2w_1^* + w_2^*)\right] + \mathcal{O}(\varepsilon Q^{*3}).$$

The perturbation equations of higher orders may be treated in exactly the same way.

The Transformed Hamiltonian. A Second Averaging

Once the transformation generated by W^* is applied to the given Hamiltonian, it becomes

$$H^* = \omega_1^*\Lambda_1^* - \varepsilon b_0\Lambda_2^* + \varepsilon M_0\sqrt{2\Lambda_2^*} + \varepsilon\left(B_0 + L_0\sqrt{2\Lambda_2^*} + \tfrac{1}{2}B_0 Q^{*2}\nu_{11}^*\right)\cos w_2^*$$
$$+\mathcal{O}(\varepsilon^{3/2}Q^{*3}) + \mathcal{O}(\varepsilon^2). \tag{8.51}$$

Because of the averaging done, the angle w_1^* is absent from H^* and, thus, Λ_1^* is a constant.

We will introduce some trivial transformations in H_5^* to put into evidence the order of magnitude of the terms. We first introduce three finite constants C_0, C_1, C_2 such that $\Lambda_1^* = C_0\sqrt{\varepsilon}$, $B_0 = C_1\sqrt{\varepsilon}$ and $L_0 = C_2\sqrt{\varepsilon}$. We also introduce a time transformation $t \to \tau = \varepsilon t$ and divide the Hamiltonian by ε. With this time scale, at the given order, the Hamiltonian becomes

$$K^* = K_0^* + K_1^*, \tag{8.52}$$

where

$$K_0^*(\Lambda_2^*) = \sqrt{-\nu_{11}^* M_0}\, C_0\sqrt[4]{2\Lambda_2^*} - b_0\Lambda_2^* + M_0\sqrt{2\Lambda_2^*}$$
$$K_1^*(w_2^*, \Lambda_2^*) = \sqrt{\varepsilon}\left(C_1 + C_2\sqrt{2\Lambda_2^*} + \tfrac{1}{2}C_1 Q^{*2}\nu_{11}^*\right)\cos w_2^*.$$

K^* is the Hamiltonian of a one-degree-of-freedom Hamiltonian system, which must be solved to complete the solution of the given system. An approximate solution may be obtained by means of an additional averaging. This second transformation is an ordinary one (non-resonant). The Hori kernel is $K_0^*(\Lambda_2^{**})$ and the first perturbation equation is

$$\{K_0^*, W_1^{**}\}_2 = K_1^{**} - K_1^*(w_2^{**}, \Lambda_2^{**}).$$

The double asterisk indicates the functions resulting from the second transformation and the subscript 2 in the Poisson bracket is there just to stress that we are working with the former second subscript (corresponding to the only remaining degree of freedom). The averaging is done over w_2^{**}:

$$K_1^{**} = \frac{1}{2\pi} \int_0^{2\pi} K_1^* dw_2^{**}$$

or

$$K_1^{**} = 0. \tag{8.53}$$

The perturbation equation may be written as

$$-\omega_2^{**} \frac{\partial W_1^{**}}{\partial w_2^{**}} = -K_1^*, \tag{8.54}$$

where

$$\omega_2^{**} = \frac{\sqrt{-\nu_{11}^* M_0 C_0}}{2(2\Lambda_2^{**})^{3/4}} - b_0 + \frac{M_0}{\sqrt{2\Lambda_2^{**}}}. \tag{8.55}$$

The solution of the perturbation equation is

$$W_1^{**} = \frac{\sqrt{\varepsilon}}{\omega_2^{**}} \left(C_1 + C_2 \sqrt{2\Lambda_2^{**}} + \frac{1}{2} C_1 Q^{*2} \nu_{11}^* \right) \sin w_2^{**}. \tag{8.56}$$

The only condition is the non-resonance condition $w_2^{**} \neq 0$. It is easy to plot the locus of $w_2 = 0$ (for $b_0 = 0$) in the frame of Fig. 8.1. The result is the dotted line r shown there. That line is very close to the curves corresponding to the pendulum separatrices showing that, in this example, this *secular* resonance may only occur for large-amplitude librations.

The Post-Pendulum Approximation

To complete this example, it is necessary to consider the actual construction of the solutions. We may, first, extend the latest canonical transformation to include the two degrees of freedom. This is easily done by just introducing $\Lambda_1^{**} = \Lambda_1^*$ and considering $W^{**}(w_2^{**}, \Lambda_1^{**}, \Lambda_2^{**})$ as the Lie generator of a complete transformation. The fact that it was determined from considerations on only one degree of freedom is irrelevant.

The Hamiltonian resulting from this transformation is (with the actual time scale t):

$$H^{**} = \omega_1^{**} \Lambda_1^{**} - \varepsilon b_0 \Lambda_2^{**} + \varepsilon M_0 \sqrt{2\Lambda_2^{**}}, \tag{8.57}$$

where

$$\omega_1^{**} = \sqrt{-\varepsilon \nu_{11}^* M_0} \sqrt[4]{2\Lambda_2^{**}}, \tag{8.58}$$

whose solutions are

$$\begin{aligned} w_1^{**} &= \omega_1^{**}(t - t_0) & \Lambda_1^{**} &= \Lambda_1^* = \text{const} \\ w_2^{**} &= \omega_2^{**} \varepsilon (t - t_0) & \Lambda_2^{**} &= \text{const.} \end{aligned} \tag{8.59}$$

The transformation of $w^{**}, \Lambda^{**} \Rightarrow w^*, \Lambda^*$, at the order of approximation of this solution, is
$$w_i^* = E_{W^{**}} w_i^{**} = w_i^{**} + \{w_i^{**}, W_1^{**}\},$$
$$\Lambda_i^* = E_{W^{**}} \Lambda_i^{**} = \Lambda_i^{**} + \{\Lambda_i^{**}, W_1^{**}\}$$
($i = 1, 2$). For instance,
$$w_1^* = w_1^{**} + \frac{\partial W_1^{**}}{\partial \Lambda_1^{**}},$$
or
$$w_1^* = w_1^{**} - \frac{\sqrt{\varepsilon} C_1 \nu_{11}^*}{2\omega_2^{**} \omega_1^{**}} \sin w_2^{**}$$
$$+ \frac{\sqrt{-\nu_{11}^* M_0}}{2\omega_2^{**2}(2\Lambda_2^{**})^{3/4}} \left(C_1 + C_2 \sqrt{2\Lambda_2^{**}} + \frac{1}{2} C_1 Q^{*2} \nu_{11}^* \right) \sin w_2^{**}.$$

It is worth noting that no singularity appears in the transformation of the first angle.

At the same order of approximation, the transformation $w^*, \Lambda^* \Rightarrow w, \Lambda$ is
$$w_i = E_{W^*} w_i^* \simeq w_i^* + \{w_i^*, W_2^*\},$$
$$\Lambda_i = E_{W^*} \Lambda_i^* \simeq \Lambda_i^* + \{\Lambda_i^*, W_2^*\}.$$
($i = 1, 2$). The sequence is a mere calculation of some partial derivatives and their products.

To obtain the post–pendulum solutions of the given Hamiltonian system, it is, now, enough to substitute the time-functions $w_i(t), \Lambda_i(t)$, thus obtained, into (8.40) and (8.42). Once more, this is only a trivial calculation that can be omitted. (In practical applications the last steps may be performed numerically.)

8.7 Example with a Separated Hori Kernel

Let us reconsider the previous example with some essential modifications. First, we discard the term $M\sqrt{-2I_3^*}\cos(\phi_1 + \phi_3)$ to get a separated Hori kernel, as discussed in Sect. 8.2. We recall that this term is the main one in the astronomical problem from which this Hamiltonian was taken. In that frame, to discard this term would be a wrong decision, but here we are only interested in having one suitable example. Second, differently of the previous example, we assume that the coefficients a, b, B, L and their derivatives are finite quantities. This assumption puts critical and long-period perturbations on an equal footing, which is essential in order to have an example significant for the given theory.

We proceed as before and expand H_0 in a Taylor series in powers of $\xi = I_1 - I_1^*$. Hence

8.7 Example with a Separated Hori Kernel

$$H_0 = H_0(I_1^*) + X_2 + X_3 + X_4 + \cdots, \tag{8.60}$$

where the X_k are the same as the previous example. In the same way, the disturbing potential εR is also expanded in powers of ξ:

$$\varepsilon R = \varepsilon \left[R^{(0)} + R^{(1)} + R^{(2)} + \cdots \right], \tag{8.61}$$

where

$$\begin{aligned} R^{(0)} &= \left(a_0 + b_0 I_3 + B_0 \cos\phi_1 + L_0 \sqrt{-2I_3} \cos\phi_3 \right) \\ R^{(1)} &= \left(a_1 + b_1 I_3 + B_1 \cos\phi_1 + L_1 \sqrt{-2I_3} \cos\phi_3 \right) \xi \\ R^{(2)} &= \tfrac{1}{2} \left(a_2 + b_2 I_3 + B_2 \cos\phi_1 + L_2 \sqrt{-2I_3} \cos\phi_3 \right) \xi^2; \end{aligned} \tag{8.62}$$

the subscripts 1 and 2 in the coefficients denote that they are, respectively, first and second derivatives with respect to I_1. All coefficients are calculated at the exact resonance value I_1^* and are, thus, constants. We also assume $B_0 > 0$ and $b_0 > 0$.

Let us select the leading terms of H. They are the first non-constant term of H_0, that is, X_2, and the leading terms of R. Then

$$\begin{aligned} H_\text{lead} &= X_2(\xi) + \varepsilon R^{(0)} \\ &= \tfrac{1}{2} \nu_{11}^* \xi^2 + \varepsilon \left(a_0 + b_0 I_3 + B_0 \cos\phi_1 + L_0 \sqrt{-2I_3} \cos\phi_3 \right). \end{aligned} \tag{8.63}$$

To have all terms of H_lead on the same footing, we assume $\xi = \mathcal{O}(\sqrt{\varepsilon})$ and $S \equiv (\xi, \sqrt{\varepsilon})$. The functions F_k ($k > 2$) are

$$\begin{aligned} F_3 &= X_3 + \varepsilon R^{(1)}, \\ F_4 &= X_4 + \varepsilon R^{(2)}, \quad \text{etc.} \end{aligned} \tag{8.64}$$

The Hori kernel $H_2^* = H_\text{lead}(\phi^*, \xi^*, I_3^*)$ may be separated into $H_{2(1)}^*(\phi_1^*, \xi^*)$ and $H_{2(2)}^*(\phi_3^*, I_3^*)$.

1. $H_{2(1)}^* = \tfrac{1}{2}\nu_{11}^* \xi^{*2} + \varepsilon B_0 \cos\phi_1^*$ is the Hamiltonian of a simple pendulum with negative mass (since $\nu_{11}^* = -6\sqrt[3]{2} < 0$). The corresponding differential equations are

$$\frac{d\phi_1^*}{du} = \frac{\partial H_2^*}{\partial \xi^*} = \nu_{11}^* \xi^* \qquad \frac{d\xi^*}{du} = -\frac{\partial H_2^*}{\partial \phi_1^*} = \varepsilon B_0 \sin\phi_1^*. \tag{8.65}$$

Limiting ourselves to the case of small oscillations about the stable equilibrium and using the pendulum angle–action variables w_1^*, Λ_1^* (see Sect. B.3.1), we have, at the first approximation,

$$\phi_1^* = \sqrt{\frac{2\nu_{11}^* \Lambda_1^*}{\omega_1}} \sin w_1^*$$

$$\xi^* = -\sqrt{\frac{2\omega_1 \Lambda_1^*}{\nu_{11}^*}} \cos w_1^*, \qquad (8.66)$$

where

$$\omega_1 = \frac{dw_1^*}{du} = \sqrt{-\varepsilon\nu_{11}^* B_0}.$$

($\Lambda_1^* < 0$ and $-\varepsilon\nu_{11}^* B_0 > 0$. By assumption, $B_0 > 0$.)

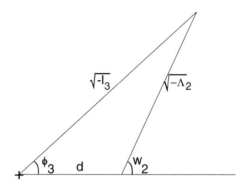

Fig. 8.2. The transformation $I_3^* \to \Lambda_2^*$. ($d = L_0/b_0$)

2. $H_{2(2)}^* = \varepsilon(b_0 I_3^* + L_0 \sqrt{-2I_3^*} \cos \phi_3^*)$ is the Hamiltonian of a planar harmonic oscillator with center away from the origin. It is enough to shift the origin to $d = L_0/b_0$ to write the oscillator in its own angle–action variables: $H_{2(2)}^* = \varepsilon b_0 \Lambda_2^* + \text{const}$. The comparison of the two expressions of $H_{2(2)}^*$ gives (see Fig. 8.2)

$$\Lambda_2^* = I_3^* - \frac{1}{2}\left(\frac{L_0}{b_0}\right)^2 + \left(\frac{L_0}{b_0}\right)\sqrt{-2I_3^*} \cos \phi_3.$$

The angle variable is, now, $w_2^* = \varepsilon b_0 u + \text{const}$.

First Perturbation Equation. The Post-Pendulum Approximation

The first perturbation equation is

$$\{F_2, W_2^*\}_1 = H_3^* - F_3. \qquad (8.67)$$

To have an explicit expression of the involved functions, we need to write F_3 with the variables (w_1^*, Λ_1^*). From (8.31) and (8.62) we have

$$F_3 = 4\sqrt[3]{4}\xi^{*3} + \varepsilon\left[a_1 + b_1 I_3^* + B_1 \cos \phi_1^* + L_1\sqrt{-2I_3^*} \cos \phi_3^*\right]\xi^* \qquad (8.68)$$

or, taking into account the approximate solution of the pendulum,

$$F_3 = C_0(-\omega_1 \Lambda_1)^{3/2}(3\cos w_1^* + \cos 3w_1^*) + \varepsilon(C_1 + C_2 I_3^* + C_3)\sqrt{-\omega_1 \Lambda_1^*}\cos w_1^*$$
$$+ \frac{\varepsilon C_3 \nu_{11}}{4\sqrt{\omega_1}}(-\Lambda_1^*)^{3/2}(\cos w_1^* - \cos 3w_1^*) + \varepsilon C_4 \sqrt{2\omega_1 \Lambda_1^* I_3^*}\cos\phi_3^* \cos w_1^*,$$

where, for simplicity, we have introduced

$$C_0 = -\sqrt[3]{4}\left(\frac{2}{-\nu_{11}^*}\right)^{\frac{3}{2}}, \quad C_1 = \frac{-a_1\sqrt{2}}{\sqrt{-\nu_{11}^*}}, \quad C_2 = \frac{-b_1\sqrt{2}}{\sqrt{-\nu_{11}^*}},$$

$$C_3 = \frac{-B_1\sqrt{2}}{\sqrt{-\nu_{11}^*}}, \quad C_4 = \frac{-L_1\sqrt{2}}{\sqrt{-\nu_{11}^*}}, \tag{8.69}$$

and discarded terms of order higher than $\mathcal{O}(\varepsilon^{3/2}\mathcal{Q}^3)$ ($\mathcal{Q}$ is proportional to the libration amplitude of ϕ_1^*). It then follows that $H_3^* = <F_3> = 0$, and the integration of the first perturbation equation is elementary giving

$$W_2^* = C_0(-\Lambda_1^*)^{3/2}\sqrt{\omega_1}(3\sin w_1^* + \frac{1}{3}\sin 3w_1^*) + \varepsilon(C_1 + C_2 I_3^* + C_3)\mathcal{Q}^*\sin w_1^*$$
$$+ \frac{\varepsilon C_3 \nu_{11}}{4}\mathcal{Q}^{*3}(\sin w_1^* - \frac{1}{3}\sin 3w_1^*) + \varepsilon C_4 \sqrt{-2I_3^*}\mathcal{Q}^*\cos\phi_3^*\sin w_1^*$$

up to order $\mathcal{O}(\varepsilon \mathcal{Q}^3)$. See (8.49).

Second Perturbation Equation. The Post-Post-Pendulum Approximation

The second perturbation equation is

$$\{F_2, W_3^*\}_1 = H_4^* - \Psi_4, \tag{8.70}$$

where

$$\Psi_4 = -10\xi^{*4} + \frac{1}{2}\varepsilon\left(a_2 + b_2 I_3^* + B_2\cos\phi_1^* + L_2\sqrt{-2I_3^*}\cos\phi_3^*\right)\xi^{*2}$$
$$+ \{F_3, W_2^*\}_1 + \{F_2, W_2^*\}_\varrho + \frac{1}{2}\{\{F_2, W_2^*\}_1, W_2^*\}_1. \tag{8.71}$$

The calculations become cumbersome and some computational help is, now, essential. We will just write down the results, as the details of the calculations are not of interest *per se*. The average of Ψ_4 is

$$H_4^* = \left(\frac{15C_0^2}{2} - \frac{15}{\nu_{11}^{*2}}\right)\omega_1^2 \Lambda_1^{*2} + \frac{\varepsilon}{2\nu_{11}^*}(b_2 I_3^* + B_2 + a_2)\omega_1 \Lambda_1^* - \frac{\varepsilon}{8}B_2 \Lambda_1^{*2}$$
$$-3\varepsilon C_0(C_1 + C_2 I_3 + C_3)\omega_1 \Lambda_1^* - 3\varepsilon\left(C_0 C_4 - \frac{L_2}{6\nu_{11}^*}\right)\omega_1 \Lambda_1^*\sqrt{-2I_3^*}\cos\phi_3^*$$
$$+ \frac{3\varepsilon}{4}C_0 C_3 \nu_{11}^* \Lambda_1^{*2} + \frac{\varepsilon^2}{4}(C_1 + C_3)^2 - \frac{\varepsilon^2}{4\omega_1}(C_1 + C_2 I_3^* + C_3)C_3 \nu_{11}^* \Lambda_1^*$$
$$+ \frac{\varepsilon^2}{4}(2C_1 + C_2 I_3^* + 2C_3)C_2 I_3^* + \frac{3\varepsilon^2 C_3^2 \nu_{11}^{*2}}{32\omega_1^2}\Lambda_1^{*2}$$
$$+ \frac{\varepsilon^2}{2}\left(C_1 + C_2 I_3^* + C_3 - \frac{C_3 \nu_{11}^*}{2\omega_1}\Lambda_1^*\right)C_4 \sqrt{-2I_3^*}\cos\phi_3^* - \frac{\varepsilon^2}{2}C_4^2 I_3^* \cos^2\phi_3^* \tag{8.72}$$

and the integration of the homological equation gives

$$\begin{aligned}W_3^* = &\left(3C_0^2 - \frac{10}{\nu_{11}^{*2}}\right)\omega_1 \Lambda_1^{*2}\sin 2w_1^* - \left(\frac{3C_0^2}{8} + \frac{5}{4\nu_{11}^{*2}}\right)\omega_1 \Lambda_1^{*2}\sin 4w_1^*\\&+ \left(\frac{a_2+B_2+b_2 I_3^*}{4\nu_{11}^*} - \frac{C_0(C_1+C_2 I_3^*+C_3)}{2}\right)\varepsilon \Lambda_1^* \sin 2w_1^* - \frac{C_0 C_3 \nu_{11}^*}{2\omega_1}\varepsilon \Lambda_1^{*2}\sin 2w_1^*\\&+ \frac{B_2+2C_0 C_3 \nu_{11}^*}{32\omega_1}\varepsilon \Lambda_1^{*2}\sin 4w_1^* + \left(\frac{L_2}{4\nu_{11}^*} - \frac{C_0 C_4}{2}\right)\varepsilon \Lambda_1^*\sqrt{-2I_3^*}\cos\phi_3^* \sin 2w_1^*\\&+ \frac{(C_1+C_2 I_3^*+C_3)C_3 \nu_{11}^*}{8\omega_1^2}\varepsilon^2 \Lambda_1^*\sin 2w_1^* - \frac{C_3^2 \nu_{11}^{*2}}{128\omega_1^3}\varepsilon^2 \Lambda_1^{*2}(8\sin 2w_1^* - \sin 4w_1^*)\\&+ \frac{L_0 C_2 - C_4 b_0}{\omega_1^{3/2}}\varepsilon^2 \sqrt{2\Lambda_1^* I_3^*}\sin\phi_3^*\cos w_1^*\\&+ \frac{C_3 C_4 \nu_{11}^*}{8\omega_1^2}\varepsilon^2 \Lambda_1^*\sqrt{-2I_3^*}\cos\phi_3^*\sin 2w_1^*\end{aligned}$$

(8.73)

where terms including the effects of higher libration harmonics were discarded[1].

We note that all functions involved in Ψ_4 have the d'Alembert property in $(w_1^*, \sqrt{\Lambda_1^*})$, that is, they have the form

$$\sqrt{\Lambda_1^{*(q+2q')}}\; {\sin \atop \cos} qw_1^* \qquad (q, q' \in \mathbf{Z}_o^+).$$

From (8.66), we see that one function that has the d'Alembert property in $(w_1^*, \sqrt{\Lambda_1^*})$ is equivalent to a polynomial in ϕ_1^*, ξ^*. Since the bracket of any two polynomials in these variables is a polynomial of the same kind, all functions obtained with the procedures described in this example have the d'Alembert property in $(w_1^*, \sqrt{\Lambda_1^*})$. This fact guarantees that the kind of results obtained so far will reproduce itself without formal modifications in higher orders.

Construction of the Solutions

To complete this example, it is necessary to consider the actual construction of the solutions. For the first subscript, the post-pendulum solutions are

$$\phi_1 = E_{W^*}\phi_1^* \simeq \phi_1^* + \{\phi_1^*, W_2^*\}_1$$

$$\xi = E_{W^*}\xi^* \simeq \xi^* + \{\xi^*, W_2^*\}_1.$$

Since the one-degree-of-freedom transformation $(\phi_1^*, \xi^*) \Rightarrow (w_1^*, \Lambda_1^*)$ is canonical, we may use, in the brackets $\{\ ,\ \}_1$, derivatives with respect to w_1^*, Λ_1^*. Hence,

$$\phi_1 = \phi_1^* + 2\sqrt{-2\nu_{11}^*}\; C_0 \Lambda_1^* \sin 2w_1^* \qquad (8.74)$$

---

[1] One should be warned against the fact that algebraic manipulators may do different branch choices according to whether the square root of a product of two negative quantities is written as $\sqrt{-a}\sqrt{-b}$ or $\sqrt{ab}$.

$$\xi = \xi^* + \frac{2\omega_1}{\sqrt{-2\nu_{11}^*}} C_0 \Lambda_1^*(3 - \cos 2w_1^*) - \frac{\varepsilon}{\sqrt{-2\nu_{11}^*}} (C_1 + C_2 I_3^* + C_3)$$

$$- \frac{\varepsilon\sqrt{-2\nu_{11}^*}}{4\omega_1} C_3 \Lambda_1^*(1 - \cos 2w_1^*) - \frac{\varepsilon}{\sqrt{-2\nu_{11}^*}} C_4 \sqrt{-2I_3^*}\cos\phi_3^*. \quad (8.75)$$

One may note that (8.75), giving ξ, is a half-order more precise than (8.74). This difference happens because ξ is itself a quantity of order $\mathcal{O}(\sqrt{\varepsilon})$. To have ϕ_1 given at the same order, it would be necessary to consider also W_3^* and to add, to the transformation equation, the terms $\{\phi_1^*, W_3^*\}_1$ and $\frac{1}{2}\{\{\phi_1^*, W_2^*\}_1, W_2^*\}_1$.

For the higher subscript, at order $\mathcal{O}(\sqrt{\varepsilon})$, we have $\phi_3 = \phi_3^*$, $I_3 = I_3^*$. The next-order results are easily obtained:

$$\phi_3 = E_{W^*}\phi_3^* \simeq \phi_3^* + \{\phi_3^*, W_2^*\}_\varrho$$

$$I_3 = E_{W^*}I_3^* \simeq I_3^* + \{I_3^*, W_2^*\}_\varrho. \quad (8.76)$$

(The partial brackets with subscript 1 are, now, equal to zero because canonical variables in a set are independent and, thus, the derivatives of ϕ_3^* and I_3^* with respect to ϕ_1^* and ξ^* are zero.) Then

$$\phi_3 = \phi_3^* + \varepsilon\sqrt{\frac{-\Lambda_1^*}{\omega_1}} C_2 \sin w_1^* - \frac{\varepsilon}{2}\sqrt{\frac{2\Lambda_1^*}{\omega_1 I_3^*}} C_4 \cos\phi_3^* \sin w_1^*$$

$$I_3 = I_3^* + \varepsilon\sqrt{\frac{2\Lambda_1^* I_3^*}{\omega_1}} C_4 \sin\phi_3^* \sin w_1. \quad (8.77)$$

In order to know $\phi_1(t), \phi_3(t), \xi(t), I_3(t)$, it is still necessary to know $\phi_1^*(t), \phi_3^*(t), \xi^*(t), I_3^*(t)$. Since $H_2^* = H_{\text{lead}}(\phi_1^*, \phi_3^*, \xi^*, I_3^*)$, the reduced system, at this order of approximation, is

$$\frac{d\phi_1^*}{dt} = \frac{\partial H_2^*}{\partial \xi^*} = \nu_{11}^*\xi^*,$$

$$\frac{d\xi^*}{dt} = -\frac{\partial H_2^*}{\partial \phi_1^*} = \varepsilon B_0 \sin\phi_1^*,$$

$$\frac{d\phi_3^*}{dt} = \frac{\partial H_2^*}{\partial I_3^*} = \varepsilon b_0 - \frac{\varepsilon L_0}{\sqrt{-2I_3^*}}\cos\phi_3^*,$$

$$\frac{dI_3^*}{dt} = -\frac{\partial H_2^*}{\partial \phi_3^*} = \varepsilon L_0\sqrt{-2I_3^*}\sin\phi_3^*.$$

The first two equations are the same as (8.65) and their solution, up to terms in $\mathcal{Q}^2$, are those given by (8.66), where, now,

$$w_1^* = \omega_1(t - t_0) \quad (8.78)$$

and Λ_1^* is an integration constant. The two other equations are elementary and their results are easily expressed with non-singular variables:

$$x^* = \sqrt{-2I_3^*}\cos\phi_3^* = \frac{L_0\sqrt{2}}{b_0} + \sqrt{-2\Lambda_2^*}\cos w_2^*$$

$$y^* = \sqrt{-2I_3^*}\sin\phi_3^* = \sqrt{-2\Lambda_2^*}\sin w_2^*,$$

where Λ_2^* is an integration constant (action) and $w_2^* = \varepsilon b_0 t + \text{const}$. Finally, the solution of the problem is obtained by substituting the solution of the transformed system into (8.74)–(8.77).

8.8 One Degree of Freedom

Let us consider the case of just one degree of freedom:

$$F_2 = H_{\text{lead}} = \frac{1}{2}\nu_{11}^* \xi^2 + \varepsilon H_2(\theta_1, J_1^*) \tag{8.79}$$

($J_1^* = \text{const}$). The Hori kernel, in this case, is

$$\frac{d\theta_1^*}{du} = \frac{\partial F_2(\theta_1^*, \xi^*)}{\partial \xi^*} = \nu_{11}^* \xi^*, \tag{8.80}$$

$$\frac{d\xi^*}{du} = -\frac{\partial F_2(\theta_1^*, \xi^*)}{\partial \theta_1^*} = -\varepsilon \frac{dH_2(\theta_1^*, J_1^*)}{d\theta_1^*} \tag{8.81}$$

or

$$\frac{d^2\theta_1^*}{du^2} = -\varepsilon \nu_{11}^* \frac{dH_2(\theta_1^*, J_1^*)}{d\theta_1^*}. \tag{8.82}$$

In these problems, to have all terms of H_{lead} on the same footing, we must assume $\xi = \mathcal{O}(\sqrt{\varepsilon})$, that is, $d = \frac{1}{2}$ or $\mathcal{S} \equiv (\xi, \sqrt{\varepsilon})$.

8.8.1 Garfinkel's Ideal Resonance Problem

As an example of a one-degree-of-freedom system, we may consider Garfinkel's Ideal Resonance Problem (see Sect. 4.4). In this example, the Hamiltonian is

$$H(\theta_1, J_1) = H_0(J_1) - \varepsilon A(J_1)\cos\theta_1 \tag{8.83}$$

and the exact resonance J_1^* is defined by $\nu_1^* = 0$. H_0 is assumed to be such that $A^*\nu_{11}^* > 0$, where $A^* = A(J_1^*)$. H is expanded as in the previous example, that is,

$$F_2(\theta_1, \xi) = \frac{1}{2}\nu_{11}^* \xi^2 - \varepsilon A^* \cos\theta_1. \tag{8.84}$$

(As before, the derivatives of A^* with respect to J_1^* will be written A_1, A_2, etc.) The Hori kernel equations are

$$\frac{d\theta_1^*}{du} = \nu_{11}^* \xi^*, \tag{8.85}$$

$$\frac{d\xi^*}{du} = -\varepsilon A^* \sin\theta_1^*. \tag{8.86}$$

Taking the energy integral $F_2 = \text{const}$ into account, it follows that

$$\frac{1}{2\nu_{11}^*}\left(\frac{d\theta_1^*}{du}\right)^2 = F_2 + \varepsilon A^* \cos\theta_1^*. \tag{8.87}$$

This is the simple pendulum equation, which can be solved as in Sect. B.2 leading to the two classical motion regimes: libration and circulation. In this section, we consider only the case of very small oscillations about the stable equilibrium and write the solutions as in the previous examples. The equations are almost the same. However, as the Ideal Resonance Problem is the paradigm of a large number of problems in Celestial Mechanics, we shall use a more accurate solution of the Hori kernel, namely[2],

$$\sin\theta_1^* = 8\sqrt{\Upsilon}\left[\left(1 - 6\Upsilon - \frac{25}{2}\Upsilon^2 - 84\Upsilon^3\right)\sin w_1\right.$$
$$+ 3\Upsilon\left(1 + \Upsilon + \frac{11}{2}\Upsilon^2\right)\sin 3w_1 + 5\Upsilon^2(1 + 7\Upsilon)\sin 5w_1$$
$$\left. + 7\Upsilon^3 \sin 7w_1\right] + \mathcal{O}(\Upsilon^{9/2}), \tag{8.88}$$

$$\xi_1^* = 8A^*\sqrt{\frac{\varepsilon\Upsilon}{\nu_{11}^* A^*}}\left[\left(1 - 2\Upsilon - \frac{17}{2}\Upsilon^2 - 62\Upsilon^3\right)\cos w_1\right.$$
$$+ \Upsilon\left(1 + 5\Upsilon + \frac{75}{2}\Upsilon^2\right)\cos 3w_1 + \Upsilon^2(1 + 11\Upsilon)\cos 5w_1$$
$$\left. + \Upsilon^3 \cos 7w_1\right] + \mathcal{O}(\sqrt{\varepsilon}\Upsilon^{9/2}), \tag{8.89}$$

where

$$\Upsilon \stackrel{\text{def}}{=} \frac{\nu_{11}^* \Lambda_1}{32\sqrt{\varepsilon\nu_{11}^* A^*}}. \tag{8.90}$$

$\sqrt{\Upsilon}$ is a quantity of the order of the libration amplitudes in θ_1^*. We also have

$$w_1 \stackrel{\text{def}}{=} \frac{dw_1}{du} = \sqrt{\varepsilon\nu_{11}^* A^*}(1 - 4\Upsilon - 12\Upsilon^2 - 80\Upsilon^3) + \mathcal{O}(\sqrt{\varepsilon}\Upsilon^4) \tag{8.91}$$

and

$$H_2^* = \int w_1 d\Lambda_1 = -\varepsilon A^* + 32\varepsilon A^*\Upsilon(1 - 2\Upsilon - 4\Upsilon^2 - 20\Upsilon^3) + \mathcal{O}(\varepsilon\Upsilon^5). \tag{8.92}$$

We note that the operations $A^*/\sqrt{A^*\nu_{11}^*}$ and $\nu_{11}^*/\sqrt{A^*\nu_{11}^*}$, appearing in the above equations, can only be completed when the signs of A^* and ν_{11}^* are known. We recall that $A^*\nu_{11}^* > 0$ and $\Lambda_1\nu_{11}^* > 0$.

The calculation scheme is almost the same as that of the previous example and it is not repeated here. It is worth emphasizing, however, that in the

---
[2] Cf. (B.63)–(B.67), putting $mk = \varepsilon A^*$ and $m^{-1} = \nu_{11}^*$.

approximations we assumed $\Upsilon = \mathcal{O}(\sqrt{\varepsilon})$, thus allowing the results to be valid for libration amplitudes (in θ_1) of order $\mathcal{O}(\sqrt[4]{\varepsilon})$.

In the post-pendulum approximation we have

$$F_3 = \frac{1}{6}\nu^*_{111}\xi^{*3} - \varepsilon A_1 \xi^* \cos\theta_1^* \qquad (8.93)$$

or, using the approximate solution of the pendulum up to $\mathcal{O}(\Upsilon^{7/2})$:

$$\begin{aligned}F_3 &= \frac{8\varepsilon^{3/2}\sqrt{\nu^*_{11}A^*}}{\nu^*_{11}}\sqrt{\Upsilon}\bigg[-A_1\cos w_1 + A_1\Upsilon(10\cos w_1 - 9\cos 3w_1) \\ &\quad + C_1\Upsilon\left(\cos w_1 + \frac{1}{3}\cos 3w_1\right) + A_1\Upsilon^2\left(\frac{1}{2}\cos w_1 + 27\cos 3w_1\right. \\ &\quad \left. - 25\cos 5w_1\right) - C_1\Upsilon^2(5\cos w_1 - \cos 5w_1)\bigg],\end{aligned} \qquad (8.94)$$

where we have introduced

$$C_1 = \frac{8\nu^*_{111}A^*}{\nu^*_{11}}$$

and discarded terms of orders higher than $\mathcal{O}(\varepsilon^{3/2}\Upsilon^{7/2})$. Hence $H_3^* = \langle F_3 \rangle = 0$, and

$$\begin{aligned}W_2^* &= \frac{8\varepsilon}{\nu^*_{11}}\sqrt{\Upsilon}\bigg[-A_1\sin w_1 + 3A_1\Upsilon(2\sin w_1 - \sin 3w_1) \\ &\quad + C_1\Upsilon\left(\sin w_1 + \frac{1}{9}\sin 3w_1\right) + A_1\Upsilon^2\left(\frac{25}{2}\sin w_1 - 3\sin 3w_1\right. \\ &\quad \left. - 5\sin 5w_1\right) - C_1\Upsilon^2\left(\sin w_1 - \frac{4}{9}\sin 3w_1 - \frac{1}{5}\sin 5w_1\right)\bigg] \end{aligned} \qquad (8.95)$$

up to order $\mathcal{O}(\varepsilon\Upsilon^{5/2})$.

In the post-post-pendulum approximation, we have to introduce

$$\Psi_4 = \frac{1}{24}\nu^*_{1111}\xi^{*4} - \frac{1}{2}\varepsilon\xi^{*2}A_2\cos\theta_1^* + \frac{1}{2}\{F_3, W_2^*\}. \qquad (8.96)$$

Hence

$$\begin{aligned}H_4^* &= \frac{\varepsilon^2}{\nu^*_{11}}\bigg[-\frac{1}{2}A_1^2 + 16(A_1^2 + \frac{1}{8}C_1A_1 - A^*A_2)\Upsilon \\ &\quad -4\left(48A_1^2 + 6C_1A_1 - 48A^*A_2 + \frac{5}{12}C_1^2 - C_2\right)\Upsilon^2\bigg]\end{aligned}$$

and

$$W_3^* = \frac{\varepsilon^{3/2}\Upsilon}{\nu_{11}^*\sqrt{\nu_{11}^*A^*}}\left[\left(-8A_1^2 + \frac{1}{3}C_1A_1 - 8A^*A_2\right)\sin 2w_1\right.$$

$$+ 8A_1^2\Upsilon(14\sin 2w_1 - 3\sin 4w_1) + \frac{1}{9}C_1A_1\Upsilon(118\sin 2w_1 - 11\sin 4w_1)$$

$$- 8A^*A_2\Upsilon(2\sin 2w_1 + 5\sin 4w_1) - \frac{1}{3}C_1^2\Upsilon\left(2\sin 2w_1 - \frac{1}{4}\sin 4w_1\right)$$

$$\left.+ \frac{1}{3}C_2\Upsilon(8\sin 2w_1 + \sin 4w_1)\right],$$

where

$$C_2 = \frac{16\nu_{1111}^*A^{*2}}{\nu_{11}^*}.$$

The solutions, in the post-post-pendulum approximation, are

$$\sin\theta_1 = \sin\theta_1^* + \{\sin\theta_1^*, W_2^*\} + \{\sin\theta_1^*, W_3^*\} + \frac{1}{2}\{\{\sin\theta_1^*, W_2^*\}, W_2^*\},$$

$$\xi = \xi^* + \{\xi^*, W_2^*\} + \{\xi^*, W_3^*\} + \frac{1}{2}\{\{\xi^*, W_2^*\}, W_2^*\},$$

or,

$$\sin\theta_1 = \sin\theta_1^* + \frac{8}{3}\sqrt{\frac{\varepsilon}{\nu_{11}^*A^*}}\left[\frac{1}{2}C_1\Upsilon\sin 2w_1 - \frac{1}{3}C_1\Upsilon^2(25\sin 2w_1 - 14\sin 4w_1)\right]$$

$$- \frac{\varepsilon\sqrt{\Upsilon}}{\nu_{11}^*A^*}\left(2A_1^2 + 2A^*A_2 - \frac{1}{4}C_1A_1\right)\sin w_1 \qquad (8.97)$$

and

$$\xi = \xi^* + \frac{\varepsilon}{\nu_{11}^*}\left[A_1(1-16\Upsilon) - 2C_1\Upsilon(1-4\Upsilon) + 16A_1\Upsilon\cos 2w_1 + \frac{2}{3}C_1\Upsilon\cos 2w_1\right.$$

$$\left.-32A_1\Upsilon^2(\cos 2w_1 - \cos 4w_1) - \frac{4}{9}C_1\Upsilon^2(19\cos 2w_1 - 7\cos 4w_1)\right]$$

$$+ \frac{\varepsilon^{3/2}\sqrt{\Upsilon}}{\nu_{11}^*\sqrt{\nu_{11}^*A^*}}\left(2A_1^2 + 2A^*A_2 - \frac{1}{4}C_1A_1\right)\cos w_1 \qquad (8.98)$$

up to terms $\mathcal{O}(\varepsilon\sqrt{\Upsilon}, \sqrt{\varepsilon}\Upsilon^2, \Upsilon^{7/2})$ in θ_1 and $\mathcal{O}(\varepsilon^{3/2}\sqrt{\Upsilon}, \varepsilon\Upsilon^2, \sqrt{\varepsilon}\Upsilon^{7/2})$ in ξ.

The solution of the Hori kernel introduces variables akin to the angle–action variables of the simple pendulum. It is noteworthy that even a first-order approximation of these variables, followed by a Lie series perturbation procedure, allows the solutions to be constructed to any order (see [55], [56]).

9

Single Resonance near a Singularity

9.1 Resonances Near the Origin: Real and Virtual

The perturbation of a regular integrable Hamiltonian may lead to bifurcations in the phase portrait. Let us consider the case of only one resonance and the simple case of one degree of freedom. Let H_0 be the Hamiltonian of a differential rotator:

$$H_0 = \frac{1}{2}\nu_1^\circ(x_1^2 + y_1^2) + \frac{1}{8}\nu_{11}^\circ(x_1^2 + y_1^2)^2 \tag{9.1}$$

and let us assume that ν_1° and ν_{11}° have opposite signs, e.g. $\nu_1^\circ < 0$ and $\nu_{11}^\circ > 0$. In that case, H_0 has a minimum on a circle of radius $\sqrt{-4\nu_1^\circ/\nu_{11}^\circ}$ on which the direction of motion changes. This is the classical case of a twist mapping and the Poincaré–Birkhoff theorem predicts that, when the rotator is perturbed, new centers and saddle points may appear in the phase portrait near the place where the frequency of the undisturbed rotator is zero (see [63]). A typical example is shown in Fig. 9.1 right. A web of separatrices with termination at the saddle points encloses the centers forming libration lobes[1] and separating

Fig. 9.1. Phase portrait of a resonant system (right). The corresponding undisturbed differential rotator is shown on the left

---

[1] Since the considered Hamiltonian system has one degree of freedom, the separatrices meet, forming a well-defined structure without the possibility of chaotic motions.

the two regions where the motions are circulations around the origin with different directions.

The given example corresponds to a *non-central* resonance happening at a finite distance from the origin. When $\sqrt{-4\nu_1^\circ/\nu_{11}^\circ}$ is small, the bifurcation happens very close to the origin and may give rise to more complex situations. For instance, in the case of a linear perturbation (e.g. the Andoyer Hamiltonian with $k = 1$), the libration lobe may enclose the origin (see the case $\alpha = 1.1$ of Fig. C.3).

When ν_1° and ν_{11}° have the same sign (e.g. both positive), $\sqrt{-4\nu_1^\circ/\nu_{11}^\circ}$ becomes imaginary. The phase portrait of the undisturbed rotator in the (x_1, y_1) plane is shown in Fig. 9.2 (left). The motions are circulations all in the same direction. However, when $\nu_1^\circ/\nu_{11}^\circ$ is small, an important phenomenon may occur when the system is perturbed. The bifurcation is now occurring in the complex continuation of the (x_1, y_1) plane, but part of the virtual libration lobes raised by a perturbation may appear in the portrait of the perturbed system in the real phase plane. We call this phenomenon a *virtual* resonance.

Fig. 9.2. Phase portrait of a perturbed systems showing libration lobes due to a virtual resonance (right). The corresponding undisturbed differential rotator is shown on the left

Examples of dynamical systems showing virtual resonances are the Andoyer Hamiltonians with $k = 2$ and $k = 3$. The example shown in Fig. 9.2 (right) corresponds to the Andoyer Hamiltonian

$$H_2 = \frac{1}{2}\nu_1^\circ(x_1^2 + y_1^2) + \frac{1}{8}\nu_{11}^\circ(x_1^2 + y_1^2)^2 + \varepsilon\tau(x_1^2 - y_1^2)$$

when $\nu_1^\circ = 0.3$, $\nu_{11}^\circ = 2$ and $\varepsilon\tau = 0.3$ (see Sect. D.2.1). The frequency change occurs in the complex continuation of the phase plane on the circle $(x_1^2 + y_1^2) \simeq -0.6$.

9.2 One Degree of Freedom

Let us consider the Hamiltonian

$$H = H_0(J_1) + \sum_{k \geq 1} \varepsilon^k H_k(\theta_1, J_1) \tag{9.2}$$

in the particular case of only one degree of freedom.

We assume that:

(a.) $J_1 \leq 0$;
(b.) H is regular in a domain about the origin (in the sense discussed in Sect. 7.4); and
(c.) $\nu_1 = \mathrm{d}H_0/\mathrm{d}J_1 = 0$ at some point $J_1 = J_1^* < 0$ close to the origin (that is, the resonance is located near the origin).

H may be expanded about the origin and, following the notation introduced in Chap. 7, we have

$$H_0 = -\frac{1}{2}\nu_1^\circ(x_1^2 + y_1^2) + \frac{1}{8}\nu_{11}^\circ(x_1^2 + y_1^2)^2 + \cdots; \qquad (9.3)$$

$$H_k = \sum_{k' \geq 1} \mathcal{V}_k^{k'}(x_1, y_1), \qquad (9.4)$$

where $\mathcal{V}_k^{k'}$ are homogeneous functions of degree k' with respect to x_1, y_1. Because of the assumption (a), the non-singular variables x_1, y_1 are those defined by (7.2) (so that $\{x_1, y_1\} = +1$).

Order of a Resonance. *The order of the resonance is the degree of the term of H_1 of least degree in x_1, y_1.*

In the one-degree-of-freedom case studied in this section, the leading term is $\mathcal{V}_1^1$ (first-order resonance).

It is worth emphasizing that, because of assumption (c), the quantity $\nu_1^\circ = \nu_i(0) \simeq -\nu_{11}^\circ J_1^*$ is small and $\nu_1^\circ \to 0$ when $J_1^* \to 0$. Therefore, the term $\frac{1}{2}\nu_i^\circ(x_1^2 + y_1^2)$ is not enough to determine the main integrable topological features of the flow in the neighborhood of the origin. The leading terms of H are

$$H_{\text{lead}} = -\frac{1}{2}\nu_1^\circ(x_1^2 + y_1^2) + \frac{1}{8}\nu_{11}^\circ(x_1^2 + y_1^2)^2 + \varepsilon\mathcal{V}_1^1. \qquad (9.5)$$

H_{lead} is the first Andoyer Hamiltonian studied in Appendix C and the bifurcation due to the resonance appears for $|J_1^*| > |J_{1\text{crit}}^*| = \mathcal{O}(\varepsilon^{2/3})$. For $|J_1^*| < |J_{1\text{crit}}^*|$, the only qualitative effect of the perturbation is to shift the center of the family of orbits away from the origin. We then assume:

$$\nu_1^\circ = \mathcal{O}(\varepsilon^{2/3}) \qquad (9.6)$$

to have a theory representing the main regimes of motion that may take place near the origin. This assumption does not introduce a real limitation on the values of ν_1°; however, if the value of this quantity is large, the neighborhood of the origin can be studied with the simpler theory of Sect. 7.6 and the resonance zone, situated far away from the origin, may be studied with the angle–action theory of the previous chapter.

In the forthcoming developments, we follow [31] and introduce the set $\mathcal{S} \equiv (x_1, y_1, \sqrt[3]{\varepsilon})$. The adopted subscripts indicate the degree of homogeneity of the function with respect to the elements of $\mathcal{S}$. Thus,

9 Single Resonance near a Singularity

$$H_0 = \sum_{k \geq 2} X_{2k}(J_1), \tag{9.7}$$

where, because of the assumption made on ν_1° (9.6), $X_2 = 0$ and

$$X_4 = -\frac{1}{2}\nu_1^\circ(x_1^2 + y_1^2) + \frac{1}{8}\nu_{11}^\circ(x_1^2 + y_1^2)^2. \tag{9.8}$$

In the continuation, we follow the same standard developments of Lie series theory in non-singular variables. We start by considering the canonical transformation $\phi_n : (x_1, y_1) \Rightarrow (x_1^*, y_1^*)$ defined by the equation

$$f(x_1, y_1) = E_{W^*} f(x_1^*, y_1^*), \tag{9.9}$$

where $E_{W^*} f$ is the Lie series expansion about the origin of the function $f(x_1^*, y_1^*)$, generated by

$$W^* = \sum_{k=3}^{n} W_k^*(x_1^*, y_1^*, \varepsilon). \tag{9.10}$$

The W_k^* are homogeneous functions of degree k with respect to the elements of $\mathcal{S}$. The Lie series expansion is the same as that shown in (7.26).

Since the transformation ϕ_n is conservative, we have

$$H(x_1, y_1) = H^*(x_1^*, y_1^*) + \mathcal{R}_n(x_1^*, y_1^*),$$

that is,

$$H^*(x_1^*, y_1^*) + \mathcal{R}_n(x_1^*, y_1^*) = E_{W^*} H(x_1^*, y_1^*). \tag{9.11}$$

In the sequence, we introduce, in these equations, the expansions already given for H and W^* as well as

$$H^* = \sum_{k \geq 4} H_k^*(x_1^*, y_1^*, \varepsilon), \tag{9.12}$$

where the $H_k^*(x_1^*, y_1^*, \varepsilon)$ are unknown homogeneous functions of degree k with respect to the elements of $\mathcal{S}$. ($H_2^* = H_3^* = 0$.) The identification of the terms having the same degree of homogeneity gives the perturbation equations

$$\begin{aligned}
H_4^* &= X_2 + X_4 + \varepsilon \mathcal{V}_1^1 & (&= H_{\text{lead}}^*) \\
H_5^* &= & \varepsilon \mathcal{V}_1^2 + \{H_4^*, W_3^*\} \\
H_6^* &= X_6 + & \varepsilon \mathcal{V}_1^3 + \tfrac{1}{2}\{H_5^* - \varepsilon \mathcal{V}_1^2, W_3^*\} + \{H_4^*, W_4^*\} \\
\cdots & & \cdots &
\end{aligned} \tag{9.13}$$

The homological equation is

$$\{H_4^*, W_{k-2}^*\} = H_k^* - \Psi_k(x_1^*, y_1^*) \tag{9.14}$$

and the corresponding Hori kernel is given by the differential equations

$$\frac{\mathrm{d}x_1^*}{\mathrm{d}u} = \frac{\partial H_4^*}{\partial y_1^*} \qquad \frac{\mathrm{d}y_1^*}{\mathrm{d}u} = -\frac{\partial H_4^*}{\partial x_1^*}. \tag{9.15}$$

We recall that the sign of these equations follows the rules stated in Sect. 7.2. In explicit form,

$$\begin{aligned}\frac{\mathrm{d}x_1^*}{\mathrm{d}u} &= -\nu_1^\circ y_1^* + \frac{1}{2}\nu_{11}^\circ y_1^*(x_1^{*2} + y_1^{*2}), \\ \frac{\mathrm{d}y_1^*}{\mathrm{d}u} &= +\nu_1^\circ x_1^* - \frac{1}{2}\nu_{11}^\circ x_1^*(x_1^{*2} + y_1^{*2}) - \varepsilon\tau_1,\end{aligned} \tag{9.16}$$

where, for simplicity, we assumed that $\mathcal{V}_1^1 = \tau_1 x_1$ ($\tau_1 > 0$).

Equations (9.16) form an autonomous differential system whose Hamiltonian is the first Andoyer Hamiltonian. Its integration is given in Appendix C. We are thus able to proceed and obtain H_k^* and W_{k-2}^* ($k \geq 5$) by means of

$$H_k^* = \, < \Psi_k(x_1^*, y_1^*) >$$

and

$$W_{k-2}^* = \int (\Psi_k - H_k^*) \, \mathrm{d}u;$$

however, as happened with the study of Garfinkel's Ideal Resonance Problem, in Chaps. 4 and 8, the integration involves elliptic functions and integrals and the actual calculations are rather complex as can be seen in the case study of Sect. 9.4.

9.3 Many Degrees of Freedom. One Single Resonance

Let us consider the general case of one single resonance in non-singular variables. Let us consider the Hamiltonian

$$H = H_0(x_1, y_1) + \sum_{k \geq 1} \varepsilon^k H_k(x_1, y_1, q, p), \tag{9.17}$$

where H_0 is assumed to depend on x_1, y_1 only through $J_1 = -\frac{1}{2}(x_1^2 + y_1^2)$. (Again, we assume $J_1 < 0$ to be close to actual Celestial Mechanics problems.)

We also assume that

$$\nu_1^\circ = \left.\frac{\mathrm{d}H_0}{\mathrm{d}J_1}\right|_{J_1=0} \tag{9.18}$$

is a small quantity. This means that the subsystem corresponding to the variables x_1, y_1 is resonant and that this singularity appears in the vicinity of the origin.

No hypothesis is made concerning the nature of the variables q_ϱ, p_ϱ ($\varrho \geq 2$). They may be non-singular, angle–action, or any other pairs of canonical variables. We just assume that they are finite quantities.

The given Hamiltonian may be expanded as follows:

$$H_0 = \sum_{k\geq 1} X_{2k} = -\frac{1}{2}\nu_1^\circ(x_1^2 + y_1^2) + \frac{1}{8}\nu_{11}^\circ(x_1^2 + y_1^2)^2 + \cdots \qquad (9.19)$$

and

$$H_k = \sum_{k'\geq 0} \mathcal{V}_k^{k'}(x_1, y_1, p, q), \qquad (9.20)$$

where $X_{k'}$ and $\mathcal{V}_k^{k'}$ are homogeneous functions of degree k' in x_1, y_1. In an rth-order resonance, the least non-zero value of k' in H_1 is r.

The leading terms of H that should be present in the Hori kernel are

$$H_{\text{lead}} = -\frac{1}{2}\nu_1^\circ(x_1^2 + y_1^2) + \frac{1}{8}\nu_{11}^\circ(x_1^2 + y_1^2)^2 + \varepsilon F_{1\text{lead}}. \qquad (9.21)$$

As discussed in Sect. 8.4, the Poisson bracket of two homogeneous functions in x_1, y_1 is not homogeneous in these variables since the terms arising from derivatives with respect to x_1, y_1 will have a loss of two units in the degree of homogeneity. Indeed, given two functions $\psi_1(x_1, y_1, q, p)$ and $\psi_2(x_1, y_1, q, p)$ homogeneous in x_1, y_1, their Poisson bracket is

$$\{\psi_1, \psi_2\} = \left(\frac{\partial \psi_1}{\partial x_1}\frac{\partial \psi_2}{\partial y_1} - \frac{\partial \psi_2}{\partial x_1}\frac{\partial \psi_1}{\partial y_1}\right) + \sum_{\varrho=2}^{N}\left(\frac{\partial \psi_1}{\partial q_\varrho}\frac{\partial \psi_2}{\partial p_\varrho} - \frac{\partial \psi_1}{\partial q_\varrho}\frac{\partial \psi_2}{\partial p_\varrho}\right). \qquad (9.22)$$

The second part of this Poisson bracket is an ordinary operation and the degree of homogeneity of the result is equal to the sum of the degrees of homogeneity of ψ_1 and ψ_2. However, in the first part of the bracket, the operations $\frac{\partial}{\partial x_1}$ and $\frac{\partial}{\partial y_1}$ subtract, each, one unit of the degree of homogeneity and the result is an homogeneous function with two degrees of homogeneity less than the rest of the terms. Therefore, as in Sect. 7.3, we introduce

$$\{\psi_1, \psi_2\}_1 = \frac{\partial \psi_1}{\partial x_1}\frac{\partial \psi_2}{\partial y_1} - \frac{\partial \psi_2}{\partial x_1}\frac{\partial \psi_1}{\partial y_1}, \qquad (9.23)$$

$$\{\psi_1, \psi_2\}_\varrho = \sum_{\varrho=2}^{N}\left(\frac{\partial \psi_1}{\partial q_\varrho}\frac{\partial \psi_2}{\partial p_\varrho} - \frac{\partial \psi_2}{\partial q_\varrho}\frac{\partial \psi_1}{\partial p_\varrho}\right) \qquad (9.24)$$

and write the Poisson bracket as

$$\{\psi_1, \psi_2\} = \{\psi_1, \psi_2\}_1 + \{\psi_1, \psi_2\}_\varrho. \qquad (9.25)$$

Equation (8.11) still holds but, now, the term with two brackets showing the subscripts 1 and ϱ has two degree of homogeneity less than the sum of the degrees of the ψ_k while the term where the subscript 1 appears twice has four degrees homogeneity less than the sum of the degrees of the ψ_k.

9.3 Many Degrees of Freedom. One Single Resonance

Taking into account these rules and proceeding along the same steps of the calculations done in Sect. 8.3 and 8.4, we obtain the Lie series of a function $f(x_1, y_1, q, p)$ homogeneous of degree L with respect to the elements of $\mathcal{S} \equiv (x_1, y_1, \varepsilon^d)$. The generating function is assumed to be

$$W^* = \sum_{k \geq 3} W_k^*(x_1^*, y_1^*, q^*, p^*; \varepsilon), \qquad (9.26)$$

where W_k^* are homogeneous functions of degree k in the elements of $\mathcal{S}$. As in other similar calculations, in order to avoid an unlimited number of terms at every order, we assume $W_1^* = W_2^* = 0$. Hence,

$$\begin{aligned}
E_{W^*} f = \, & f \\
& + \{f, W_3^*\}_1 \\
& + \{f, W_4^*\}_1 + \tfrac{1}{2}\{\{f, W_3^*\}_1, W_3^*\}_1 \\
& + \{f, W_5^*\}_1 + \{f, W_3^*\}_\varrho + \tfrac{1}{2}\{\{f, W_3^*\}_1, W_4^*\}_1 \\
& \quad + \tfrac{1}{2}\{\{f, W_4^*\}_1, W_3^*\}_1 + \tfrac{1}{6}\{\{\{f, W_3^*\}_1, W_3^*\}_1, W_3^*\}_1 \\
& + \{f, W_6^*\}_1 + \{f, W_4^*\}_\varrho + \tfrac{1}{2}\{\{f, W_3^*\}_\varrho, W_3^*\}_1 + \tfrac{1}{2}\{\{f, W_3^*\}_1, W_3^*\}_\varrho \\
& \quad + \tfrac{1}{2}\{\{f, W_3^*\}_1, W_5^*\}_1 + \tfrac{1}{2}\{\{f, W_5^*\}_1, W_3^*\}_1 + \cdots,
\end{aligned} \qquad (9.27)$$

where we have put terms of degree L in the first row, terms of degree $L+1$ in the second row, etc. At the given order, the only difference with respect to the expansion given by (7.26) is the subscript 1 in all brackets and additional terms like $\{f, W_3^*\}_\varrho$ in the order $L+2$ row and others brackets with subscript ϱ in higher orders. The order $L+2$ is enough for the practical applications presented in this chapter.

We now introduce the canonical transformation $\phi_n : (x_1, y_1, q, p) \Rightarrow (x_1^*, y_1^*, q^*, p^*)$ defined by the equation

$$f(x_1, y_1, q, p) = E_{W^*} f(x_1^*, y_1^*, q^*, p^*). \qquad (9.28)$$

Since the transformation is conservative, we have

$$H(x_1, y_1, q, p) = H^*(x_1^*, y_1^*, q^*, p^*) + \mathcal{R}_n(x_1^*, y_1^*, q^*, p^*), \qquad (9.29)$$

that is,

$$H^*(x_1^*, y_1^*, q^*, p^*) + \mathcal{R}_n(x_1^*, y_1^*, q^*, p^*) = E_{W^*} H(x_1^*, y_1^*, q^*, p^*). \qquad (9.30)$$

The perturbation equations are obtained by introducing in this equation the expansions already given for E_{W^*} and H, as well as

$$H^* = \sum_{k \geq 4} H_k^*(x_1^*, y_1^*, q^*, p^*; \varepsilon). \qquad (9.31)$$

They are

$$
\begin{aligned}
H_4^* &= X_2 + X_4 + \varepsilon \mathcal{V}_1^1 \\
H_5^* &= \varepsilon \mathcal{V}_1^2 + \{H_4^*, W_3^*\}_1 \\
H_6^* &= X_6 + \varepsilon \mathcal{V}_1^3 + \varepsilon^2 \mathcal{V}_2^0 + \{H_4^*, W_4^*\}_1 + \tfrac{1}{2}\{\{H_4^*, W_3^*\}_1, W_3^*\}_1 + \varepsilon\{\mathcal{V}_1^2, W_3^*\}_1 \\
H_7^* &= \varepsilon \mathcal{V}_1^4 + \varepsilon^2 \mathcal{V}_2^1 + \{H_4^*, W_5^*\}_1 + \{H_4^*, W_3^*\}_\varrho + \tfrac{1}{2}\{\{H_4^*, W_3^*\}_1, W_4^*\}_1 \\
&\quad + \tfrac{1}{2}\{\{H_4^*, W_4^*\}_1, W_3^*\}_1 + \tfrac{1}{6}\{\{\{H_4^*, W_3^*\}_1, W_3^*\}_1, W_3^*\}_1 \\
&\quad + \varepsilon\{\mathcal{V}_1^2, W_4^*\}_1 + \tfrac{\varepsilon}{2}\{\{\mathcal{V}_1^2, W_3^*\}_1, W_3^*\}_1 + \{X_6 + \varepsilon \mathcal{V}_1^3 + \varepsilon^2 \mathcal{V}_2^0, W_3^*\}_1 \\
&\cdots \quad \cdots .
\end{aligned}
$$
(9.32)

(Brackets with the subscript ϱ only appear in the equation for H_7^*.)

In this case, the homological equation is

$$\{H_4^*, W_{k-2}^*\}_1 = H_k^* - \Psi_k(x_1^*, y_1^*, q^*, p^*) \tag{9.33}$$

and the corresponding Hori kernel is formed by the differential equations

$$\frac{dx_1^*}{du} = \frac{\partial H_4^*}{\partial y_1^*} \qquad \frac{dy_1^*}{du} = -\frac{\partial H_4^*}{\partial x_1^*}. \tag{9.34}$$

On Examples and Case Studies

In previous chapters, we have always presented theories followed by examples. However, when motions near singularities of the angle–action variables are considered, it becomes more convenient to adopt a different approach. In this chapter, instead of an example, we present a case study where the given theory is not simply applied, but adapted. Indeed, to define a general resonant system including all features found in real problems would introduce many unnecessary complications. For instance, it may happen that some of the q, p are also small quantities, and other parts of the partial Poisson bracket denoted as $\{\ ,\ \}_\varrho$ may have a loss of degrees of homogeneity. In addition, it is not possible to make general assumptions on the order of magnitude of x_1, y_1 and ν_1° putting all terms in H_{lead} on the same footing in all cases. In the case of a first-order resonance, such as the one studied below, x_1, y_1 are assumed to be of order $\mathcal{O}(\sqrt[3]{\varepsilon})$. For a second-order resonance, we should assume x_1, y_1 of order $\mathcal{O}(\sqrt{\varepsilon})$ and, for a third-order resonance, we should assume x_1, y_1 of order $\mathcal{O}(\varepsilon)$.[2]

9.4 A First-Order Resonance Case Study

Let us reconsider the example studied in Sects. 7.8 and 8.6:

---

[2] These assumptions are justified by properties of the Andoyer Hamiltonians. See Appendices C and D.

9.4 A First-Order Resonance Case Study

$$H = -\frac{1}{2I_1^2} - 2I_1 + \varepsilon \left(a(I_1) + b' I_3 + B \cos \phi_1 + L \sqrt{-2I_3} \cos \phi_3 \right.$$
$$\left. + M \sqrt{-2I_3} \cos(\phi_1 + \phi_3) \right), \quad (9.35)$$

where we changed bI_3 into $b'I_3$ to avoid confusion with the b of the Andoyer Hamiltonian. We assume, here, that all coefficients are constants, except a (taken as $a = a_1 I_1$). In addition, we assume that M, b', a_1 are finite and that B, L are small (and of the same order). In the vicinity of the resonance of ϕ_1, the leading terms of the Hamiltonian are

$$H_{\text{lead}} = \frac{1}{2}\nu_{11}^\circ (I_1 - I_1^\circ)^2 + \varepsilon B \cos \phi_1 + \varepsilon M \sqrt{-2I_3} \cos(\phi_1 + \phi_3), \quad (9.36)$$

where

$$I_1^\circ = 2^{-1/3} \qquad \nu_{11}^\circ = -6\sqrt[3]{2}.$$

The resonance in ϕ_1 is of first order and to have all terms on an equal footing, we assume $B = \mathcal{O}(\sqrt{-I_3})$, $\xi = I_1 - I_1^\circ = \mathcal{O}(\varepsilon^{2/3})$ and $I_3 = \mathcal{O}(\varepsilon^{2/3})$. With these assumptions, the Poisson bracket of two homogeneous functions in ξ, I_3 is homogeneous, and we do not have to split it in the way discussed in the previous section. The Lie series expansion of a function of $(\phi_1, \phi_3, \xi, I_3)$ is obtained from (9.27) by deleting the brackets with subscript ϱ, changing the brackets with subscript 1 into true Poisson brackets, and assuming that the terms are grouped according to their degrees of homogeneity in the elements of $\mathcal{S} \equiv (\sqrt{\xi}, \sqrt{-I_3}, \varepsilon^{1/3})$.

A different approach was adopted by Message [70] in the study of this problem and its extension to higher-order resonances where the first approximation solutions were obtained assuming that $\mathcal{S} \equiv (\sqrt{\xi}, \sqrt{-I_3}, \sqrt{\varepsilon})$.

The integrability of H_{lead}, necessary for the application of Hori theory, is trivially proved when we use the Sessin transformation (see Sect. 9.5.1)[3] to introduce the new set of canonical variables

$$X = x_3 + \beta \cos \phi_1$$
$$Y = y_3 + \beta \sin \phi_1 \qquad (9.37)$$
$$\vartheta_2 = \phi_1$$
$$\mathcal{G} = \xi - I_3 + \beta(x_3 \cos \phi_1 + y_3 \sin \phi_1) + \frac{1}{2}\beta^2,$$

where

$$x_3 = \sqrt{-2I_3} \cos(\phi_1 + \phi_3), \qquad y_3 = \sqrt{-2I_3} \sin(\phi_1 + \phi_3)$$

and $\beta = B/M$ is a constant. The definition of β is such that the two periodic terms of H_{lead} are merged into only one term. With these new variables, we have

---

[3] In addition, the (canonical) analogy $w_1 = \phi_1 + \phi_3$, $w_2 = \phi_1$, $J_1 = I_3 < 0$, $J_2 = \xi - I_3$ was used.

$$H_{\text{lead}} = \frac{1}{2}\nu_{11}^\circ \left(\mathcal{G} - \frac{1}{2}(X^2 + Y^2)\right)^2 + \varepsilon MX. \tag{9.38}$$

The angle ϑ_2 is absent from the transformed Hamiltonian. Therefore, $\mathcal{G} = $ const is a first integral of H_{lead} (Sessin integral) and this Hamiltonian is integrable.

We may, also, introduce the angle–action variables corresponding to X, Y:

$$J = \frac{1}{2}(X^2 + Y^2) \qquad \vartheta_1 = \arctan \frac{Y}{X} \tag{9.39}$$

and their inverses

$$\begin{aligned} X &= \sqrt{2J}\cos\vartheta_1 \\ Y &= \sqrt{2J}\sin\vartheta_1. \end{aligned} \tag{9.40}$$

The invariance of Poisson brackets to canonical transformations allows us to indifferently use ϑ_1, J or Y, X (in this order because $J > 0$) as the first pair of variables. We may use, in each operation, those variables appearing to be the most convenient.

9.4.1 The Hori Kernel

The Hori kernel is

$$H_4^* = H_{\text{lead}}^*(\phi^*, I^*) = \frac{1}{2}\nu_{11}^\circ \left(\mathcal{G}^* - \frac{1}{2}(X^{*2} + Y^{*2})\right)^2 + \varepsilon MX^*. \tag{9.41}$$

This Hamiltonian is the first Andoyer Hamiltonian. To use the solutions given in Appendix C, it is convenient to convert it into the standard $b > 0, \tau > 0$ case. To do this, we initially consider, instead of H_4^*, the Hamiltonian

$$\mathcal{F}_4 = \frac{1}{2}\nu_{11}^\circ \mathcal{G}^{*2} - H_4^* = \frac{1}{2}\nu_{11}^\circ \mathcal{G}^*(X^{*2} + Y^{*2}) - \frac{1}{8}\nu_{11}^\circ (X^{*2} + Y^{*2})^2 - \varepsilon MX^*.$$

Comparing to (C.12), the coefficients of this Andoyer Hamiltonian are

$$\begin{aligned} a &= \nu_{11}^\circ \mathcal{G}^* \\ b &= -\frac{1}{2}\nu_{11}^\circ \\ \tau &= -M. \end{aligned}$$

The coefficient of $(X^{*2} + Y^{*2})^2$ is, now, positive, since $\nu_{11}^\circ < 0$. In order to have the same behavior shown in Fig. C.3, we assume $M < 0$ (that is, $\tau > 0$; otherwise, a trivial angle transformation should be made beforehand).

Limiting ourselves to the case of small oscillations about the stable equilibrium, the solution of H_4^* is (see Sect. C.9)

$$J^* = J_c + \frac{\varepsilon\tau h_0}{w_0}\gamma\cos w + \mathcal{O}(\gamma^2) \tag{9.42}$$

$$\vartheta_1^* = \pi + \gamma\sin w + \mathcal{O}(\gamma^2), \tag{9.43}$$

where

$$h_0 = \sqrt{2\mathcal{J}_c} \qquad (9.44)$$

$$\omega_0 = \sqrt{\frac{\varepsilon\tau(a+3bh_0^2)}{h_0}} + \mathcal{O}(\gamma^2) \qquad (9.45)$$

$$w = \omega_0(u - u_0). \qquad (9.46)$$

$\mathcal{J}_c$ is the center of the oscillation and $\gamma > 0$ is a constant of the order of the oscillation amplitude of ϑ_1. $\mathcal{J}^*, \vartheta_1^*$ are the transformed $\mathcal{J}, \vartheta_1$. u is the independent variable of the Hori kernel equations and u_0 is chosen such that when $w = 0$, we have $\vartheta_1 = \pi$ and $\mathcal{J} = \mathcal{J}_0$ is maximum. It is worth emphasizing that the opposite directions of the motion in the solutions of the Hamiltonians H_4^* and $\mathcal{F}_4$ have already been taken into account through a transformation of w into $-w$ in the solutions of the Andoyer Hamiltonian.

To complete the solution of the Hori kernel, we still need to integrate the last pair of canonical equations:

$$\frac{d\vartheta_2^*}{du} = \frac{\partial H_4^*}{\partial \mathcal{G}^*} = \nu_{11}^\circ(\mathcal{G}^* - \mathcal{J}^*)$$

$$\frac{d\mathcal{G}^*}{du} = 0$$

whose solutions are $\mathcal{G}^* = \mathcal{G}_0$ (const) and

$$\vartheta_2^* = \chi_0 - \mathcal{P}_1 \gamma \sin w + \mathcal{O}(\gamma^2), \qquad (9.47)$$

where

$$\chi_0 = \vartheta_{20} + \nu_{11}^\circ(\mathcal{G}_0 - \mathcal{J}_c)u \qquad (9.48)$$

is a uniformly varying angle and

$$\mathcal{P}_1 = \frac{\varepsilon\tau h_0 \nu_{11}^\circ}{\omega_0^2}. \qquad (9.49)$$

It is worth noting that, because of the integration with respect to w, the order of every term decreases and $\mathcal{P}_1$ is finite[4]. This characteristic will repeat itself in the next terms of the series giving ϑ_2^* and the numerical convergence may arise only from the decreasing value of the powers of γ.

9.4.2 First Perturbation Equation

The first perturbation equation is

$$\{H_4^*, W_3^*\} = H_5^* - \varepsilon \mathcal{V}_1^2, \qquad (9.50)$$

---

[4] In the control of the orders recall that $h_0 = \mathcal{O}(\varepsilon^{1/3})$ and $\omega_0 = \mathcal{O}(\varepsilon^{2/3})$.

where
$$\mathcal{V}_1^2(\phi, I) = a_1\xi + b'I_3 + L(x_3\cos\phi_1 + y_3\sin\phi_1)$$

(the same $\mathcal{V}_1^2$ of Sect. 7.8). Let us write $\mathcal{V}_1^2$ with the new variables $\vartheta, \mathcal{J}, \mathcal{G}$. To do this, it is useful to have the inverses

$$\begin{aligned}\xi &= \mathcal{G} - \mathcal{J} \\ I_3 &= -\mathcal{J} + \beta\sqrt{2\mathcal{J}}\cos(\vartheta_2 - \vartheta_1) - \tfrac{1}{2}\beta^2.\end{aligned} \quad (9.51)$$

We also have

$$x_3\cos\phi_1 + y_3\sin\phi_1 = \sqrt{2\mathcal{J}}\cos(\vartheta_2 - \vartheta_1) - \beta.$$

Hence, at the point (ϕ^*, I^*), with the new variables, we have

$$\mathcal{V}_1^2 = a_1\mathcal{G}^* - (a_1 + b')\mathcal{J}^* + (L + b'\beta)\sqrt{2\mathcal{J}^*}\cos(\vartheta_2^* - \vartheta_1^*) - L\beta - \frac{1}{2}b'\beta^2.$$

9.4.3 Averaging

The next step in the application of Hori theory is the averaging of $\mathcal{V}_1^2$ to determine H_5^* in such a way that the average of the right-hand side of (9.50) is zero. That is,

$$H_5^* = \varepsilon < \mathcal{V}_1^2 > = \lim_{\hat{u}\to\infty}\frac{1}{\hat{u}}\int_0^{\hat{u}}\mathcal{V}_1^2(\vartheta^*, \mathcal{J}^*, \mathcal{G}^*)du. \quad (9.52)$$

The only cumbersome term in the integrand is the periodic one. It may be expanded as a trigonometric series in w with coefficients expanded, themselves, in powers of γ. After some manipulation, we obtain (at the considered order)

$$\begin{aligned}\sqrt{2\mathcal{J}^*}\cos(\vartheta_2^* - \vartheta_1^*) &= h_0\cos\chi_0 + \frac{\varepsilon\tau}{\omega_0}\gamma\cos\chi_0\cos w \\ &\quad + h_0(1 + \mathcal{P}_1)\gamma\sin\chi_0\sin w.\end{aligned} \quad (9.53)$$

Hence,

$$H_5^* = \varepsilon < \mathcal{V}_1^2 > = \varepsilon a_1\mathcal{G}_0 - \varepsilon(a_1 + b')\mathcal{J}_c - \varepsilon L\beta - \frac{1}{2}\varepsilon b'\beta^2. \quad (9.54)$$

The solution of the first perturbation equation is completed with the integration giving W_3^*:

$$W_3^* = \int(\varepsilon\mathcal{V}_1^2 - H_5^*)du,$$

that is,

$$\begin{aligned}W_3^* &= -\varepsilon(a_1 + b')\frac{\varepsilon\tau h_0}{\omega_0^2}\gamma\sin w \\ &\quad -\varepsilon(L + b'\beta)\left(\frac{h_0\sin\chi_0}{\nu_{11}^\circ(\mathcal{G}_0 - \mathcal{J}_c)} + D_1\gamma\sin(\chi_0 + w) + D_2\gamma\sin(\chi_0 - w)\right),\end{aligned} \quad (9.55)$$

where

$$D_1 = \frac{1}{2w_0}\frac{\varepsilon\tau - w_0 h_0(1+\mathcal{P}_1)}{\nu_{11}^\circ(\mathcal{G}_0 - \mathcal{J}_c) + w_0} \qquad D_2 = \frac{1}{2w_0}\frac{\varepsilon\tau + w_0 h_0(1+\mathcal{P}_1)}{\nu_{11}^\circ(\mathcal{G}_0 - \mathcal{J}_c) - w_0}.$$

The transformation $\vartheta, \mathcal{J}, \mathcal{G} \Rightarrow \vartheta^*, \mathcal{J}^*, \mathcal{G}^*$ is formally defined by the Lie series $f(\vartheta, \mathcal{J}, \mathcal{G}) = E_{W^*} f(\vartheta^*, \mathcal{J}^*, \mathcal{G}^*)$; at the first post-identity approximation we have $f(\vartheta, \mathcal{J}, \mathcal{G}) = f^* + \{f^*, W_3^*\}$, where, for simplicity, we have written $f^* = f(\vartheta^*, \mathcal{J}^*, \mathcal{G}^*)$.

The problem is, now, that we have W_3^* given as a function of u and some integration constants and we need to know $W_3^*(\vartheta^*, \mathcal{J}^*, \mathcal{G}^*)$ to calculate the Poisson brackets appearing in the Lie transformations. This is not, in general, an easy task, as it involves the construction of the inverse transformation $w_1, \chi_0, \gamma, \mathcal{G}_0 \Rightarrow \vartheta^*, \mathcal{J}^*, \mathcal{G}^*$.

The inversion of the variables, giving $\vartheta_1^*, \mathcal{J}^*$, is, in general, cumbersome. However, when only the harmonic approximation of the solutions of the Hori kernel is used, they are easily obtained:

$$\gamma \cos w = \frac{w_0}{\varepsilon\tau h_0}(\mathcal{J}^* - \mathcal{J}_c)$$
$$\gamma \sin w = -\sin \vartheta_1^*. \tag{9.56}$$

The inversion of the second pair of equations is trivial. In particular, we have

$$\sin \chi_0 = \sin \vartheta_2^* - \mathcal{P}_1 \sin \vartheta_1^* \cos \vartheta_2^* + \mathcal{O}(\gamma^2)$$
$$\cos \chi_0 = \cos \vartheta_2^* + \mathcal{P}_1 \sin \vartheta_1^* \sin \vartheta_2^* + \mathcal{O}(\gamma^2). \tag{9.57}$$

We note that in the terms factored by coefficients of order $\mathcal{O}(\gamma)$, the approximations $\sin \chi_0 = \sin \vartheta_2^*$ and $\cos \chi_0 = \cos \vartheta_2^*$ are enough. We thus obtain

$$W_3^* = \varepsilon(a_1 + b')\frac{\varepsilon\tau h_0}{w_0^2}\sin\vartheta_1^* - \frac{\varepsilon(L + b'\beta)h_0}{\nu_{11}^\circ(\mathcal{G}^* - \mathcal{J}_c)}\sin\vartheta_2^*$$
$$- \varepsilon(L + b'\beta)(D_1 + D_2)\frac{w_0}{\varepsilon\tau h_0}(\mathcal{J}^* - \mathcal{J}_c)\sin\vartheta_2^*$$
$$+ \varepsilon(L + b'\beta)\left(D_1 - D_2 + \frac{h_0 \mathcal{P}_1}{\nu_{11}^\circ(\mathcal{G}^* - \mathcal{J}_c)}\right)\sin\vartheta_1^* \cos\vartheta_2^*.$$

9.4.4 The Post-Harmonic Solution

To complete this example, it is necessary to consider the actual construction of the solutions. At this point, we have to consider one particularity of our example. Since $\mathcal{J} > 0$, we have $\{\vartheta_1, \mathcal{J}\} = -1$. This means that in the Poisson brackets with respect to the variables $\vartheta^*, \mathcal{J}^*, \mathcal{G}^*$, the derivatives concerning the first pair of conjugate canonical variables must be taken in the order $\partial/\partial\mathcal{J}^*, \partial/\partial\vartheta_1^*$.

The post-harmonic solution is, then, given by

$$\dot{\vartheta}_1 = E_{W^*}\vartheta_1^* \simeq \vartheta_1^* + \{\vartheta_1^*, W_3^*\} = \vartheta_1^* - \frac{\partial W_3^*}{\partial \mathcal{J}^*}$$

$$\dot{\vartheta}_2 = E_{W^*}\vartheta_2^* \simeq \vartheta_2^* + \{\vartheta_2^*, W_3^*\} = \vartheta_2^* + \frac{\partial W_3^*}{\partial \mathcal{G}^*}$$

$$\mathcal{J} = E_{W^*}\mathcal{J}^* \simeq \mathcal{J}^* + \{\mathcal{J}^*, W_3^*\} = \mathcal{J}^* + \frac{\partial W_3^*}{\partial \vartheta_1^*}$$

$$\mathcal{G} = E_{W^*}\mathcal{G}^* \simeq \mathcal{G}^* + \{\mathcal{G}^*, W_3^*\} = \mathcal{G}^* - \frac{\partial W_3^*}{\partial \vartheta_2^*}.$$

The derivatives are easy to calculate yielding

$$\vartheta_1 = \vartheta_1^* + \varepsilon(L + b'\beta)(D_1 + D_2)\frac{\omega_0}{\varepsilon\tau h_0}\sin\vartheta_2^*,$$

$$\vartheta_2 = \vartheta_2^* + \frac{\varepsilon(L+b'\beta)h_0}{\nu_{11}^\circ(\mathcal{G}^* - \mathcal{J}_c)^2}\sin\vartheta_2^*$$
$$- \varepsilon(L+b'\beta)\frac{\partial(D_1+D_2)}{\partial\mathcal{G}^*}\frac{\omega_0}{\varepsilon\tau h_0}(\mathcal{J}^* - \mathcal{J}_c)\sin\vartheta_2^*$$
$$+ \varepsilon(L+b'\beta)\left(\frac{\partial(D_1-D_2)}{\partial\mathcal{G}^*} - \frac{h_0\mathcal{P}_1}{\nu_{11}^\circ(\mathcal{G}^* - \mathcal{J}_c)^2}\right)\sin\vartheta_1^*\cos\vartheta_2^*.$$

Similarly, for the actions, we have

$$\mathcal{J} = \mathcal{J}^* + \varepsilon(a_1 + b')\frac{\varepsilon\tau h_0}{\omega_0^2}\cos\vartheta_1^*$$
$$+ \varepsilon(L+b'\beta)\left(D_1 - D_2 + \frac{h_0\mathcal{P}_1}{\nu_{11}^\circ(\mathcal{G}^* - \mathcal{J}_c)}\right)\cos\vartheta_1^*\cos\vartheta_2^*,$$

$$\mathcal{G} = \mathcal{G}^* + \frac{\varepsilon(L+b'\beta)h_0}{\nu_{11}^\circ(\mathcal{G}^* - \mathcal{J}_c)}\cos\vartheta_2^*$$
$$+ \varepsilon(L+b'\beta)(D_1+D_2)\frac{\omega_0}{\varepsilon\tau h_0}(\mathcal{J}^* - \mathcal{J}_c)\cos\vartheta_2^*$$
$$+ \varepsilon(L+b'\beta)\left(D_1 - D_2 + \frac{h_0\mathcal{P}_1}{\nu_{11}^\circ(\mathcal{G}^* - \mathcal{J}_c)}\right)\sin\vartheta_1^*\sin\vartheta_2^*.$$

In the last set of equations, $\vartheta_1^*, \vartheta_2^*, \mathcal{J}^*, \mathcal{G}^*$ are the solutions of the averaged Hamiltonian

$$H_4^* + H_5^* = \frac{1}{2}\nu_{11}^\circ(\mathcal{G}^* - \mathcal{J}^*)^2 + \varepsilon M\sqrt{2\mathcal{J}^*}\cos\vartheta_1^* + \varepsilon a_1\mathcal{G}^*, \qquad (9.58)$$

where we discarded the constant terms of H_5^*. These solutions are the same as those of the Hori kernel, with a small modification due to the term $\varepsilon a_1\mathcal{G}^*$. $\mathcal{J}^*$ and ϑ_1^* are the same functions given by (9.42) and (9.43), just replacing, there, $w(u)$ by

$$\widehat{w} = \omega_0(t - t_0). \qquad (9.59)$$

In addition, $\mathcal{G}^* = \text{const}$ and

$$\vartheta_2^* = \vartheta_{20} + \nu_{11}^\circ(\mathcal{G}^* - \mathcal{J}_c)t + \varepsilon a_1 t - \mathcal{P}_1 \gamma \sin \widehat{w}.$$

(The only modification with respect to (9.47) is a correction in the progressive part of this angle.)

From a strict point of view, the canonical transformation $\mathcal{J}, \vartheta, \mathcal{G} \Rightarrow \mathcal{J}^*, \vartheta^*, \mathcal{G}^*$ is not a solution of Bohlin's problem in its original form, because the critical angle ϑ_1^* was not eliminated from the Hamiltonian. However, in terms of the angle–action variables $\widehat{w}$ and

$$\Lambda = \frac{\omega_0 h_0^2 \gamma^2}{4\pi(a + 3bh_0^2)} + \mathcal{O}(\gamma^4), \tag{9.60}$$

(see C.74) we have, at the considered order of approximation,

$$H_{(5)}^*(\widehat{w}, \vartheta_2^*, \Lambda, \mathcal{G}^*) = -\mathcal{H}_4(\Lambda) + \varepsilon a_1 \mathcal{G}^*, \tag{9.61}$$

where $\mathcal{H}_4(\Lambda)$ is a series in Λ whose leading terms are given by (C.77)[5]. It is worth noting that the transformation $\mathcal{J}^*, \theta^*, \mathcal{G}^* \to \Lambda, \widehat{w}, \theta_2^*, \mathcal{G}^*$ is canonical because the transformation $\mathcal{J}^*, \theta_1^* \to \Lambda, \widehat{w}$ is independent of θ_2^* and $\mathcal{G}^*$.

Therefore, when the problem is considered with the variables $\Lambda, \widehat{w}$ instead of $\mathcal{J}^*, \vartheta_1^*$, the Bohlin's problem is solved, provided only that the non-resonance condition is satisfied.

9.4.5 Secular Resonance

In the previous sections, the averaging followed the general Hori-theory averaging rule: $H_5^* = \varepsilon < \mathcal{V}_1^2 >$ and all periodic terms of $\varepsilon \mathcal{V}_1^2$ were integrated in u and included in W_3^*. The non-resonance condition, in this case, is that the frequencies of the combinations of the angles χ_0 and w present in $\mathcal{V}_1^2$ are not small.

One important case, in which the non-resonance condition is not satisfied, happens when $\mathcal{G}_0 - \mathcal{J}_c$ is small and the angle ϑ_2^* is critical. In this case, the adoption of the averaging rule of Sect. 9.4.3 leads to a small divisor. It is worth emphasizing that $\mathcal{G}_0$ and $\mathcal{J}_c$ are quantities of order $\mathcal{O}(\varepsilon^{2/3})$, but this smallness was already considered in the perturbation equations where a perturbation of order $\mathcal{O}(\varepsilon^{5/3})$ gives a generating function of order $\mathcal{O}(\varepsilon)$. The smallness to which we refer now arises from the fact that their values become close to one another, making their difference yet smaller than themselves.

When this happens, we need to use the von Zeipel averaging rule: we split $\mathcal{V}_1^2$ into its secular, long-period and short-period parts (considering ϑ_2 as a slow angle and w as a fast one)[6], and average only on the fast angle:

---

[5] The $-$ sign in front of $\mathcal{H}_4(\Lambda)$ comes from the fact that $\{\widehat{w}, \Lambda\} = \{\vartheta_1, \mathcal{J}\} = -1$ and that we restored the angle-action order of the variables in $H_{(5)}^*$.

[6] We need to have in mind that χ_0 is just an auxiliary angle and not one of the variables of the problem.

$$H_5^* = \frac{1}{2\pi} \int_0^{2\pi} \varepsilon \mathcal{V}_1^2 dw. \qquad (9.62)$$
$$(\vartheta_2 = \text{const})$$

The periodic term of the integrand may be written, in this case,

$$\sqrt{2\mathcal{J}^*}\cos(\vartheta_2^* - \vartheta_1^*) = -h_0 \cos\vartheta_2^* - \frac{\varepsilon\tau}{\omega_0}\gamma\cos\vartheta_2^*\cos w - h_0\gamma\sin\vartheta_2^*\sin w,$$

hence

$$H_5^* = \varepsilon a_1 \mathcal{G}_0 - \varepsilon(a_1 + b')\mathcal{J}_c - \varepsilon L\beta - \frac{1}{2}\varepsilon b'\beta^2 - \varepsilon(L + b'\beta)h_0 \cos\vartheta_2^*. \qquad (9.63)$$

The integration of $(\varepsilon \mathcal{V}_1^2 - H_5^*)$ gives, now,

$$W_3^* = -\varepsilon(a_1 + b')\frac{\varepsilon\tau h_0}{\omega_0^2}\gamma\sin w \qquad (9.64)$$
$$-\varepsilon(L + b'\beta)\frac{\gamma}{\omega_0^2}\left(\varepsilon\tau\cos\vartheta_2^*\sin w - \omega_0 h_0 \sin\vartheta_2^*\cos w\right).$$

The continuation is done exactly as before, and the results are similar. The most important difference appears in the equations giving the time variation of the averaged variables $\vartheta_1^*, \vartheta_2^*, \mathcal{J}^*, \mathcal{G}^*$, which are now solutions of the Hamiltonian

$$H_4^* + H_5^* = \frac{1}{2}\nu_{11}^\circ(\mathcal{G}^* - \mathcal{J}^*)^2 + \varepsilon M\sqrt{2\mathcal{J}^*}\cos\vartheta_1^* + \varepsilon a_1 \mathcal{G}^* - \varepsilon(L + b'\beta)h_0 \cos\vartheta_2^* \qquad (9.65)$$

or, introducing the variables $\widehat{w}, \Lambda$ (in which Bohlin's problem appears as solved):

$$H_{(5)}^* = -\mathcal{H}_4(\Lambda) + \varepsilon a_1 \mathcal{G}^* - \varepsilon(L + b'\beta)h_0 \cos\vartheta_2^*. \qquad (9.66)$$

9.4.6 Secondary Resonances

Another instance in which the non-resonance condition is not satisfied, appears when one of the divisors $\nu_{11}^\circ(\mathcal{G}_0 - \mathcal{J}_c) \pm \omega_0$ is small and the corresponding angles $\vartheta_2^* \pm w$ becomes critical; that is, a resonance occurs between the proper angle of the Andoyer Hamiltonian, w, and the slow angle ϑ_2^*. (We recall that ω_0 is also a quantity of order $\mathcal{O}(\varepsilon^{2/3})$.) In this case, we shall follow a von Zeipel averaging rule exactly as in the previous section. Once more, we divide $\mathcal{V}_1^2$ into its secular, long-period and short-period parts, but now the slow angle is the critical combination of ϑ_2^* and w, and the average is done over all angles which do not reduce themselves to a multiple of the critical one. Let us consider, for instance, that the critical angle is $\vartheta_2^* - w$. In this case, the averaged Hamiltonian is

$$H_5^* = \frac{1}{2\pi}\int_0^{2\pi} \varepsilon \mathcal{V}_1^2 dw, \qquad (9.67)$$
$$(\vartheta_2 - w = \text{const})$$

that is,

$$H_5^* = \varepsilon a_1 \mathcal{G}_0 - \varepsilon(a_1+b')\mathcal{J}_c - \varepsilon L\beta - \frac{1}{2}\varepsilon b'\beta^2 - \varepsilon(L+b'\beta)\frac{\varepsilon\tau + \omega_0 h_0}{2\omega_0}\gamma\cos(\vartheta_2^* - w). \tag{9.68}$$

The integration of $(\varepsilon \mathcal{V}_1^2 - H_5^*)$ gives[7]

$$W_3^* = -\varepsilon(a_1+b')\frac{\varepsilon\tau h_0}{\omega_0^2}\gamma\sin w - \varepsilon(L+b'\beta)\frac{h_0}{\omega_0}\sin\vartheta_2^*$$
$$+\varepsilon(L+b'\beta)\frac{\varepsilon\tau - h_0\omega_0}{4\omega_0^2}\gamma\sin(\vartheta_2^* + w).$$

The continuation is, again, done exactly as before, and the results are similar. The averaged variables $\vartheta_1^*, \vartheta_2^*, \mathcal{J}^*, \mathcal{G}^*$ are, now, solutions of the Hamiltonian

$$H_4^* + H_5^* = \frac{1}{2}\nu_{11}^\circ(\mathcal{G}^* - \mathcal{J}^*)^2 + \varepsilon M\sqrt{2\mathcal{J}^*}\cos\vartheta_1^* + \varepsilon a_1 \mathcal{G}^*$$
$$+\varepsilon(L+b'\beta)\frac{\varepsilon\tau + h_0\omega_0}{2\omega_0}\left(\frac{\omega_0}{\varepsilon\tau h_0}(\mathcal{J}^* - \mathcal{J}_c)\cos\vartheta_2^* - \sin\vartheta_1^*\sin\vartheta_2^*\right),$$

where (9.56) was used to obtain the last term. In terms of $\widehat{w}, \Lambda$, we have

$$H_{(5)}^* = -\mathcal{H}_4(\Lambda) + \varepsilon a_1 \mathcal{G}^* - \varepsilon(L+b'\beta)\frac{\varepsilon\tau + h_0\omega_0}{2\omega_0}\gamma\cos(\vartheta_2^* - w). \tag{9.69}$$

The problem is, thus, reduced to one degree of freedom (only the critical angle of the secondary resonance remains in $H_{(5)}^*$). An identical procedure may be followed when the angle is $\vartheta_2^* + w$ or another combination of ϑ_2^* and w present in a more complete representation of $\mathcal{V}_1^2$ becomes critical.

9.4.7 Initial Conditions Diagram

Let us draw the main features of the studied example on a diagram. This diagram is called the "initial conditions diagram" since the axes are the initial conditions $\mathcal{G}_0$ and $\mathcal{J}_0$, where $\mathcal{J}_0$ is the value of $\mathcal{J}^*$ at $w = 0$. In fact, to have a figure with the same aspect of those found in applications to actual problems, the x-axis in Fig. 9.3 represents the difference $\mathcal{G}_0 - \mathcal{J}_0$ instead of just $\mathcal{G}_0$.

To avoid having to give numbers to the coefficients of the given problem, we adopt $A_1^2 = (4\varepsilon\tau/b)^{2/3}$ as unit for both $\mathcal{G}_0$ and $\mathcal{J}_0$ (see Sect. C.2.1). With these units, $\mathcal{J} = \xi^2/2$ and $\mathcal{G} = 3\alpha/8$ (where ξ and α are the parameters defined in Sect. C.2.1).

---

[7] In the integration in w, it is necessary to use as independent angles $w, \phi = \vartheta_2^* - w$ instead of w, ϑ_2^*.

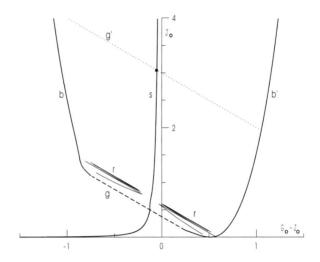

Fig. 9.3. Initial conditions diagram

Law of Structure

Curve s in Fig. 9.3 is the locus of the center of libration $\mathcal{J}_c$ calculated with (C.19). We also plot its continuation for $\alpha < 1$ calculated with (C.20) and (C.21).

Libration Limits

Curves b and b' in Fig. 9.3 show the limits of the libration zone given by ξ_2 and ξ_3 of Sect. C.4. The line g is the lower limit of the libration zone, which corresponds to the catastrophe value $\alpha = 1$ of the Andoyer Hamiltonian. In this case, $\mathcal{G}_0 = 3/8$ and, given our choice of the abcissas, g is a straight line inclined by 45 degrees (when equal scales are used for both axes).

Secondary Resonance

The secondary resonances occur for $\mathcal{G}_0 - \mathcal{J}_c \pm \omega_0 = 0$. In the case of the Hori kernel, the calculation shows that, because of the smallness of $\mathcal{G}_0 - \mathcal{J}_c$, these resonances do not occur in this problem. However, when the initial conditions diagram is drawn to interpret results of numerical experiments, we have to look for the critical lines of the full averaged Hamiltonian and not only those of the Hori kernel. We have, thus, to consider the contributions coming from εa_1 to the motion of the angle ϑ_2. When a_1 is considered, $\mathcal{G}_0 - \mathcal{J}_c + \varepsilon a_1$ is no longer a small quantity and the locus of the resonance may occur inside the libration zone. The curves r in Fig. 9.3 show the locus of the initial conditions for which the angles $\vartheta_2 \pm w$ become critical.

Secular Resonance

The secular resonance occurs for $\mathcal{G}_0 - \mathcal{J}_c = 0$. The value of this difference is given by the law of structure showing that, in this problem, it is never equal to zero, but always small. When the full averaged Hamiltonian is considered, the motion of the angle ϑ_2^* becomes equal to $\mathcal{G}_0 - \mathcal{J}_c + \varepsilon a_1$. If $a_1 > 0$, it may happen that the line $\mathcal{G}_0 = -\varepsilon a_1$ crosses the law of structure at some point. Let it be the point shown by a dot in Fig. 9.3. In this case, the angle ϑ_2 becomes critical over all solutions on the line g' of the figure.

The initial conditions diagram is an important tool for understanding the results of numerical experiments. In that case, we have also to fix the initial value of the angles. In the given example we could take, for instance, $\vartheta_{10} = \pi, \vartheta_{20} = 0$; we discard, for interpretation purposes only, the differences between these angles and the averaged ones and consider $\vartheta_{10}^* = \pi, \vartheta_{20}^* = 0$, instead of the actual given values. The condition $\vartheta_{10}^* = \pi$ is satisfied by both $w = 0$ and $w = \pi$. Then, each oscillation will be represented in the initial conditions diagram by two points: one when $\mathcal{J}^*$ is a maximum and another when it is a minimum (the two borders of the oscillation). Since $\mathcal{G}^*$ is a constant, these two points may be over a line with the same inclination as g and g', one on each side of the curve giving the law of structure.

9.5 Sessin Transformation and Integral

The integrability of the Hamiltonian

$$H = A(C + J_1 + J_2)^2 + D_1\sqrt{2J_1}\cos w_1 + D_2\sqrt{2J_2}\cos w_2, \qquad (9.70)$$

where $w \in \mathbf{T}^2$ and $J \in \mathbf{R}_+^2$ are angle–action variables and $A, C, D_j \in \mathbf{R}$ are constants, has been proved by Sessin [86], [87] through the introduction of a new set of variables.

The Sessin transformation becomes trivial if we introduce, beforehand, the non-singular Poincaré variables associated with w, J:

$$x_i = \sqrt{2J_i}\cos w_i$$
$$y_i = \sqrt{2J_i}\sin w_i. \qquad (9.71)$$

In non-singular variables, the Hamiltonian becomes

$$H = AQ^2 + D_1 x_1 + D_2 x_2, \qquad (9.72)$$

where

$$Q = C + \frac{1}{2}(x_1^2 + y_1^2 + x_2^2 + y_2^2). \qquad (9.73)$$

The surfaces $H = \text{const}$ are fourth-degree four-dimensional surfaces that are reduced to a 4-sphere when $D_1 = D_2 = 0$. The corresponding differential equations are

$$\frac{dx_i}{dt} = -2AQy_i$$
$$\frac{dy_i}{dt} = 2AQx_i + D_i. \tag{9.74}$$

Sessin has shown that the canonical transformation[8]

$$X_1 = \frac{D_1}{D}x_1 + \frac{D_2}{D}x_2 \qquad X_2 = -\frac{D_2}{D}x_1 + \frac{D_1}{D}x_2$$
$$Y_1 = \frac{D_1}{D}y_1 + \frac{D_2}{D}y_2 \qquad Y_2 = -\frac{D_2}{D}y_1 + \frac{D_1}{D}y_2, \tag{9.75}$$

where $D = \sqrt{D_1^2 + D_2^2}$, reduces the Hamiltonian to

$$H = A\widehat{Q}^2 + D_1 X_1, \tag{9.76}$$

where

$$\widehat{Q} = C + \frac{1}{2}(X_1^2 + Y_1^2 + X_2^2 + Y_2^2), \tag{9.77}$$

which may be easily integrated. Indeed, in this case, we obtain for the equations corresponding to the subscript $i = 2$:

$$\frac{dX_2}{dt} = -2A\widehat{Q}Y_2$$
$$\frac{dY_2}{dt} = 2A\widehat{Q}X_2$$

and, eliminating $\widehat{Q}$ from them,

$$X_2 \frac{dX_2}{dt} + Y_2 \frac{dY_2}{dt} = 0$$

or

$$\mathcal{G} = \frac{1}{2}(X_2^2 + Y_2^2) = \text{const} \tag{9.78}$$

(the *Sessin integral*).

In terms of the given variables w, J, the Sessin integral is written

$$D^2 \mathcal{G} = D_2^2 J_1 + D_1^2 J_2 - 2 D_1 D_2 \sqrt{J_1 J_2} \cos(w_1 - w_2). \tag{9.79}$$

The Sessin transformation may be easily interpreted as a rotation in the four-dimensional phase space. Indeed, H is formed by the symmetric part $x_1^2 + y_1^2 + x_2^2 + y_2^2$, which is invariant to rotations, and the linear term $\sum D_i x_i$, which is one vector in the (x_1, x_2) plane and is affected by rotations. It is then enough to consider one rotation on the (x_1, x_2) plane that brings the vector $\sum D_i x_i$ to one of the the principal axes (e.g. $\sum D_i x_i = DX_1$) and to introduce a rotation on the (y_1, y_2) plane such that the resulting four-dimensional rotation is canonical.

---

[8] Linear transformations of this kind were called *orthogonal* by Poincaré [80], who used them to diagonalize quadratic Hamiltonians.

9.5 Sessin Transformation and Integral

Exercise 9.5.1. Show that the transformation $(x, y) \Rightarrow (X, Y)$ defined by

$$\begin{pmatrix} X_1 \\ X_2 \end{pmatrix} = \mathsf{R} \begin{pmatrix} x_1 \\ x_2 \end{pmatrix} \qquad \begin{pmatrix} Y_1 \\ Y_2 \end{pmatrix} = \mathsf{R} \begin{pmatrix} y_1 \\ y_2 \end{pmatrix},$$

where R is the planar rotation matrix

$$\mathsf{R} = \begin{pmatrix} \cos\alpha & \sin\alpha \\ -\sin\alpha & \cos\alpha \end{pmatrix},$$

is canonical.

Exercise 9.5.2 ([32]). Show the integrability of the Hamiltonian

$$H = \mathcal{F}\left(\sum_1^N J_i\right) + \sum_1^N D_i \sqrt{2J_i} \cos w_i, \qquad (9.80)$$

where $\mathcal{F}$ is a differentiable function, $w \in \mathbf{T}^N$ and $J \in \mathbf{R}_+^N$ are angle–action variables and $D_i \in \mathbf{R}$ are constants.

9.5.1 The Restricted (Asteroidal) Case

The Hamiltonian (9.70) comes from the study of first-order resonance in a system of two planets and its symmetry certainly played a role in the discovery of the Sessin transformation. In restricted systems, in which one planet is replaced by one asteroid, the symmetry is broken by the fact that the attraction of the asteroid on the planet is neglected. Instead of (9.70), we have, as in the asteroidal case studied in Sect. 9.4,

$$H = A(C + J_1 + J_2)^2 + D_1\sqrt{2J_1}\cos w_1 + D_2 \cos w_2, \qquad (9.81)$$

It is easy to see that a restricted form of the Sessin transformation can be applied to change this Hamiltonian into a trivially integrable case [100]. The Sessin transformation in this case is written

$$\begin{aligned} Y_1 &= y_1 + \beta \sin w_2 & \vartheta_2 &= w_2 \\ X_1 &= x_1 + \beta \cos w_2 & \Lambda_2 &= \mathcal{G}(y_1, x_1, J_2), \end{aligned} \qquad (9.82)$$

where $\beta = \dfrac{D_2}{D_1}$ and the function $\mathcal{G}$ is defined by the relations $\{\vartheta_2, \Lambda_2\} = 1$ and $\{X_1, \Lambda_2\} = \{Y_1, \Lambda_2\} = 0$, so that the transformation $(y_1, x_1; w_2, J_2) \Rightarrow (Y_1, X_1; \vartheta_2, \Lambda_2)$ is canonical[9]. Hence,

$$\frac{\partial \mathcal{G}}{\partial J_2} = 1$$

$$\frac{\partial \mathcal{G}}{\partial y_1} = -\frac{\partial \mathcal{G}}{\partial J_2} \beta \sin w_2$$

$$\frac{\partial \mathcal{G}}{\partial x_1} = -\frac{\partial \mathcal{G}}{\partial J_2} \beta \cos w_2,$$

---
[9] Since $J_1 > 0$, the order of the variables in the brackets is $(y_1, x_1; w_2, J_2)$.

that is,
$$\mathcal{G} = J_2 - \beta(x_1 \cos w_2 + y_1 \sin w_2) + C', \tag{9.83}$$
where C' is an arbitrary integration constant. The new Hamiltonian is
$$H = A\left(C'' - \frac{\beta^2}{2} + \frac{1}{2}(X_1^2 + Y_1^2) + \mathcal{G}\right)^2 + D_1 X_1, \tag{9.84}$$
where $C'' = C - C'$. Since the new Hamiltonian does not depend on ϑ_2, we have $\mathcal{G} = \text{const}$ (*Sessin integral*).

Exercise 9.5.3. Show that in the case $J_1 < 0$, that is, when the Hamiltonian is
$$H = A(C + J_1 + J_2)^2 + D_1\sqrt{-2J_1}\cos w_1 + D_2 \cos w_2,$$
the derivation is analogous, but the Sessin integral is
$$\mathcal{G} = J_2 + \beta(x_1 \cos w_2 + y_1 \sin w_2) + C' \tag{9.85}$$
and the new Hamiltonian is
$$H = A\left(C'' + \frac{\beta^2}{2} - \frac{1}{2}(X_1^2 + Y_1^2) + \mathcal{G}\right)^2 + D_1 X_1. \tag{9.86}$$

10

Nonlinear Oscillators

10.1 Quasiharmonic Hamiltonian Systems

A quasiharmonic Hamiltonian system is a system whose unperturbed part $H_0(J)$ is a linear function of the actions. That is,

$$H = \sum_{i=1}^{N} \omega_i J_i + R(\theta, J, \varepsilon), \qquad (10.1)$$

where $\omega_i = \text{const}$ and R is a 2π-periodic function of the angles θ_i, analytical in ε and vanishing for $\varepsilon = 0$. This system is a paradigm of many conservative nonlinear systems.

In terms of coordinates and momenta, these systems arise from the study of systems of differential equations which are reduced to separate harmonic oscillators when the small parameter ε is zero. In appropriate variables, the Hamiltonian is written

$$H = \frac{1}{2} \sum_{i=1}^{N} \omega_i (x_i^2 + y_i^2) + R_1(x, y, \varepsilon), \qquad (10.2)$$

where $R_1(x, y, \varepsilon) = R(\theta, J, \varepsilon)$. The corresponding differential equations are[1]

$$\dot{x}_i = -\omega_i y_i - \frac{\partial R_1}{\partial y_i} \qquad \dot{y}_i = +\omega_i x_i + \frac{\partial R_1}{\partial x_i} \qquad (10.3)$$

or

$$\ddot{x}_i + \omega_i^2 x_i = \phi_i(x, y, \dot{x}, \dot{y}, \varepsilon), \qquad (10.4)$$

where the function ϕ_i vanishes when $\varepsilon = 0$. The angle–action variables associated with

$$H_0 = \frac{1}{2} \sum_{i=1}^{N} \omega_i (x_i^2 + y_i^2) \qquad (10.5)$$

---

[1] The sign of these equations correspond to the choice $J_i > 0$. See Sect. 7.2.

are
$$\theta_i = \arctan\sqrt{\frac{y_i}{x_i}} \tag{10.6}$$

and
$$J_i = \frac{1}{2}(x_i^2 + y_i^2). \tag{10.7}$$

We notice that, in this problem, all elements of the Hessian of H_0, $\det\left(\frac{\partial^2 H_0}{\partial J_i \partial J_j}\right)$, are equal to zero.

10.2 Formal Solutions. General Case

Let us consider the application of the Hori theory to the canonical system defined by the Hamiltonian

$$H = \sum_{i=1}^{N} \omega_i J_i + \sum_{k=1}^{\infty} \varepsilon^k F_k(\theta, J). \tag{10.8}$$

Let us assume that, for $\varepsilon = 0$, there is no trivial degeneracy (in Schwarzschild's sense), that is, $\omega_i \neq 0$ for all $i = 1, \cdots, N$.

Let us introduce the canonical transformation $\phi_n : (\theta, J) \Rightarrow (\theta^*, J^*)$ defined by
$$f(\theta, J) = E_{W^*} f(\theta^*, J^*),$$

where
$$W^* = \sum_{k=1}^{n} W_k^*(\theta^*, J^*, \varepsilon) \tag{10.9}$$

with W_k^* of order ε^k. Following the steps of Sect. 6.2, we obtain the same set of equations as there. However, the high-order derivatives of H_0 now vanish because H_0 is a linear function of the actions.

The homological equation is
$$H_k^* = \{H_0, W_k^*\} + \Psi_k(\theta^*, J^*) \tag{10.10}$$

or
$$\sum_{i=1}^{N} \omega_i \frac{\partial W_k^*}{\partial \theta_i^*} = \Psi_k(\theta^*, J^*) - H_k^*,$$

where H_k^* is the term of order ε^k of the new Hamiltonian and Ψ_k is a function that is completely known when the equations for the subscripts smaller than k are solved. Since (θ, J) are the angle–action variables of the undisturbed Hamiltonian H_0, we may use the averaging operation of Sect. 6.2 in the form

$$H_k^*(J^*) = <\Psi_k(\theta^*, J^*)>,$$

where $<\ldots>$ stands for the average over the angles θ_i^* ($i = 1, \cdots, N$) from 0 to 2π. If Ψ_k is written as

$$\Psi_k = \sum_{h \in D_k} \varepsilon^k A_{kh}(J^*) \cos(h \mid \theta^*), \qquad (10.11)$$

where $D_k \subset \mathbf{Z}^N$ is a finite set, we obtain

$$H_k^* = \varepsilon^k A_{k0}(J^*) \qquad (10.12)$$

and

$$\sum_{i=1}^{N} \omega_i \frac{\partial W_k^*}{\partial \theta_i^*} = \sum_{h \in D_k \setminus \{0\}} \varepsilon^k A_{kh}(J^*) \cos(h \mid \theta^*). \qquad (10.13)$$

A particular solution of this equation is

$$W_k^*(\theta^*, J^*) = \sum_{h \in D_k \setminus \{0\}} \frac{\varepsilon^k A_{kh}(J^*) \sin(h \mid \theta^*)}{(h \mid \omega)}. \qquad (10.14)$$

This solution introduces the divisors $(h \mid \omega)$. The condition for the application of the method under consideration is the *non-resonance condition* $(h \mid \omega) \neq 0$ for all integer vectors $h \in D_k \setminus \{0\}$ and all $k \leq n$. When this non-resonance condition is satisfied, we obtain the Lie generator of the canonical transformation ϕ_n, which transforms the Hamiltonian H into a function depending only on the actions J_i^*, except for a remainder R_n divisible by ε^{n+1}.

The resulting Hamiltonian is

$$H^* = \sum_{i=1}^{N} \omega_i J_i^* + \sum_{k=1}^{n} H_k^*(J^*) \qquad (10.15)$$

and the differential equations spanned by H^* have the solutions

$$\theta_i^* = \omega_i t + \sum_{k=1}^{n} \frac{\partial H_k^*(J^*)}{\partial J_i^*} t + \text{const} \qquad J_i^* = \text{const}. \qquad (10.16)$$

(Note that Hori's formal integral of Sect. 6.7 is, here, simply, $H_0^* = \sum_{i=1}^{N} \omega_i J_i^* = \text{const}$.)

The formal solutions of the problem stated in Sect. 10.1 are, then,

$$\theta_i = E_{W^*} \theta_i^*, \qquad J_i = E_{W^*} J_i^*, \qquad (10.17)$$

where θ_I^*, J_i^* are the solutions of the resulting (averaged) system and W^* is the Lie generator defined by (10.9) and (10.14).

10.3 Exact Commensurability of Frequencies (Resonance)

Let us assume that the coefficients ω_i satisfy, simultaneously, L independent commensurability relations

$$(\bar{h}_\varrho \mid \omega) = 0 \qquad (\varrho = N - L + 1, \cdots, N), \tag{10.18}$$

where $\bar{h}_\varrho \in \mathbf{Z}^N$.

To study this case, we introduce, beforehand, the Lagrangian point transformation

$$\begin{aligned}\phi_\mu &= \theta_\mu & (\mu = 1, \cdots, M = N - L) \\ \phi_\varrho &= (\bar{h}_\varrho \mid \theta) & (\varrho = M + 1, \cdots, N),\end{aligned} \tag{10.19}$$

which is completed by the introduction of new actions by means of the Jacobian canonical condition in the particular form

$$\sum_{i=1}^N J_i \, \delta\theta_i = \sum_{i=1}^N I_i \, \delta\phi_i$$

or

$$\sum_{i=1}^N J_i \, \delta\theta_i = \sum_{\mu=1}^M I_\mu \, \delta\theta_\mu + \sum_{\varrho=M+1}^N I_\varrho \, (\bar{h}_\varrho \mid \delta\theta).$$

The identification of both sides leads to

$$\begin{aligned} J_\mu &= I_\mu + \sum_\varrho I_\varrho \bar{h}_{\varrho\mu} \\ J_\sigma &= \sum_\varrho I_\varrho \bar{h}_{\varrho\sigma} \end{aligned} \tag{10.20}$$

($\sigma = M+1, \cdots, N$), where $\bar{h}_{\varrho\nu}$ means the νth integer component of the vector $\bar{h}_\varrho$.

It then follows that

$$H_0(J) = \sum_{i=1}^N \omega_i J_i = \sum_{\mu=1}^M \omega_\mu I_\mu + \sum_{\varrho=M+1}^N I_\varrho \, (\bar{h}_\varrho \mid \omega) \tag{10.21}$$

or, because of (10.18),

$$H_0(J) = \sum_{\mu=1}^M \omega_\mu I_\mu. \tag{10.22}$$

The complete Hamiltonian is, now,

$$\widehat{H} = \sum_{\mu=1}^M \omega_\mu I_\mu + \widehat{R}(\phi, I, \varepsilon). \tag{10.23}$$

It is current in the literature on nonlinear mechanics to refer to this system as a resonant oscillator involving L simultaneous resonances. However,

10.3 Exact Commensurability of Frequencies (Resonance)

because of the linearity of $\widehat{H}_0(J)$, the L commensurabilities given by (10.18) are *essential*: They give rise to L *degenerate* degrees of freedom.

Since resonances and essential commensurabilities are different concepts in perturbation theories, we rather say that the Hamiltonian system under consideration in this section is *degenerate* with $M = N - L$ short-period and L long-period variables. Therefore, we adopt the von Zeipel averaging rule, which takes into due account the fact that $\widehat{H}_0$ is independent of the actions I_ϱ. We use

$$H_k^*(\phi_\varrho^*, I^*) = <\tilde{\Psi}_k(\phi^*, I^*)>, \qquad (10.24)$$

where $<\cdots>$ stands for the average over the angles ϕ_μ ($\mu = 1, \cdots, M$), from 0 to 2π. If $\tilde{\Psi}_k$ has the form

$$\tilde{\Psi}_k = \sum_{h \in D_k} \varepsilon^k A_{kh}(I^*) \cos(h \mid \phi^*), \qquad (10.25)$$

where $D_k \subset \mathbf{Z}^N$, it follows that

$$H_k^*(\phi_\varrho^*, I^*) = \tilde{\Psi}_{k(\mathrm{S})}(I^*) + \tilde{\Psi}_{k(\mathrm{LP})}(\phi_\varrho^*, I^*) \qquad (10.26)$$

and

$$\sum_{\mu=1}^{M} \omega_\mu \frac{\partial W_k^*}{\partial \phi_\mu^*} = \tilde{\Psi}_{k(\mathrm{SP})}(\phi^*, I^*), \qquad (10.27)$$

where the subscripts S, LP, SP stand for the different parts of $\tilde{\Psi}_k$ (secular, long-period and short-period).

The result of the averaging operation is a system of canonical equations whose Hamiltonian is

$$H^* = \sum_{\mu=1}^{M} \omega_\mu I_\mu^* + \sum_{k=1}^{n} H_k^*(\phi_\varrho^*, I^*). \qquad (10.28)$$

This Hamiltonian is equal to $\widehat{H}$, except for a remainder divisible by ε^{n+1}.

The system defined by the new Hamiltonian leads to

$$I_\mu^* = \mathrm{const},$$

the reduced system

$$\dot{\phi}_\varrho^* = \sum_{k=1}^{n} \frac{\partial}{\partial I_\varrho^*} H_k^*(\phi_\varrho^*, I^*), \qquad \dot{I}_\varrho^* = -\sum_{k=1}^{n} \frac{\partial}{\partial \phi_\varrho^*} H_k^*(\phi_\varrho^*, I^*) \qquad (10.29)$$

and the quadratures

$$\phi_\mu = \int \frac{\partial H^*}{\partial I_\mu^*} dt.$$

The averaging reduces the given system to a system with $L = N - M$ degrees of freedom. The continuation depends on the complete integrability, or not, of the system defined by the leading part, $\widehat{H}_1(\phi_\varrho^*, I^*)$, of the new Hamiltonian.

It is worthwhile mentioning that the linearity of $\widehat{H}_0(I_\mu)$ does not influence these results, which are the same as those already obtained with the more general Hamiltonians in Sect. 10.2. However, the point transformation $(\theta, J) \Rightarrow (\phi, I)$ does not preserve the d'Alembert property of the given function. When this property is an essential feature in the given problem, we may use (10.19) only to separate the secular, long-period and short-period parts of Ψ_k, as follows:

(a.) $\Psi_{k(S)}(J^*)$ is the average of Ψ_k over all angles;
(b.) $\Psi_{k(LP)}(\theta^*, J^*)$ is the collection of periodic terms of Ψ_k independent of ϕ_μ; they have arguments $(h|\theta)$, where h is a linear combination of the $\bar{h}_\varrho$ ($\Psi_{k(LP)}$ does not depend on ϕ_μ, but may depend on θ_μ);
(c.) $\Psi_{k(SP)}(\theta^*, J^*)$ is the collection of remaining periodic terms of Ψ_k, dependent on ϕ_μ.

The averaging rule is now written

$$<\Psi_k(\theta^*, J^*)> = \Psi_{k(S)} + \Psi_{k(LP)} \tag{10.30}$$

and the homological equation is separated into

$$H_k^*(\theta^*, J^*) = \Psi_{k(S)}(J^*) + \Psi_{k(LP)}(\theta^*, J^*) \tag{10.31}$$

and

$$\sum_{i=1}^{N} \omega_i \frac{\partial W_k^*}{\partial \theta_i^*} = \Psi_{k(SP)}(\theta^*, J^*). \tag{10.32}$$

10.4 Birkhoff Normalization

Let us consider a regular dynamical system with N degrees of freedom, in the variables q, p, and let us assume the existence of a stable equilibrium point $P_0 \equiv (q_0, p_0)$.

The motion in the neighborhood of P_0 is governed by the Hamiltonian

$$H = \sum_{k=2}^{\infty} X_k(q - q_0, p - p_0), \tag{10.33}$$

where X_k are the components of the Taylor series expansion of the Hamiltonian. The subscript k means the degree of homogeneity of the functions with respect to the components $(q_i - q_{0i})$ and $(p_i - p_{0i})$. The functions X_0 and X_1 are missing in H: X_0 is a constant and does not contribute to the equations of motion and $X_1 = 0$ because the coefficients of the linear terms are zero at an equilibrium point. X_2 is a sign-definite quadratic form and we assume that a rotation of the axes was done, bringing the coordinate system to the principal axes of X_2. We may also change the scales along these axes to have

$$X_2 = \sum_{i=1}^{N} \frac{1}{2} \omega_i (x_i^2 + y_i^2), \tag{10.34}$$

where x_i, y_i are the coordinates along the scaled principal axes of the quadratic form (centered at P_0).

Birkhoff's normalization is the name given to the reduction of H to a polynomial in $(x_i^2 + y_i^2)$ – the so-called Birkhoff normal form. Before stating the main result on this topic, we introduce some usual nomenclature concerning resonance. We say that a commensurability relation of order k holds when there exist $\bar{h} \in \mathbf{Z}^N$ such that

$$(\bar{h} \mid \omega) = 0$$

and

$$||\bar{h}||_1 = \sum_{i=1}^{N} |\bar{h}_i| = k,$$

where $\bar{h} \equiv (\bar{h}_1, \bar{h}_2, \cdots, \bar{h}_N)$ and $\omega \equiv (\omega_1, \omega_2, \cdots, \omega_N)$ [2]. The following result is well known.

Theorem 10.4.1 (Birkhoff). *If the frequencies ω_i do not satisfy any commensurability relation of order n or smaller, there is a canonical transformation such that H is reduced to a Birkhoff normal form of degree n except for terms of degree higher than n.*

The normal form of degree n may be constructed by means of the perturbation theory for the neighborhood of the origin described in Sect. 7.6, with only a few modifications. Indeed, the given Hamiltonian is regular at the origin but has no rotational symmetry (the terms with odd subscripts: $X_3, X_5, \cdots$ are not equal to zero as in Sect. 7.6). Moreover, it is not perturbed (that is, $H_k = 0$ for all $k > 0$). In order to have a more straightforward approach of the proof of Birkhoff's theorem, it is convenient to transform the variables to angle–action variables, by means of

$$x_i = \sqrt{2J_i} \cos \theta_i \qquad y_i = \sqrt{2J_i} \sin \theta_i \tag{10.35}$$

and the inverse relation giving the actions

$$J_i = \frac{1}{2}(x_i^2 + y_i^2). \tag{10.36}$$

We thus have

---

[2] Other definitons of order are usual. In the study of nonlinear oscillators, the order is often defined as $k = ||\bar{h}||_1 - 2$. In the study of planetary motions, when the ω are mean motions, it is defined as $k = |\sum_{i=1}^{N} \bar{h}_i|$.

$$X_2 = \sum_{i=1}^{N} \omega_i J_i \tag{10.37}$$

and

$$X_k = \sum_{h \in D_k} A_{kh} \exp \mathrm{i}(h \mid \theta), \tag{10.38}$$

where the $A_{kh}(J)$ are homogeneous functions of degree k in $\sqrt{J_i}$. $D_k \subset \mathbf{Z}^N$ is the set of integers associated with the powers of $\sqrt{J_i}$ in A_{kh} by d'Alembert property rules. This means that, for all $h \in D_k$, $||h||_1 = k, k-2, k-4, \cdots$ (see Sect. 7.3). It is worth mentioning that we have introduced complex A_{kh} and resorted to exponential functions only to avoid separating sine and cosine terms in X_k.

The conservation of the Hamiltonian is written

$$H^*(J^*) + \mathcal{R}(\theta^*, J^*) = E_{W^*} \sum_{k=2}^{\infty} X_k(\theta^*, J^*) \tag{10.39}$$

and the comparison of the terms with the same degree of homogeneity in the elements of $\mathcal{S} \equiv (\sqrt{J_1}, \sqrt{J_2}, \cdots, \sqrt{J_N})$ leads to the equations

$$
\begin{aligned}
H_2^* &= X_2 \\
H_3^* &= X_3 + \{X_2, W_3^*\} \\
H_4^* &= X_4 + \{X_3, W_3^*\} + \{X_2, W_4^*\} + \tfrac{1}{2}\{\{X_2, W_3^*\}, W_3^*\} \\
H_5^* &= X_5 + \{X_4, W_3^*\} + \{X_3, W_4^*\} + \{X_2, W_5^*\} + \tfrac{1}{2}\{\{X_3, W_3^*\}, W_3^*\} \\
&\quad + \tfrac{1}{2}\{\{X_2, W_3^*\}, W_4^*\} + \tfrac{1}{2}\{\{X_2, W_4^*\}, W_3^*\} + \tfrac{1}{6}\{\{\{X_2, W_3^*\}, W_3^*\}, W_3^*\} \\
H_6^* &= X_6 + \{X_5, W_3^*\} + \{X_4, W_4^*\} + \{X_3, W_5^*\} + \{X_2, W_6^*\} \\
&\quad + \tfrac{1}{2}\{\{X_4, W_3^*\}, W_3^*\} + \tfrac{1}{2}\{\{X_3, W_4^*\}, W_3^*\} + \tfrac{1}{2}\{\{X_3, W_3^*\}, W_4^*\} \\
&\quad + \tfrac{1}{2}\{\{X_2, W_3^*\}, W_5^*\} + \tfrac{1}{2}\{\{X_2, W_4^*\}, W_4^*\} + \tfrac{1}{2}\{\{X_2, W_5^*\}, W_3^*\} \\
&\quad + \tfrac{1}{6}\{\{\{X_3, W_3^*\}, W_3^*\}, W_3^*\} + \tfrac{1}{6}\{\{\{X_2, W_3^*\}, W_3^*\}, W_4^*\} \\
&\quad + \tfrac{1}{6}\{\{\{X_2, W_3^*\}, W_4^*\}, W_3^*\} + \tfrac{1}{6}\{\{\{X_2, W_4^*\}, W_3^*\}, W_3^*\} \\
&\quad + \tfrac{1}{24}\{\{\{\{X_2, W_3^*\}, W_3^*\}, W_3^*\}, W_3^*\} \\
&\cdots \\
H_n^* &= X_n + \sum_{k=2}^{n-1}\{X_k, W_{n-k+2}^*\} + \cdots + \tfrac{1}{(n-2)!}\{\{\cdots\{X_2, W_3^*\},\cdots\}, W_3^*\}.
\end{aligned}
\tag{10.40}$$

We recall that the degree of homogeneity of the Poisson bracket of two functions, in the elements of $\mathcal{S}$, is two units lower than the sum of the degrees of homogeneity of the two functions.

These equations may be written in a generic form (homological equation) as

$$\{X_2, W_k^*\} = H_k^* - \Psi_k(\theta^*, J^*) \tag{10.41}$$

and the equations of the associated Hori kernel are

$$\frac{\mathrm{d}\theta_i^*}{\mathrm{d}u} = \frac{\partial X_2}{\partial J_i^*} = \omega_i, \qquad \frac{\mathrm{d}J_i^*}{\mathrm{d}u} = 0, \qquad (10.42)$$

whose general solution is

$$\theta_i^* = \omega_i u + \gamma_i, \qquad J_i^* = C_i, \qquad (10.43)$$

where γ_i, C_i are integration constants[3].

The first perturbation equation is

$$\{X_2, W_3^*\} = H_3^* - \Psi_3(\theta^*, J^*), \qquad (10.44)$$

where $\Psi_3 = X_3$. Because of the rules of formation of the set D_3, all $h \in D_3$ are such that $||h||_1$ is equal to 1 or 3. Thus, there is no secular term in Ψ_3. Since, by hypothesis, there is no commensurability relation of order 3, it follows $H_3^* = \ = 0$.

The second perturbation equation is

$$\{X_2, W_4^*\} = H_4^* - \Psi_4(\theta^*, J^*), \qquad (10.45)$$

where $\Psi_4 = X_4 + \{X_3, W_3^*\} + \frac{1}{2}\{\{X_2, W_3^*\}, W_3^*\}$.

Since the d'Alembert property is preserved by Poisson brackets, Ψ_4 may be written

$$\Psi_4 = \sum_{h \in D_4} B_{4h}(J^*) \exp \mathrm{i}(h \mid \theta^*). \qquad (10.46)$$

Because of the rules of formation of the set D_4, all $h \in D_4$ are such that $||h||_1$ is equal to 0, 2 or 4. Thus, if there is no commensurability relation of order 4,

$$H_4^* = \ = B_{40}(J^*). \qquad (10.47)$$

Similar reasoning may be done at all orders $k \leq n$ if no commensurability of order n or smaller exist. The result is

$$H^* = \sum_{k=1}^{[n/2]} H_{2k}^*(J^*) + \mathcal{R}_{n+1}(\theta^*, J^*), \qquad (10.48)$$

and this proves Birkhoff's theorem since H_{2k}^* are functions only of the $J_i = \frac{1}{2}(x_i^{*2} + y_i^{*2})$. We note that no infinite series expansion was used in

---

[3] Some simplified expressions with fewer brackets to calculate are

$$\begin{aligned}
\Psi_4 &= X_4 + \tfrac{1}{2}\{H_3^* + X_3, W_3^*\} \\
\Psi_5 &= X_5 + \tfrac{1}{2}\{H_4^* + X_4, W_3^*\} + \tfrac{1}{2}\{H_3^* + X_3, W_4^*\} - \tfrac{1}{12}\{\{H_3^* - X_3, W_3^*\}, W_3^*\} \\
\Psi_6 &= X_6 + \tfrac{1}{2}\{H_5^* + X_5, W_3^*\} + \tfrac{1}{2}\{H_4^* + X_4, W_4^*\} + \tfrac{1}{2}\{H_3^* + X_3, W_5^*\} - \\
&\quad \tfrac{1}{12}\{\{H_4^* - X_4, W_3^*\}, W_3^*\} - \tfrac{1}{12}\{\{H_3^* - X_3, W_4^*\}, W_3^*\} - \\
&\quad \tfrac{1}{12}\{\{H_3^* - X_3, W_3^*\}, W_4^*\}.
\end{aligned}$$

the construction of the canonical transformation since, by construction, we adopted

$$W^* = \sum_{k=3}^{n} W_k^*. \tag{10.49}$$

A convergence analysis is thus only necessary if we want to know the asymptotic behavior of $\mathcal{R}$ for $n \to \infty$.

10.4.1 A Formal Extension Including One Single Resonance

From the viewpoint of perturbation theory, a Birkhoff normalization can, generally[4], be done even when one commensurability of order $5 \leq k \leq n$ exists (except for a remainder of degree higher than n, of course).

We may proceed exactly as in the previous section up to obtaining the homological equation

$$\{X_2, W_k^*\} = H_k^* - \Psi_k(\theta^*, J^*) \tag{10.50}$$

and the associated Hori kernel defined by

$$\begin{aligned} \theta_i^* &= \omega_i u + \gamma_i \\ J_i^* &= C_i. \end{aligned} \tag{10.51}$$

The function Ψ_k may be written

$$\Psi_k = \sum_{h \in D_k} B_{k,h}(J^*) \exp \mathrm{i}(h \mid \theta^*) \tag{10.52}$$

and, because of the formation rules of the set D_k, all $h \in D_k$ are such that $||h||_1$ is equal to $k, k-2, k-4, \cdots$.

When *one* commensurability relation

$$(\bar{h} \mid \omega) = 0 \tag{10.53}$$

exists, the averaging rule should be modified to become:

$$H_k^* = <\Psi_k> = B_{k,0}(J^*) + \sum_{h \in D_k'} B_{k,h}(J^*) \exp \mathrm{i}(h \mid \theta^*), \tag{10.54}$$

where $D_k' \subset D_k$ is the subset formed by all multiples of $\bar{h}$ of order at most equal to k. If no other commensurability of order n, or smaller, exists, the Hamiltonian resulting from the nth-order perturbation equation is

$$H^* = \sum_{k=1}^{[n/2]} H_{2k}^*(\bar{h} \mid \theta^*, J^*) + \mathcal{R}_{n+1}(\theta^*, J^*). \tag{10.55}$$

---

[4] That is, apart from the case of some exceptional conditions to be discussed later.

We may now introduce the Lagrangian point transformation

$$\begin{aligned}
\phi_\mu &= \theta_\mu^* \quad (\mu = 1, 2, \cdots, N-1) \\
\phi_N &= (\bar{h} \mid \theta^*) \\
I_\mu &= J_\mu^* - J_N^* \bar{h}_\mu / \bar{h}_N \\
I_N &= J_N^* / \bar{h}_N,
\end{aligned} \quad (10.56)$$

where $\bar{h} \equiv (\bar{h}_1, \bar{h}_2, \ldots, \bar{h}_N) \in \mathbf{Z}^N$. The only angle present in the resulting Hamiltonian $\widehat{H}$ is ϕ_N. The corresponding canonical equations are

$$\frac{d\phi_N}{dt} = \frac{\partial \widehat{H}}{\partial I_N}, \qquad \frac{dI_N}{dt} = -\frac{\partial \widehat{H}}{\partial \phi_N},$$

$$\frac{d\phi_\mu}{dt} = \frac{\partial \widehat{H}}{\partial I_\mu} = G_\mu(\phi_N, I), \qquad \frac{dI_\mu}{dt} = -\frac{\partial \widehat{H}}{\partial \phi_\mu} = 0, \quad (10.57)$$

$(\mu = 1, 2, \ldots, N-1)$, where

$$\widehat{H} = \sum_{k=2}^n \widehat{B}_{k,0}(I) + \sum_{k=2}^n \sum_\ell \widehat{B}_{k,\ell\bar{h}}(I) \exp i\ell\phi_N. \quad (10.58)$$

The Hamiltonian $\widehat{H}$ may be reduced to one degree of freedom and, therefore, is integrable. The I_μ ($\mu = 1, 2, \ldots, N-1$) are constants and the angles ϕ_μ may be obtained by integrating the corresponding equations (when the functions $I_N(t)$ and $\phi_N(t)$ are known):

$$\phi_\mu = \int G_\mu[\phi_N(t), I_\mu, I_N(t)] dt.$$

If it is possible to construct angle–action variables associated with the Hamiltonian $\widehat{H}$, the composition of the canonical transformations $(\theta, J) \Rightarrow (\theta^*, J^*) \Rightarrow (\phi, I)$ with the canonical transformation defining the angle–action variables associated with $\widehat{H}$ transforms H into a normal form (up to order n).

The conditions for the existence of angle–action variables of $\widehat{H}(\phi_N, I)$ are almost the same of the stability theorem of Arnold (see [71], Sect. IX.E). To find them, let us analyze the structure of $\widehat{H}$. Let $\widehat{B}_{k_1,\ell_0\bar{h}} \exp i\ell_0\phi_N$ be the lowest order periodic term in $\widehat{H}$ and let $C_{k_2,0}$ be the lowest-order non-periodic term of $\widehat{H}$ actually dependent on I_N. Because of the d'Alembert property, $\widehat{B}_{k_1,\ell_0\bar{h}}(I)$ is at least of order $\ell_0\bar{h}$. If $k_2 < k_1$, $C_{k_2,0}$ will lead the expansion of $\widehat{H}$ in the neighborhood of the origin of the (ϕ_N, I_N) plane. Then, in the limit $I_N \to 0$, the solutions of the system are circular motions of radius $I_N = \text{const}$ and frequency $\Omega_n = \partial C_{k_2,0}/\partial I_N$. Therefore, there is a neighborhood of the origin where periodic solutions exist and, then, angle–action variables may be constructed.

10.4.2 The Comensurabilities of Lower Order

The cases $||\bar{h}||_1 \leq 4$ were excluded from the above discussion for obvious reasons. In general, the lowest-order terms $\widehat{B}_{k_1,\ell\bar{h}}(I) \exp i\ell\phi_N$ have $\ell = \pm 1$. Since $k_2 \geq 4$, we have $k_1 = ||\bar{h}_1|| \leq k_2$. Therefore, $C_{4,0}(I)$ may no longer lead the expansion. These cases are thus distinct and need to be considered one by one. Let us first recall that commensurabilities $||\bar{h}||_1 = 2$ can only appear in $\widehat{H}$ through the average of Ψ_{2k} ($k \geq 2$) (X_2 does not contain periodic terms) and are, thus, at least of fourth order.

In the case $||\bar{h}||_1 = 3$, the lowest-order possible periodic terms in $\widehat{H}$ are $\widehat{B}_{3,\pm\bar{h}}(I) \exp(\pm i\phi_N)$. If $\widehat{B}_{3,\pm\bar{h}} \neq 0$, these periodic terms will lead the new Hamiltonian. Let us write their sum as $A_{3,\bar{h}}(I) \sin(\phi_N - \alpha_N)$, where the amplitude $A_{3,\bar{h}}(I)$ and the phase α_N are elementary functions of $\widehat{B}_{3,\pm\bar{h}}(I)$. This sum is 0 at the origin and also on the rays $\phi_N = \alpha_N (\mod \pi)$. The phase portrait of the leading term is similar to that shown in Fig. 10.2. No periodic orbits exist in this case. Therefore, one necessary condition for having periodic solutions in the case $||\bar{h}||_1 = 3$ is $\widehat{B}_{3,\pm\bar{h}} = 0$, that is, that no resonant term exist in X_3.

In the cases $||\bar{h}||_1 = 2$ and $||\bar{h}||_1 = 4$, the lowest-order possible secular and periodic terms are $C_{4,0}(I)$ and $\widehat{B}_{4,\pm\bar{h}}(I) \exp(\pm i\phi_N)$, respectively. If $C_{4,0} \neq 0$ and $B_{4,\pm\bar{h}} \neq 0$, these terms will lead the new Hamiltonian. Let us write their sum as $A_{(}4,0)(I) + A_{4,\bar{h}}(I) \sin(\phi_N - \alpha_N)$, where the coefficients $A_{4,0}(I)$, $A_{4,\bar{h}}(I)$ and the phase α_N are simple functions of $C_{4,0}$ and $\widehat{B}_{4,\pm\bar{h}}$. The phase portrait of the leading terms depends on the values of $A_{4,0}(I)$ and $A_{4,\bar{h}}(I)$. If $|A_{4,0}| > |A_{4,\bar{h}}|$, in a neighborhood $0 < I_N < I_{\lim}$, the isoenergetic curves in the (I_N, ϕ_N) plane are closed and are loci of periodic orbits that may be used to construct an angle-action pair of variables. Otherwise, $\widehat{H}$ will be constant along the rays $\phi_N = \alpha_N + \arcsin(-A_{4,0}/A_{4,\bar{h}})$ going away from the origin and no periodic orbits circling the origin can exist.

10.5 The Restricted Three-Body Problem

An important two-degrees-of-freedom gyroscopic system, in Celestial Mechanics, is the restricted three-body problem. It is the planar problem of the motion of a particle of negligible mass in the gravitational field created by two bodies (*primaries*) moving around their common center of gravity in circular Keplerian orbits. If we consider one reference system rotating with the two primaries so that, in this system, they remain fixed, we have the same gyroscopic system studied in Sect. 1.7.1. W is, now, the potential energy of the gravitational interaction between the third body and the two primaries. To avoid artificial difficulties due to the negligible mass of the third particle, we use the Hamiltonian per unit mass. In this case, (1.66) becomes

$$H = \frac{p^2}{2} - [\boldsymbol{\Omega}, \boldsymbol{r}, \boldsymbol{p}] + V(\boldsymbol{r}), \tag{10.59}$$

where $\boldsymbol{r} \equiv (\xi_1, \xi_2)$, $\boldsymbol{p} \equiv (p_1, p_2) = \dot{\boldsymbol{r}} + \boldsymbol{\Omega} \times \boldsymbol{r}$ and $\boldsymbol{\Omega}$ is the angular velocity vector of the reference frame; V is the gravitational potential created by the two primaries at the point $\boldsymbol{r}$:

$$V = -\frac{Gm_1}{d_1} - \frac{Gm_2}{d_2},$$

where m_1, m_2 are the masses of the primaries, d_1, d_2 the distances of the primaries to the third body and G the gravitational constant. For simplicity in the calculations, we introduce units such that the distance between the primaries is $R = 1$, the angular velocity of the rotating frame is $\Omega = |\boldsymbol{\Omega}| = 1$ and the sum of the masses is $m_1 + m_2 = 1$. The third Kepler law, $R^3 \Omega^2 = G(m_1 + m_2)$, shows that, with these units, $G = 1$. We write the masses of the primaries as $m_1 = 1 - \mu$ and $m_2 = \mu$ and obtain d_1, d_2 immediately from the geometry of the problem (Fig. 10.1):

$$d_1^2 = (\mu + \xi_1)^2 + \xi_2^2 \qquad d_2^2 = (1 - \mu - \xi_1)^2 + \xi_2^2.$$

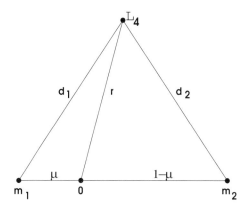

Fig. 10.1. Lagrangian point $\mathcal{L}_4$

This system has five equilibrium solutions. Three equilibrium solutions (*Euler collinear solutions*) lie on the axis passing through the two primaries (one of each side and one between them). The other two equilibrium solutions (*Lagrange equilateral solutions*) lie on the vertices of two equilateral triangles having the two primaries as the two other vertices[5]. The Lagrange solutions, designated as $\mathcal{L}_4$ and $\mathcal{L}_5$, are given by $\boldsymbol{r}^0 \equiv (\xi_1^0, \xi_2^0)$ with

$$\xi_1^0 = \frac{1}{2} - \mu \qquad \xi_2^0 = \pm \frac{\sqrt{3}}{2}. \tag{10.60}$$

---

[5] For a thorough discussion of these solutions, see [66], [95]. For a description of the motions around the Lagrangian points, see [24].

10.5.1 Equations of the Motion Around the Lagrangian Point $\mathcal{L}_4$

The theories presented in this book may be used to study the oscillations about the Lagrangian points. To do this, the first step is to expand the potential V in powers of $\boldsymbol{x} = \boldsymbol{r} - \boldsymbol{r}^0$. The transformation $\boldsymbol{x} = \boldsymbol{r} - \boldsymbol{r}^0$ is also introduced in the kinetic part of the Hamiltonian, along with the transformation $\boldsymbol{y} = \boldsymbol{p} - \boldsymbol{p}^0$, where $\boldsymbol{p}^0$ is the vector momentum of the equilibrium solution. We recall that, at the equilibrium points, $\dot{\boldsymbol{r}} = 0$ but $\boldsymbol{p} \neq 0$. A simple calculation, using the definition of $\boldsymbol{p}$, shows that $p_1^0 = -\xi_2^0$, $p_2^0 = \xi_1^0$.

The result of these operations is

$$H = H_2 + H_3 + H_4 + \cdots,$$

where

$$H_2 = \frac{1}{2} y_1^2 + \frac{1}{2} y_2^2 + x_2 y_1 - x_1 y_2 + \frac{1}{8} x_1^2 - \frac{1}{4} \gamma x_1 x_2 - \frac{5}{8} x_2^2$$

$$H_3 = -\frac{7\sqrt{3}}{144} \gamma x_1^3 + \frac{3\sqrt{3}}{16} x_1^2 x_2 + \frac{11\sqrt{3}}{48} \gamma x_1 x_2^2 + \frac{3\sqrt{3}}{16} x_2^3$$

$$H_4 = \frac{37}{128} x_1^4 + \frac{25}{96} \gamma x_1^3 x_2 - \frac{123}{64} x_1^2 x_2^2 - \frac{15}{32} \gamma x_1 x_2^3 - \frac{3}{128} x_2^4$$

and

$$\gamma \overset{\text{def}}{=} 3\sqrt{3}(1 - 2\mu).$$

The constant term $-\frac{11}{8} - \frac{1}{216}\gamma^2$ was discarded since it does not contribute to the differential equations.

The linear approximation of the equations of the motion around $\mathcal{L}_4$ is obtained from H_2. It is the same Hamiltonian as that studied in Sect. 2.9 (see 2.122) with $a = \frac{1}{4}$, $b = -\frac{5}{4}$ and $d = -\frac{\gamma}{4}$. The eigenvalues are

$$\lambda_j = \mp \frac{1}{2} \sqrt{-2 \pm \sqrt{\gamma^2 - 23}} \overset{\text{def}}{=} \mp i\omega_k \tag{10.61}$$

($k = 1, 2$). The necessary conditions for which the four eigenvalues are imaginary and different is $\sqrt{23} < \gamma < \sqrt{27}$. The lower limit corresponds to having μ equal to the Routh critical mass ratio $\mu_1 = \frac{1}{2}(1 - \sqrt{\frac{23}{27}}) = 0.03852\ldots$ and the upper limit to the degenerate cases $\mu = 0$ and $\mu = 1$. The system is linearly stable for $0 < \mu < \mu_1$.

The components of the two first eigenvectors $\tilde{A}_k$ ($k = 1, 2$) are

$$\begin{aligned} A_{1,k} &= 4\lambda_k^2 - 9 \\ A_{2,k} &= -8\lambda_k + \gamma \\ A_{3,k} &= 4\lambda_k^3 - \lambda_k - \gamma \\ A_{4,k} &= -4\lambda_k^2 + \gamma\lambda_k - 9. \end{aligned} \tag{10.62}$$

We recall that $|\tilde{A}_k|$ is arbitrary and therefore the choice is not unique (here, the choice is not the same as that made in Sect. 2.9.1).

10.5 The Restricted Three-Body Problem

To get the Hamiltonian in normal form, we have to use the transformation given by (2.121). The functions $\varrho_{N+k,k}$ (see 2.127) are

$$\varrho_{31} = 8\,\mathrm{i}\,\omega_1(\omega_1^2 - \omega_2^2)(9 + 4\omega_1^2)$$
$$\varrho_{42} = 8\,\mathrm{i}\,\omega_2(\omega_2^2 - \omega_1^2)(9 + 4\omega_2^2).$$

Adopting the convention $0 < \omega_1 < \omega_2$, we have $-\mathrm{i}\varrho_{31} < 0$ and $-\mathrm{i}\varrho_{42} > 0$. This means that the resulting action variables are such that $J_1 < 0$ and $J_2 > 0$.

The canonical transformation to the angle–action variables is

$$z_j = \sum_{k=1}^{2} \sqrt{\frac{J_k}{-\mathrm{i}\varrho_{N+k,k}}}\,(\tilde{A}_k \mathrm{e}^{-\mathrm{i}w_k} + \overline{\tilde{A}_k}\mathrm{e}^{\mathrm{i}w_k}), \qquad (10.63)$$

where the overline indicates complex conjugation.

The lower order part of the Hamiltonian is

$$X_2 = \omega_1 J_1 + \omega_2 J_2. \qquad (10.64)$$

In terms of the corresponding Poincaré non-singular variables $\widehat{x}_k, \widehat{y}_k$, we have

$$X_2 = -\frac{1}{2}\omega_1(\widehat{x}_1^2 + \widehat{y}_1^2) + \frac{1}{2}\omega_2(\widehat{x}_2^2 + \widehat{y}_2^2). \qquad (10.65)$$

The difference of signs in the two terms comes from the fact that the the rules for transformation into non-singular variables are not the same for $J_k > 0$ and $J_k < 0$ (see Sect. 7.2).

With the help of an algebraic processor, it is possible to obtain the higher-order parts in terms of the angle–action variables. However, the results are enormous expressions that we refrain from reproducing here. In schematic form, we may write,

$$X_k = \sum_{h \in D_k} B_{k,h}(J)\,\mathrm{e}^{\mathrm{i}(h|w)},$$

where D_n is the set of all h such that $||h||_1$ is n or $n-2$ or $n-4$, etc. and $B_{n,h}$ are homogeneous functions of degree $n/2$ in the actions J_1, J_2. (The X_n and the functions Ψ_n formed in the construction of the perturbation equations have the d'Alembert property.)

The first perturbation equation is

$$\{X_2, W_3^*\} = X_3^* - X_3(w^*, J^*). \qquad (10.66)$$

Because of the rules of formation of the set D_3, there are no secular terms in X_3. If there is no commensurability between the proper frequencies ω_1 and ω_2, then $X_3^* = <X_3> = 0$.

The second perturbation equation is

$$\{X_2, W_4^*\} = X_4^* - \Psi_4(\theta^*, J^*), \qquad (10.67)$$

where $\Psi_4 = X_4 + \{X_3, W_3^*\} + \frac{1}{2}\{\{X_2, W_3^*\}, W_3^*\}$. Ψ_4 is given by a summation over $h \in D_4$, where D_4 is the set of all h such that $||h||_1$ is 0, 2 or 4. When commensurabilities are excluded, X_4^* is equal to the constant term of the trigonometric polynomial Ψ_4.

10.5.2 Internal 2:1 Resonance

Let us consider the case $\omega_2 = 2\omega_1$. A simple calculation using the fact that $\omega_1^2 + \omega_2^2 = 1$ and the relationship between γ and the frequencies in symmetrical form, $\gamma = \sqrt{27 - 16\omega_1^2\omega_2^2}$, easily gives:

$$\omega_1 = \frac{1}{\sqrt{5}} \qquad \omega_2 = \frac{2}{\sqrt{5}} \qquad \gamma = \frac{\sqrt{611}}{5} \tag{10.68}$$

and the value of μ is the critical value

$$\mu_2 = 0.0242938970\cdots.$$

The transformation discussed above allows the angle–action variables to be introduced. We get, for instance,

$$X_2 = \frac{1}{\sqrt{5}} J_1 + \frac{2}{\sqrt{5}} J_2 \tag{10.69}$$

and

$$\begin{aligned}
X_3 = &\; J_1\sqrt{-J_1}\;(3.339604686\cos w_1 - 9.096749597\sin w_1 \\
&\qquad +6.212220054\cos 3w_1 - 10.32983562\sin 3w_1) \\
&+ J_1\sqrt{J_2}\;[-9.171392587\cos w_2 + 23.06026831\sin w_2 \\
&\qquad -20.71122172\cos(2w_1 + w_2) + 28.04734005\sin(2w_1 + w_2) \\
&\qquad -1.322197100\cos(2w_1 - w_2) + 0.298260493\sin(2w_1 - w_2)] \\
&+ \sqrt{-J_1}J_2\;[-3.061304295\cos w_1 + 6.830954032\sin w_1 \\
&\qquad -21.96976280\cos(w_1 + 2w_2) + 24.35104283\sin(w_1 + 2w_2) \\
&\qquad -5.667138826\cos(w_1 - 2w_2) - 11.57701846\sin(w_1 - 2w_2)] \\
&+ J_2\sqrt{J_2}\;(1.940102278\cos w_2 - 4.329118636\sin w_2 \\
&\qquad +7.534561805\cos 3w_2 - 6.840066733\sin 3w_2).
\end{aligned}$$

The integration of X_3 may take into account the existence of the critical angle $2w_1 - w_2$ among the arguments. The Von Zeipel averaging rule can be used to easily obtain the average of X_3. All terms are averaged over w_1, w_2 except the secular (which is absent from X_3) and the terms depending on the critical angle. That is

$$\begin{aligned}
<X_3> \;=\; &J_1\sqrt{J_2}\,[\,-1.322197100\cos(2w_1 - w_2) \\
&+ 0.2982604928\sin(2w_1 - w_2)].
\end{aligned}$$

The two leading terms of the transformed Hamiltonian are $X_2^* = X_2(J_1^*, J_2^*)$ and $X_3^* = <X_3(J_1^*, J_2^*)>$. We may use the theory presented in the previous section to obtain some higher-order terms of the new Hamiltonian and of the generating function of the canonical transformation $(w, J) \to (w^*, J^*)$.

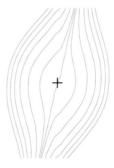

Fig. 10.2. 2:1 resonance. Phase portrait of $H^*_{(3)}$ in the plane $I_1 = 0$. Polar coordinates: $\sqrt{I_2}, \phi$.

Nevertheless, this last step is not necessary to get the main result about this internal resonance.

We may reduce this system to one degree of freedom through the point transformation

$$\begin{aligned} \phi' &= w_1^* & I_1 &= J_1^* + 2J_2^* \\ \phi &= w_2^* - 2w_1^* & I_2 &= J_2^* \end{aligned} \quad (10.70)$$

and the leading part of the Hamiltonian becomes

$$H^*_{(3)} = \frac{1}{\sqrt{5}} I_1 + (2I_2 - I_1)\sqrt{I_2}\,(1.322197100 \cos\phi + 0.2982604928 \sin\phi)$$

Since ϕ' is absent from $H^*_{(3)}$, I_1 is a constant. Figure 10.2 shows the resulting curves $H^*_{(3)} = \text{const}$ on the manifold defined by $I_1 = 0$. It shows asymptotic branches (rays) going away from the origin (shown with a plus sign). Therefore, the equilibrium solution $I_2 = 0$ of the reduced transformed system is unstable [67], [68] (see also [27] and references therein).

10.5.3 Internal 3:1 Resonance

Let us consider, now, the case $\omega_2 = 3\omega_1$. In this case,

$$\omega_1 = \frac{1}{\sqrt{10}} \qquad \omega_2 = \frac{3}{\sqrt{10}} \qquad \gamma = \frac{\sqrt{639}}{5} \quad (10.71)$$

and the value of μ is the critical value

$$\mu_3 = 0.0135160160\ldots.$$

We proceed as before and obtain

$$X_2 = \frac{1}{\sqrt{10}} J_1 + \frac{3}{\sqrt{10}} J_2. \quad (10.72)$$

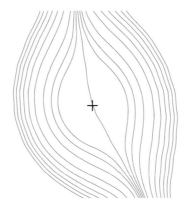

Fig. 10.3. 3:1 resonance. Phase portrait of $H^*_{(4)}$ in the plane $I_1 = 0$. Polar coordinates: $\sqrt{I_2}, \phi$

The expression for X_3 is similar to the previous one, but with different numerical coefficients. Since it does not contain, now, any critical term, its integration is trivial, giving $<X_3> = 0$ and W_3^*. We have, then, to consider the second perturbation equation:

$$\{X_2, W_4^*\} = X_4^* - X_4 - \frac{1}{2}\{X_3, W_3^*\}. \tag{10.73}$$

The integration of X_4 may take into account the existence of the critical angle $3w_1 - w_2$ among the arguments. The Von Zeipel averaging rule can be used to easily obtain the average of X_4. All terms are averaged over w_1, w_2 except the secular and the terms depending on the critical angle. These calculations are cumbersome and we show only the result.

The new Hamiltonian depends on only one angle, and we use the existence of one first integral to reduce it to one degree of freedom. The point transformation is, now,

$$\begin{array}{ll} \phi' = w_1^* & I_1 = J_1^* + 3J_2^* \\ \phi = w_2^* - 3w_1^* & I_2 = J_2^* \end{array} \tag{10.74}$$

and the leading part of the transformed Hamiltonian is

$$H^*_{(4)} = \frac{1}{\sqrt{10}} I_1 + 0.246875 I_1^2 + 0.6955357143 I_1 I_2 - 4.1705357141 I_2^2$$
$$+ (3I_2 - I_1)^{3/2}\sqrt{I_2}\,(12.70709081\cos\phi + 4.384418361\sin\phi)$$
$$+ (3I_2 - I_1)I_1\sqrt{I_2}\,(4.235696938\cos\phi + 1.461472787\sin\phi).$$

As in the previous case, since ϕ' is absent from $H^*_{(4)}$, I_1 is a constant. Figure 10.3 shows the resulting curves $H^*_{(4)} = $ const on the manifold defined by $I_1 = 0$. It shows asymptotic branches (rays) going away from the origin (shown with a plus sign). Therefore, as in the case of the 2:1 internal resonance, the equilibrium solution $I_2 = 0$ of the reduced transformed system is unstable.

10.5.4 Other Internal Resonances

We have already seen that I_4 is unstable at the 2:1 and 3:1 internal resonances. The other resonance, which may occur when H_4^* is considered, is the 1:1 resonance ($\omega_1 = \omega_2$). However, the solutions in this case cannot be studied in the same way as the previous two resonances because, in this case, the transformation to angle–action variables introduced in Sect. 10.5 can no longer be used (the divisors $\varrho_{31}, \varrho_{42}$ become equal to zero). Because of the d'Alembert property of the functions X_k and Ψ_k, these three resonances are the only ones that may appear in $H_{(4)}^*$. For other resonances, $||h||_1 > 4$ and they only appear in higher orders. Let us consider, as an example, the case $3\omega_1 = 2\omega_3$. In that case,

$$\omega_1 = \frac{2}{\sqrt{13}} \qquad \omega_2 = \frac{3}{\sqrt{13}} \qquad \gamma = \frac{3\sqrt{443}}{13} \qquad (10.75)$$

and the value of μ is

$$\mu_2 = 0.0326224067\ldots.$$

If we proceed as before, we obtain

$$X_2^* = \frac{2}{\sqrt{13}} J_1^* + \frac{3}{\sqrt{13}} J_2^*,$$

$$X_3^* = 0$$

and

$$X_4^* = -984.3080179 J_1^{*2} + 2765.829446 J_1^* J_2^* - 494.3119688 J_2^{*2}.$$

If we use variables (ϕ, I) defined in the same way as in the previous sections, with the multiplier $3/2$ instead of 2 or 3, we obtain for the leading part of the Hamiltonian:

$$H_{(4)}^* = \frac{2}{\sqrt{13}} I_1 - 984.3080179 I_1^{*2} + 5718.753500 I_1 I_2 - 6857.749179 I_2^{*2}.$$

Since no angle appears in $H_{(4)}^*$, the curves $H_{(4)}^* = \text{const}$ are concentric circles in the manifolds $I_1 = \text{const}$. The terms depending on the angle $\phi = 3w_1 - 2w_2$, which will appear at the next order, will somewhat distort these circles far from the origin, but will not change the fact that the origin is, now, a center. This situation is different of those found in the two previous cases. In those cases, the existence of a solution along rays asymptotic to the origin was enough to say that the equilibrium was unstable. In the case of the 3:2 internal resonance, the origin is a stable equilibrium solution of the Hamiltonian H^* (reduced to one degree of freedom). This is a necessary condition for the stability of the Lagrange equilibrium solutions $\mathcal{L}_4$ and $\mathcal{L}_5$, but it is not sufficient. From Arnold's theorem [72], we need also to have the condition $X_4^*(\omega_1, \omega_2) = 2765.829446 \omega_1 \omega_2 - 494.3119688 \omega_2^2 - 984.3080179 \omega_1^2 \neq 0$, which is satisfied.

This situation will reproduce itself for all other high-order resonances. In all cases $H^*_{(4)}$ will be a quadratic from in I_1, I_2 differing from the above $H^*_{(4)}$ only in the numerical coefficients.

10.6 The Hénon–Heiles Hamiltonian

One important example of a quasiharmonic system with one commensurability is the system defined by the Hénon–Heiles Hamiltonian [46]

$$H = \frac{1}{2}(x_1^2 + y_1^2) + \frac{1}{2}(x_2^2 + y_2^2) + x_1^2 x_2 - \frac{1}{3}x_2^3 \tag{10.76}$$

or

$$H = X_2 + X_3,$$

where

$$X_2 = \frac{1}{2}(x_1^2 + y_1^2) + \frac{1}{2}(x_2^2 + y_2^2) = J_1 + J_2 \tag{10.77}$$

$$X_3 = x_1^2 x_2 - \frac{1}{3}x_2^3$$

$$= \frac{1}{6} J_2 \sqrt{2 J_2} \, (3 \cos\theta_2 + \cos 3\theta_2)$$

$$+ \frac{1}{2} J_1 \sqrt{2 J_2} \, [2 \cos\theta_2 + \cos(2\theta_1 + \theta_2) + \cos(2\theta_1 - \theta_2)]$$

and θ_j, J_j are the angle–action variables associated with the Poincaré pairs x_j, y_j. This system has the order-2 commensurability

$$\omega_1 - \omega_2 = 0.$$

We may perform a formal averaging following the theory given in Sect. 10.4.1 and reduce this Hamiltonian to one degree of freedom up to order n.

The first perturbation equation is

$$\{X_2, W_3^*\} = H_3^* - X_3(\theta^*, J^*). \tag{10.78}$$

Since there is no critical term in X_3, the solution of this equation is

$$H_3^* \;=\; <X_3> \;=\; 0$$

and

$$W_3^* = \int X_3 du, \tag{10.79}$$

where u is related to θ_1^* and θ_2^* through the solution of the Hori kernel associated with the homological equation:

$$\theta_1^* = u + \gamma_1 \qquad \theta_2^* = u + \gamma_2 \tag{10.80}$$

10.6 The Hénon–Heiles Hamiltonian

(J_1^* and J_2^* are constants in the Hori kernel). When we introduce these equations into (10.79), perform the integration and use the inverse transformation to go back to the variables θ_i^*, we obtain

$$W_3^* = -\frac{1}{6}J_2^*\sqrt{2J_2^*}\left(3\sin\theta_2^* + \frac{1}{3}\sin 3\theta_2^*\right)$$
$$+J_1^*\sqrt{2J_2^*}\left[\sin\theta_2^* + \frac{1}{6}\sin(2\theta_1^* + \theta_2^*) + \frac{1}{2}\sin(2\theta_1^* - \theta_2^*)\right].$$

The next perturbation equation is

$$\{X_2, W_4^*\} = H_4^* - \Psi_4,$$

where

$$\Psi_4 = X_4(\theta^*, J^*) + \frac{1}{2}\{H_3^* + X_3, W_3^*\},$$

or, taking into account the fact that $X_4 = 0$ and $H_3^* = 0$,

$$\Psi_4 = \frac{1}{2}\{X_3, W_3^*\}.$$

After some calculations, we obtain

$$\Psi_4 = -\frac{1}{12}J_1^{*2}\left(5 + 4\cos 2\theta_1^* - \cos 4\theta_1^*\right) - \frac{1}{12}J_2^{*2}\left(5 + 4\cos 2\theta_2^* - \cos 4\theta_2^*\right)$$
$$+\frac{1}{6}J_1^*J_2^*\left[2 - 2\cos 2\theta_1^* - 2\cos 2\theta_2^* + \cos\left(2\theta_1^* + 2\theta_2^*\right) - 7\cos\left(2\theta_1^* - 2\theta_2^*\right)\right]$$

and

$$H_4^* = <\Psi_4> = -\frac{5}{12}(J_1^* + J_2^*)^2 + \frac{7}{6}J_1^*J_2^*[1 - \cos(2\theta_1^* - 2\theta_2^*)]. \quad (10.81)$$

The expression for W_4^* is obtained in exactly the same way as W_3^*. With the help of an algebraic manipulator, we may easily compute higher order terms and obtain $H_5^* = 0$ and

$$H_6^* = \frac{1}{432}\left(101J_1^{*3} - 235J_2^{*3}\right) - \frac{1}{16}J_1^*J_2^*\left(65J_1^* - 47J_2^*\right)$$
$$-\frac{1}{72}J_1^*J_2^*\left(161J_1^* - 175J_2^*\right)\cos(2\theta_1^* - 2\theta_2^*).$$

In terms of the variables

$$\phi_1 = \theta_1^* \qquad\qquad I_1 = J_1^* + J_2^*$$
$$\phi_2 = \theta_2^* - \theta_1^* + \frac{\pi}{2} \qquad\qquad I_2 = J_2^*, \qquad (10.82)$$

we have, to order 6,

$$\widehat{H}^*_{(6)} = H^*_2 + H^*_4 + H^*_6$$
$$= I_1 - \frac{5}{12}I_1^2 + \frac{101}{432}I_1^3 - 7I_1^2 I_2 + \frac{56}{3}I_1 I_2^2$$
$$- \frac{112}{9}I_2^3 + \frac{7}{3}I_2(I_1 - I_2)(1 + \frac{23}{12}I_1 - 4I_2)\sin^2\phi_2.$$

$\widehat{H}^*_{(6)}$ is an integrable Hamiltonian; it depends on only one angle and has the integral $I_1 = J_1^* + J_2^* = \text{const}$.

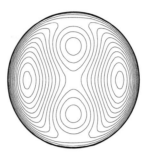

Fig. 10.4. Phase portrait of $H^*_{(6)}$ for $I_1 = 0.2$

The additive constant $\pi/2$ in the definition of ϕ_2 was introduced to have the phase portrait with the same orientation seen in the published surfaces of section of the Hénon–Heiles Hamiltonian. Those surfaces are defined as intersections with the plane $x_1 = 0$ (that is, $\theta_1 = 90$ degrees). The given phase portrait is not constructed with the variables x, y but with the averaged ones.

The structure of the singular points and the phase portrait of $H^*_{(k)}$ becomes more intricate as the order k of the formal solution increases (see [63], Sect. 1.4). For instance, the phase portrait of $H^*_{(4)}$ shows only the two centers at $\phi_2 = 0 \pmod{\pi}$. The phase portrait of $H^*_{(6)}$ (Fig. 10.4) shows four centers (and three saddle points). The thick border corresponds to $J_1^* = 0$.

10.6.1 The Toda Lattice Hamiltonian

A similar example of a quasiharmonic system with one commensurability is the Toda lattice Hamiltonian:

$$H = \frac{1}{2}(y_1^2 + y_2^2) + \frac{1}{24}e^{2x_2}\left[e^{2\sqrt{3}x_1} + e^{-2\sqrt{3}x_1}\right] + \frac{1}{24}e^{-4x_2} - \frac{1}{8} \quad (10.83)$$

(see [7]). After the Taylor expansion of the exponentials about the origin, we obtain

$$H = \frac{1}{2}(x_1^2 + x_2^2 + y_1^2 + y_2^2) + x_1^2 x_2 - \frac{1}{3}x_2^3$$
$$+ \frac{1}{2}(x_1^2 + x_2^2)^2 + x_1^4 x_2 + \frac{2}{3}x_1^2 x_2^3 + \frac{1}{3}x_2^5 + \cdots. \quad (10.84)$$

The first part of this Hamiltonian is the Hénon–Heiles Hamiltonian. Therefore, the construction of a formal solution is similar to that above, up to the fourth order perturbation equation, where, in contrast to the previous case, $X_4, X_5, \cdots$ are no longer equal to zero and need to be considered in calculating Ψ_k ($k \geq 4$). They are

$$X_4(\theta^*, J^*) = 2(J_1^* \cos^2 \theta_1^* + J_2^* \cos^2 \theta_2^*)^2$$

$$X_5(\theta^*, J^*) = 4J_1^{*2}\sqrt{2J_2^*}\,\cos\theta_2^*\cos^4\theta_1^* - \frac{4}{3}J_2^{*2}\sqrt{2J_2^*}\cos^5\theta_2^*$$

$$+ \frac{8}{3} J_1^* J_2^* \sqrt{2J_2^*}\,\cos^3\theta_2^* \cos^2\theta_1^*$$

$$X_6(\theta^*, J^*) = \frac{8}{5} J_1^{*3} \cos^6\theta_1^* + 8J_1^{*2} J_2^* \cos^2\theta_2^* \cos^4\theta_1^* + \frac{88}{45} J_2^{*3} \cos^6\theta_2^*$$

$$+ \frac{8}{3} J_1^* J_2^{*2} \cos^4\theta_2^* \cos^2\theta_1^*.$$

Proceeding exactly as before, we obtain

$$\widehat{H}_{(6)}^* = \frac{1}{27} I_1(27 + 9I_1 + 5I_1^2) + \frac{4}{9} I_2(3 - 7I_1)(I_1 - I_2) \sin^2 \phi_2. \quad (10.85)$$

Again, $I_1 = J_1^* + J_2^*$ is a constant. The phase portrait of $H_{(6)}^*$ is shown in Fig. 10.5. It shows two centers and does not differ significantly from the phase portrait of $H_{(4)}^*$. The Toda lattice Hamiltonian is integrable and surfaces of section of the full Hamiltonian show that the next orders will not change the topology of the phase portrait.

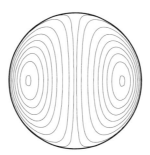

Fig. 10.5. Phase portrait of $H_{(6)}^*$ for $I_1 = 0.2$

10.7 Systems with Multiple Commensurabilities

In the previous sections, it was seen that when the basic frequencies of one quasiharmonic system satisfies one commensurability relation, it is possible to proceed with the formal reduction of this system to one degree of freedom and thus obtain a formal solution of the given system. When there are L commensurability relations, the system may only be reduced to L degrees of freedom and, very particular cases excepted, the new Hamiltonian H^* depends on all variables $J_\varrho^*, \theta_\varrho^*$ ($\varrho = N - L + 1, \ldots, N$). It is not a near-integrable system and, in general, it is not possible to separate a lower degree integrable or reducible part.

The wildest cases are those with N degrees of freedom and $N-1$ commensurability relations ($N \geq 3$). A well-studied example is the N-dimensional three-phonon interaction Hamiltonian, defined by the following properties:

(a.) $\omega_j = j\omega_1$ ($j = 1, \ldots, N$);
(b.) $R(\theta, J, \varepsilon)$ is homogeneous of degree $3/2$ with respect to the variables J_k and has the d'Alembert property; it is non-singular at the origin (see Sect. 7.3).

This Hamiltonian may be written as $H = X_2 + X_3$, where

$$X_2 = \sum_{j=1}^{N} j J_j;$$

$$X_3 = \varepsilon \sum_{h \in D_3} a_h(J) \cos(h \mid \theta).$$

D_3 is the set formed by all $h \in \mathbf{Z}^N$ such that $||h||_1 = 1, 3$.

We may apply to this system the algorithm described in Sect. 10.3 and reduce the Hamiltonian to $N-1$ degrees of freedom The first perturbation equation is

$$\{X_2, W_3^*\} = H_3^* - X_3 \tag{10.86}$$

and the averaging operation is $H_3^* = <X_3>$. H_3^* will be formed only by the long-period part of H_1, that is, by terms of two different kinds:

(a.) terms whose arguments include three different angles with coefficients $+1, +1, -1$, e.g. $\theta_i + \theta_j - \theta_k$ with $i + j - k = 0$;
(b.) terms whose arguments include two different angles with coefficients $+2, -1$, e.g. $2\theta_i - \theta_j$ with $2i - j = 0$.

In the next order,

$$\Psi_4 = \frac{1}{2}\{H_3^* + X_3, W_3^*\}.$$

By construction, the terms in W_3^* are such that $||h||_1 = 1, 3$; that is, $\sum h_j$ may be equal to $-3, -1, +1, +3$. The bracket $\{H_3^* + X_3, W_3^*\}$ generates terms with $||h||_1$ even, in the interval $[-6, +6]$. It is also worth noting that $\Psi_{3(\mathrm{SP})} = 0$, but $\Psi_{4(\mathrm{SP})} \neq 0$.

10.7.1 The Ford–Lunsford Hamiltonian. 1:2:3 Resonance

An example is the Hamiltonian

$$H = J_1 + 2J_2 + 3J_3 + \varepsilon\alpha J_1\sqrt{J_2}\cos(2\theta_1 - \theta_2) + \varepsilon\beta\sqrt{J_1 J_2 J_3}\cos(\theta_1 + \theta_2 - \theta_3) \tag{10.87}$$

thoroughly studied by Ford and Lunsford [35]. It is also known as the 1:2:3 resonance [83].

In this case, both periodic terms are such that $\sum jh_j = 0$. It is the Hamiltonian of an irreducible system with two degrees of freedom and it is already as averaged as possible. The impossibility of construction of formal solutions for this system is related to the high chaoticity of the solutions revealed by Poincaré maps.

It is easy to show, using the equations of Sect. 7.6, that this system is irreducible even in the neighborhood of the origin (that is, for $J_j \simeq 0$).

10.8 Parametrically Excited Systems

In this section, we consider systems in which an excitation appears in coefficients of the differential equation. One example is the linear oscillator with a time-dependent frequency, whose Hamiltonian is

$$H(q,p,t) = \frac{p^2}{2} + \frac{1}{2}A(t)q^2. \tag{10.88}$$

The differential equations spanned by this Hamiltonian may be combined to give the well-known Hill equation $\ddot{q} + A(t)q = 0$. One particular case of the Hill equation is the Mathieu equation, obtained when we introduce

$$A(t) = \omega^2(1 + \varepsilon\cos kt). \tag{10.89}$$

The Mathieu equation is a linear ordinary differential equation with periodic coefficients for which powerful instruments, such as Floquet's theory exist ([78], Sect. 5.2). We could add a nonlinear term, say, αq^4 and transform the harmonic oscillator into a nonlinear oscillator. In this case, it would no longer be possible to use linear theories, but, when ε is small, it is possible to deal with the problem using perturbation theories. This will be done in this section, but, initially, to keep the development limited, only the linear Mathieu equation will be considered. The linear case is enough to show the main features of these systems, namely, that a small parametric oscillation can produce a large response when the frequency of the excitation is close to twice the proper frequency of the system.

If we introduce the angle–action variables of the harmonic oscillator $\ddot{q} + \omega^2 q = 0$,

$$J_1 = \frac{1}{2\omega}(\omega^2 q^2 + p^2) \qquad \sin\theta_1 = \sqrt{\frac{\omega}{2J_1}}q, \qquad (10.90)$$

the Hamiltonian becomes

$$H = \omega J_1 + \frac{1}{2}\varepsilon\omega J_1 \cos kt - \frac{1}{4}\varepsilon\omega J_1[\cos(2\theta_1 + kt) + \cos(2\theta_1 - kt)] \quad (10.91)$$

or, in the extended phase space,

$$H = \omega J_1 + kJ_2 + \frac{1}{2}\varepsilon\omega J_1 \cos\theta_2 - \frac{1}{4}\varepsilon\omega J_1[\cos(2\theta_1 + \theta_2) + \cos(2\theta_1 - \theta_2)], \quad (10.92)$$

where $\theta_2 = kt$ and J_2 is the action canonically conjugate to θ_2.

The resulting Hamiltonian for $\varepsilon = 0$, $H_0 = \omega J_1 + kJ_2$, is linear in the two actions and completely degenerate. It is thus convenient to use the extended point transformation $(\theta, J) \to (w, I)$ defined by the equations

$$\theta_1 = \omega w_1 \qquad J_1 = \frac{1}{\omega}(I_1 + kI_2)$$
$$\theta_2 = kw_1 - w_2 \qquad J_2 = -I_2$$

to put into evidence the degenerate degree of freedom (the action I_2 is absent from H_0). Hence

$$H = I_1 + \frac{\varepsilon}{2}(I_1 + kI_2)\left(\cos(kw_1 - w_2) - \frac{1}{2}\cos[(2\omega + k)w_1 - w_2]\right.$$
$$\left. - \frac{1}{2}\cos[(2\omega - k)w_1 + w_2]\right). \qquad (10.93)$$

To continue, we have to choose one canonical perturbation theory and obtain a canonical transformation $(w, I) \to (w^*, I^*)$ that eliminates the short-period terms. In this very particular case, the von Zeipel–Brouwer theory is more expeditious. We follow closely the procedures described in Sect. 3.4 for the case $N = 2, M = 1$.

The first von Zeipel–Brouwer perturbation equations are

$$H_0(I^*) = H_0^*(I^*)$$

$$\nu_1^* \frac{\partial S_1(w, I^*)}{\partial w_1} + H_1(w, I^*) = H_1^*(w, I^*), \qquad (10.94)$$

where $H_0(I^*) = I_1^*$, $\nu_1^* = 1$ and

$$H_1(w, I^*) = \frac{1}{2}(I_1^* + kI_2^*)\left(\cos(kw_1 - w_2) - \frac{1}{2}\cos[(2\omega + k)w_1 - w_2]\right.$$
$$\left. - \frac{1}{2}\cos[(2\omega - k)w_1 + w_2]\right).$$

H_1^* and S_1 are functions to be determined. Using the von Zeipel averaging rule, we obtain

$$H_1^* = <H_1(w, I^*)> = \frac{1}{2\pi}\int_0^{2\pi} H_1 dw_1 = 0 \qquad (10.95)$$

and

$$\frac{\partial S_1}{\partial w_1} = -\tfrac{1}{2}(I_1^* + kI_2^*)\left(\cos(kw_1 - w_2) - \frac{1}{2}\cos[(2\omega + k)w_1 - w_2]\right.$$
$$\left. - \frac{1}{2}\cos[(2\omega - k)w_1 + w_2]\right)$$

or

$$S_1 = -\tfrac{1}{2}(I_1^* + kI_2^*)\left(\frac{1}{k}\sin(kw_1 - w_2) - \frac{\sin[(2\omega + k)w_1 - w_2]}{2(2\omega + k)}\right.$$
$$\left. - \frac{\sin[(2\omega - k)w_1 + w_2]}{2(2\omega - k)}\right).$$

This equation introduces the divisors $2\omega \pm k$ and can only be considered when none of these divisors is equal to zero. In this case, we may proceed and consider the next von Zeipel–Brouwer equation (when the divisors $2\omega \pm 2k$ will appear). And so on.

In the absence of resonance, the transformed Hamiltonian is independent of the angle w_1^*. However, when one of the divisors $2\omega \pm k$ is close to zero, the term with the critical angle cannot be included in S_1 and should be included in H_1^*. If, for instance, $2\omega - k \simeq 0$, equation (10.94) is split into

$$H_1^* = -\frac{1}{4}(I_1^* + kI_2^*)\cos[(2\omega - k)w_1 + w_2] \qquad (10.96)$$

and

$$\frac{\partial S_1}{\partial w_1} = -\frac{1}{2}(I_1^* + kI_2^*)\left(\cos(kw_1 - w_2) - \frac{1}{2}\cos[(2\omega + k)w_1 - w_2]\right)$$

or

$$S_1 = -\frac{1}{2}(I_1^* + kI_2^*)\left(\frac{1}{k}\sin(kw_1 - w_2) - \frac{\sin[(2\omega + k)w_1 - w_2]}{2(2\omega + k)}\right).$$

The next von Zeipel–Brouwer perturbation equation is the third of (3.40). However, since both angles w_1 and w_2 will remain in H^*, we have to consider this fact when writing the equations. (In this case, the von Zeipel–Brouwer theory eliminates the short-period perturbations but does not reduce the number of degrees of freedom of the given Hamiltonian.) Writing that equation in an explicit way, we get

$$\frac{\partial S_2}{\partial w_1} + \frac{\partial H_1}{\partial I_1^*}\frac{\partial S_1}{\partial w_1} + \frac{\partial H_1}{\partial I_2^*}\frac{\partial S_1}{\partial w_2} + H_2 = H_2^* + \frac{\partial H_1^*}{\partial w_1}\frac{\partial S_1}{\partial I_1^*} + \frac{\partial H_1^*}{\partial w_2}\frac{\partial S_1}{\partial I_2^*},$$

where, to avoid writing too much, we have introduced $\nu_1^* = 1$ and $\nu_2^* = \nu_{ij}^* = 0\ (i,j = 1,2)$. The solution of this equation is easy.

The second and third terms of the left-hand side are products of periodic terms. They will give rise to new periodic terms with arguments $2\omega w_1$, $4\omega w_1$, $2(kw_1 - w_2)$, $2(\omega + k)w_1 - 2w_2$, $2(\omega - k)w_1 + 2w_2$ and $2(2\omega + k)w_1 - 2w_2$ and to the non-periodic term

$$-\frac{\omega}{16(2\omega + k)}(I_1^* + kI_2^*).$$

The periodic terms will contribute ordinary periodic terms to S_2; the second and third terms of the right-hand side are also products of periodic terms, but they give rise only to periodic terms with arguments $2\omega w_1$, $4\omega w_1$, $2(kw_1 - w_2)$ and $2(\omega - k)w_1 + 2w_2$.

Then, the averaging rule gives

$$H_2^* = -\frac{\omega}{16(2\omega + k)}(I_1^* + kI_2^*)$$

and the resulting partial differential equation can be solved to obtain $S_2(w, I^*)$. The uprise of new small divisors is ruled out by the fact that we are, now, analyzing the neighborhood of $2\omega - k = 0$ and the new periodic terms generated by the series products are not critical there. This integration is elementary and is not shown here.

The resulting second-order Hamiltonian is

$$H_{(2)}^*(w*, I^*) = I_1^* - \frac{\varepsilon}{4}(I_1^* + kI_2^*)\cos\left[(2\omega - k)w_1^* + w_2^*\right] - \varepsilon^2\omega\frac{I_1^* + kI_2^*}{16(2\omega + k)}.$$

It is worth noting that all angles appearing in the above calculations have the form $a\omega w_1 + b(kw_1 - w_2)\ (a,b \in \mathbb{Z})$ and that no angles with a different form can be produced by the product of trigonometric functions of two of them. Thus, no term in w_2 alone can be formed, and the only angle that may appear in H^* is the critical angle $(2\omega - k)w_1^* + w_2^*$ or an integer multiple of it. The transformed system has thus, actually, only one degree of freedom and can be solved. In order to do this in a ordered way, we perform one more change of variables and write $H_{(2)}^*$ as a one-degree-of-freedom Hamiltonian. The transformation is

$$\begin{aligned}\alpha_1 &= (2\omega - k)w_1^* + w_2^* & \Lambda_1 &= I_2^* \\ \alpha_2 &= w_1^* & \Lambda_2 &= I_1^* - (2\omega - k)I_2^*.\end{aligned} \qquad (10.97)$$

In terms of the new variables, the second-order Hamiltonian becomes

$$\widehat{H}_{(2)}^* = (2\omega - k)\Lambda_1 + \Lambda_2 - \frac{\varepsilon}{4}(2\omega\Lambda_1 + \Lambda_2)\cos\alpha_1 - \varepsilon^2\omega\frac{2\omega\Lambda_1 + \Lambda_2}{16(2\omega + k)}.$$

$$(10.98)$$

The angle α_2 is ignorable and this gives the first integral

$$\Lambda_2 = I_1^* - (2\omega - k)I_2^* = \text{const}. \tag{10.99}$$

The system is reduced to three differential equations:

$$\begin{aligned}
\dot{\alpha}_1 &= (2\omega - k) - \frac{\varepsilon\omega}{2}\cos\alpha_1 - \frac{\varepsilon^2\omega^2}{8(2\omega + k)}, \\
\dot{\Lambda}_1 &= -\frac{\varepsilon}{4}(2\omega\Lambda_1 + \Lambda_2)\sin\alpha_1, \\
\dot{\alpha}_2 &= 1 + -\frac{\varepsilon}{4}\cos\alpha_1 - \frac{\varepsilon^2\omega}{16(2\omega + k)}.
\end{aligned} \tag{10.100}$$

The two first equations define one one-degree-of-freedom Hamiltonian system. The first equation may be written as

$$\frac{d\alpha_1}{\cos\alpha_1 + \gamma} = \frac{\varepsilon\omega dt}{2}, \tag{10.101}$$

where, for simplicity, we have introduced

$$k' = k + \frac{\varepsilon^2\omega^2}{8(2\omega + k)}$$

and

$$\gamma = -\frac{2(2\omega - k')}{\varepsilon\omega}.$$

We see that there are two different kinds of solutions according to $|\gamma| < 1$ or $|\gamma| > 1$. When $|\gamma| > 1$, the angle α_1 circulates monotonically. The period of circulation is easily calculated from the integral over one cycle. It is

$$T = \frac{4\pi}{\varepsilon\omega\sqrt{\gamma^2 - 1}}. \tag{10.102}$$

It is easy to write the equation for $d\Lambda_1/d\alpha_1$ and see that in one cycle Λ_1 oscillates and returns to the initial point. The only condition to be satisfied is that $\omega\Lambda_1 + \Lambda_2 > 0$, but this quantity is the averaged J_1, which is positive by its definition.

When $|\gamma| < 1$, the situation is more complicated as the denominator becomes equal to zero when $\alpha_1 = \arccos(-\gamma)$ and the integral of the left-hand side of (10.101) goes to infinity logarithmically, with $\dot{\alpha}_1$ positive in one side and negative in the other. Thus, α_1 has two equilibrium points (one stable and one unstable). In turn, Λ_1 also has one equilibrium point corresponding to $\omega\Lambda_1 + \Lambda_2 = 0$. However, the Hessian of the corresponding one-degree-of-freedom system is always negative indicating that the equilibrium points where $\dot{\alpha}_1 = \dot{\Lambda}_1 = 0$ are unstable. The phase portrait of the system is shown in Fig. 10.6. It shows the main characteristic of parametrically excited systems in the neighborhood of $\omega = k/2$. No matter how small ε is, the oscillations grow without limit.

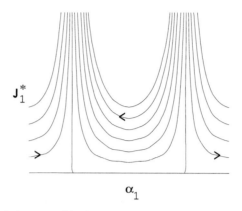

Fig. 10.6. Solutions $J_1^*(\alpha_1)$ of the Mathieu equation when $|\gamma| < 1$.

10.8.1 A Nonlinear Extension

Let us consider the case in which the parametrically perturbed oscillator has a cubic perturbation, with the nonlinearity of the same order as the parametric excitation ($\mu = \mathcal{O}(\varepsilon)$). Let the Hamiltonian be

$$H(q,p,t) = \frac{p^2}{2} + \frac{1}{2}\omega^2(1 + \varepsilon \cos kt)q^2 + \mu\omega^4 q^4. \tag{10.103}$$

When we consider μ and ε on the same footing, we may proceed exactly as before and obtain the second-order Hamiltonian

$$\begin{aligned}H^*_{(2)} &= (2\omega - k)\Lambda_1 + \Lambda_2 + \frac{3}{2}\mu(2\omega\Lambda_1 + \Lambda_2)^2 + \frac{\varepsilon}{4}(2\omega\Lambda_1 + \Lambda_2)\cos\alpha_1 \\ &\quad - \frac{\varepsilon^2\omega}{16(2\omega+k)}(2\omega\Lambda_1 + \Lambda_2) + \frac{17}{4}\mu^2(2\omega\Lambda_1 + \Lambda_2)^3 \\ &\quad - \frac{(9k+14\omega)}{16(2\omega+k)}\mu\varepsilon(2\omega\Lambda_1 + \Lambda_2)^2 \cos\alpha_1.\end{aligned} \tag{10.104}$$

The phase portrait of this Hamiltonian for the resonant case $2\omega - k \simeq 0$ is shown in Fig. 10.7. It shows that the main characteristic of the parametrically excited systems is changed when a small anharmonicity of the oscillator is considered. The oscillations no longer grow unbounded, but oscillate about one center. If we remember that harmonic oscillators appear in Physics generally as approximations of real nonlinear oscillators, the portrait given here is, in many cases, more suitable than the unbounded increases of the linear approximation.

If we neglect the terms of second-order in μ and ε and one trivial constant term ($k\Lambda_2/2\omega$), the above Hamiltonian may be written

$$\widehat{H} = aJ_1^* + bJ_1^{*2} + 2\varepsilon\tau J_1^* \cos 2\sigma, \tag{10.105}$$

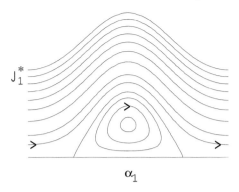

Fig. 10.7. Solutions $J_1^*(\alpha_1)$ of the Mathieu equation with a small nonlinear term ($\mu = 0.1\,\varepsilon$). Same scale and parameters as Fig. 10.6

where

$$a = \frac{2\omega - k}{2}, \qquad b = \frac{3}{2}\mu\omega^2, \qquad \tau = \frac{\omega}{8}, \qquad \sigma = \frac{\alpha_1}{2}. \qquad (10.106)$$

Recall that Λ_2 is a constant and, therefore, in the one-degree-of-freedom phase plane, the transformation $\sigma = \alpha_1/2$; $J_1^* = 2\Lambda_1 + \Lambda_2/\omega$ is canonical. $\widehat{H}$ is the second Andoyer Hamiltonian (see Sect. D.2) whose morphogenesis is dependent on the parameter

$$\gamma = -\frac{a}{2\varepsilon\tau} = -\frac{2(2\omega - k)}{\varepsilon\omega} \qquad (10.107)$$

(which is the same γ as above, if we neglect the higher-order correction embedded into k', and the same as the parameter α used in Sect. D.2). Inspection of Fig. D.2 shows that, for $|\gamma| < 1$, the origin is an unstable equilibrium point. However, in contrast to the linear case, there is, now, a stable equilibrium point off the origin. It is located at

$$\alpha_1 = \pi \qquad J_1^* = J_1^0 = \frac{\varepsilon(\gamma + 1)}{12\mu\omega}. \qquad (10.108)$$

The location of the center and the maximum amplitude of the oscillations around it are proportional to ε/μ. Therefore, the center is as close to the origin as the parametric excitation ε is small. As in the linear case, the origin $J_1^* = 0$ is a stable equilibrium point when $|\gamma| > 1$.

Karl Petrus Teodor Bohlin
© Center for History of Sciences, The Royal Swedish Academy of Sciences

A

Bohlin Theory

A.1 Bohlin's Resonance Problem

Karl Bohlin is one of the founding fathers of Hamiltonian perturbation theories. In his lone paper on the subject [8], several of the ideas extensively used since then in these theories can be found:

(a.) the search for a solution of the Hamilton–Jacobi equation as a power series in the small parameter;
(b.) the derivation of a homological equation for the generic term of the generating function of a canonical transformation;
(c.) the introduction of the square root of the perturbation as a small parameter to study the neighborhood of a resonance.

It is no wonder that Poincaré wrote three chapters of his *Méthodes Nouvelles* [80] developing Bohlin's theory. Moreover, the direct influence of Bohlin's work is felt in other chapters in the second volume of Poincaré's work. For instance, what Poincaré called "Lindstedt Method" – and that we have called, here, "Poincaré Theory" (see Sect. 3.2) – is, essentially, the same method introduced by Bohlin for the study of resonant systems, but using ε as the small parameter, instead of $\sqrt{\varepsilon}$.

The problem studied by Bohlin in his paper is that of finding series solutions of the second-order differential equation corresponding to the Hamiltonian

$$H = \frac{1}{2}p_\zeta^2 - p_\omega + \varepsilon \sum_{h \in \mathbf{Z}^2} B_h(p) \cos(h_1 \zeta - h_2 \omega), \tag{A.1}$$

where p_ζ, p_ω are actions conjugate to the angles ζ, ω, near initial conditions where the angle $h_1^\circ \zeta - h_2^\circ \omega$ leads to a null divisor.

We introduce a new set of canonical variables defined by

$$\begin{aligned} \theta_1 &= \zeta - \frac{h_2^\circ}{h_1^\circ}\omega & J_1 &= p_\zeta \\ \theta_2 &= \frac{1}{h_1^\circ}\omega & J_2 &= h_2^\circ p_\zeta + h_1^\circ p_\omega \end{aligned} \tag{A.2}$$

and Bohlin's Hamiltonian becomes

$$H = \frac{1}{2}J_1^2 + \frac{h_2^\circ}{h_1^\circ}J_1 - \frac{1}{h_1^\circ}J_2 + \varepsilon \sum_{h\in\mathbf{Z}^2} B_h \cos(h_1\theta_1 + k_h\theta_2), \qquad (A.3)$$

where $k_h = h_1 h_2^\circ - h_2 h_1^\circ$.

This is a typical two-degrees-of-freedom Hamiltonian showing no essential degeneracy in Schwarzschild's sense (see Sect. 2.7). Then, von Zeipel–Brouwer theory may be used to eliminate the non-critical terms, for which $k_h \neq 0$, and to obtain the reduced Hamiltonian

$$H^* = \frac{1}{2}J_1^{*2} + \frac{h_2^\circ}{h_1^\circ}J_1^* + \varepsilon \sum_{h\in\mathbf{Z}} B'_h(J_1^*) \cos h\theta_1^*. \qquad (A.4)$$

This one-degree-of-freedom Hamiltonian is resonant in the neighborhood of $J_1^* = -h_2^\circ/h_1^\circ$. It is more general than Garfinkel's Ideal Resonance Problem, and may be studied in a similar way.

The problem initially proposed by Bohlin included other angles. The arguments of the periodic perturbations were $(h_1\zeta - h_2\omega + \gamma_h)$, where the angles γ_h were slowly varying angles. They were considered as constant because, in the approximation adopted, the perturbation coefficients B_h were independent of the actions conjugate to γ_h. In the astronomical problem considered by Bohlin, this meant that long-term variations were discarded, and thus that the results had a short-time validity only.

Since non-critical short-period perturbations may always be considered as having been eliminated by means of an application of von Zeipel–Brouwer theory, the most general Bohlin Hamiltonian is

$$H = H_0(J_1) + \varepsilon R(\theta, J; \varepsilon), \qquad (A.5)$$

where $J \equiv (J_1, \cdots, J_N)$, $\theta \equiv (\theta_1, \cdots, \theta_N)$. The $N-1$ degrees of freedom corresponding to the subscripts $\varrho = 2, \cdots, N$ are degenerate and the actions J_ϱ are absent from H_0; the undisturbed frequencies $\nu_\varrho = \partial H_0/\partial J_\varrho$ are identically equal to zero. H_0 only depends on the action J_1 and the sought solution may be valid in a neighborhood of the action value J_1^* for which the frequency $\nu_1 = \partial H_0/\partial J_1$ vanishes.

The problem of finding a formal canonical transformation able to eliminate the critical angle from the corresponding Hamiltonian is called *Bohlin's problem*. Poincaré considered Bohlin's problem in its complete form as given by the Hamiltonian (A.5) and showed that the search for a series solution of the corresponding Hamilton–Jacobi is impaired by the rise of a singularity at the second-order of approximation (see Sect. A.3). In Chaps. 8 and 9, we have shown that, in a large number of circumstances (which include even the presence of secular and secondary resonances), a Lie series perturbation theory allows the solution of Bohlin's resonance problem to be obtained, at the sought order, without the formation of Poincaré's singularity.

A.2 Bohlin's Perturbation Equations

What is often called Bohlin theory appeared at the eve of space age as a tool to study resonant motions of artificial satellites (such as artificial satellites moving with a critical inclination [52] and geosynchronous satellites [75].) It incorporated the key idea of von Zeipel–Brouwer theory: the search for successive transformations, each able to eliminate one or more degrees of freedom, instead of searching for one transformation able to solve, at one stroke, the Hamilton–Jacobi equation. In artificial satellite theories, the starting point is the averaged Hamiltonian obtained by Brouwer [14] where the non-critical short-period angles have been eliminated. Therefore, the theory consists of the search for of one canonical transformation $(\theta, J) \Rightarrow (\theta^*, J^*)$ defined by the (Jacobian) generating function

$$S = (J^* \mid \theta) + \sum_{k=1}^{n} \varepsilon^{k/2} S_k(\theta, J^*), \tag{A.6}$$

able to eliminate the critical angle θ_1 and to reduce the number of degrees of freedom of the system. The small parameter is the square root of ε and the subscripts were accordingly adopted, indicating the order of the terms in $\sqrt{\varepsilon}$. This transformation may be able to eliminate the critical degree of freedom and the new Hamiltonian will be independent of θ_1^*.

The transformation is conservative and we have

$$H(\theta, J) = H^*(\theta_\varrho^*, J^*) + \mathcal{R}_{n+1}(\theta^*, J^*) \tag{A.7}$$

or, taking into account the canonical transformation generated by S,

$$H\left(\theta, \frac{\partial S}{\partial \theta}\right) = H^*\left(\frac{\partial S}{\partial J_\varrho^*}, J^*\right) + \mathcal{R}_{n+1}\left(\frac{\partial S}{\partial J^*}, J^*\right). \tag{A.8}$$

To identify both sides of (A.8) according to the powers of $\sqrt{\varepsilon}$, we need the power series expansions of H_k and H^*. These expansions are the same as those used in Poincaré and von Zeipel–Brouwer theories (see Sects. 3.2.1 and 3.2.2) with just a change in the small parameter. We have

$$H_0 = G_{0,0} + \sqrt{\varepsilon} G_{0,1} + \varepsilon G_{0,2} + \cdots + \varepsilon^{n/2} G_{0,n} + \cdots, \tag{A.9}$$

$$H_k = G_{k,k} + \sqrt{\varepsilon} G_{k,k+1} + \varepsilon G_{k,k+2} + \cdots + \varepsilon^{n/2} G_{k,n} + \cdots \tag{A.10}$$

and (see Sect 3.4.1)

$$H^*\left(\frac{\partial S}{\partial J_\varrho^*}, J^*\right) = H_0^*(J^*) + \sum_{k=1}^{n} \varepsilon^{k/2} [H_k^*(\theta_\varrho, J^*) + G_k'^*(\theta_\varrho, J^*)] + \cdots. \tag{A.11}$$

The functions G_{ij} and G_i' are those defined by (3.15), (3.22) and (3.39). The identification of both sides of (A.8) follows the same steps described in Sect.

3.4.2 (for $M = 1$) except that, now, $\nu_1^* = \partial H_0/\partial J_1^*$ is assumed to be a quantity of the order of the adopted small parameter ($\sqrt{\varepsilon}$), and the disturbing function εR has only the even components ($\varepsilon R = \varepsilon H_2 + \varepsilon^2 H_4 + \cdots$). Equating the parts of the same order in (A.8), we obtain the perturbation equations

$$H_0 = H_0^*,$$

$$0 = H_1^*,$$

$$\frac{\nu_1^*}{\sqrt{\varepsilon}} \frac{\partial S_1}{\partial \theta_1} + \frac{1}{2} \nu_{11}^* \left(\frac{\partial S_1}{\partial \theta_1} \right)^2 + H_2(\theta, J^*) = H_2^* + G_2'^*,$$

$$\frac{\nu_1^*}{\sqrt{\varepsilon}} \frac{\partial S_2}{\partial \theta_1} + \nu_{11}^* \frac{\partial S_1}{\partial \theta_1} \frac{\partial S_2}{\partial \theta_1} + G_{2,3} + \mathcal{E}_3' = H_3^* + G_3'^*, \qquad (A.12)$$

$$\cdots\cdots$$

$$\frac{\nu_1^*}{\sqrt{\varepsilon}} \frac{\partial S_k}{\partial \theta_1} + \nu_{11}^* \frac{\partial S_1}{\partial \theta_1} \frac{\partial S_k}{\partial \theta_1} + G_{2,k+1} + G_{4,k+1} + \cdots + \mathcal{E}_{k+1}' = H_{k+1}^* + G_{k+1}'^*,$$

$$\cdots\cdots$$

$$\frac{\nu_1^*}{\sqrt{\varepsilon}} \frac{\partial S_{n-1}}{\partial \theta_1} + \nu_{11}^* \frac{\partial S_1}{\partial \theta_1} \frac{\partial S_n}{\partial \theta_1} + G_{2,n} + G_{4,n} + \cdots + \mathcal{E}_n' = H_n^* + G_n'^*.$$

All remaining terms are of order $\mathcal{O}(\varepsilon^{(n+1)/2})$, at least, and are supposed to be grouped with the remainder $\mathcal{R}_{n+1}$. The functions $\mathcal{E}_k'$ are those defined implicitly by (3.20).

The analysis of these equations is similar to the analysis of others done in classical theories. The first equation gives H_0^* and says that it is equal to H_0 at the point $J_1 = J_1^*$. The second equation says that $H_1^* = 0$. As a consequence, $G_2'^* = 0$ and

$$G_3'^* = \sum_\varrho \frac{\partial H_2^*}{\partial \theta_\varrho} \frac{\partial S_1}{\partial J_\varrho^*}$$

(see 3.39).

The next equation is the fundamental equation of Bohlin theory, or *Bohlin's equation*:

$$\frac{\nu_1^*}{\sqrt{\varepsilon}} \frac{\partial S_1}{\partial \theta_1} + \frac{1}{2} \nu_{11}^* \left(\frac{\partial S_1}{\partial \theta_1} \right)^2 + H_2 = H_2^* \qquad (A.13)$$

or

$$\frac{\partial S_1}{\partial \theta_1} = \frac{1}{\nu_{11}^* \sqrt{\varepsilon}} \left(-\nu_1^* \pm \sqrt{\nu_1^{*2} - 2\varepsilon \nu_{11}^*(H_2 - H_2^*)} \right). \qquad (A.14)$$

($G_2'^* = 0$ since $H_1^* = 0$.) This nonlinear equation is almost the same as that which appeared as the fundamental equation (4.34) of Delaunay's theory. It is indeterminate while H_2^* is not fixed. This indeterminacy was overcome by the introduction of the *weak* averaging rule

A.2 Bohlin's Perturbation Equations

$$H_2^*(\theta_\varrho, J^*) = <H_2(\theta, J^*)>, \tag{A.15}$$

where $<\cdots>$ stands for the average over the angle θ_1. This is not the only choice found in the literature. The so-called "minimum principle":

$$H_2^*(\theta_\varrho, J^*) = \min_{\theta_1} H_2(\theta, J^*)$$

was also used. However, none of these rules was able to solve Bohlin's problem and they brought, as a consequence, the singularity discussed in the next section. If we adopt the averaging rule defined by (A.15), we have

$$\frac{\nu_1^*}{\sqrt{\varepsilon}}\frac{\partial S_1}{\partial \theta_1} + \frac{1}{2}\nu_{11}^*\left(\frac{\partial S_1}{\partial \theta_1}\right)^2 + H_{2(K)} = 0, \tag{A.16}$$

where

$$H_{2(K)}(\theta, J^*) = H_2(\theta, J^*) - <H_2(\theta, J^*)>. \tag{A.17}$$

Bohlin's equation defines several regimes of motion (as in Garfinkel's Ideal Resonance Problem), but their study can only be done when the function $H_2(\theta, J^*)$ is explicitly given.

The remaining equations may be generically written as

$$\left(\frac{\nu_1^*}{\sqrt{\varepsilon}} + \nu_{11}^*\frac{\partial S_1}{\partial \theta_1}\right)\frac{\partial S_k}{\partial \theta_1} + \Psi_{k+1} = H_{k+1}^* \tag{A.18}$$

(*homological equation*). Taking into account that the functions $G_{2,k+1}, G_{4,k+1}, \cdots, \mathcal{E}'_{k+1}$ and G'^*_{k+1} are completely known when the functions $S_1, S_2, \cdots, S_{k-1}$ are known, the term Ψ_{k+1} in the homological equation (A.18) represents a known function. In contrast with Bohlin's equation, the homological equation is linear.

In addition to the practical difficulties in the integration of (A.14), involving an elliptic integral, there are other difficulties to consider. As discussed by Garfinkel et al. [36], $\nu_1^* = \mathcal{O}(\sqrt{\varepsilon})$ while its derivative with respect to J_1^* is $\nu_{11}^* = \mathcal{O}(1)$. Thus, differentiation with respect to J_1^* reduces the order of every function having ν_1^* as a factor[1]. As a consequence, when the explicit equations of the transformation are calculated, the term S_k contributes parts of order $(k-1)$ in the expression for $\theta_1 = \partial S/\partial J_1^*$.

Since θ_1 is given by the derivative of the Jacobian generating function with respect to J_1^*, the dependence of the S_k on J_k^* must be determined. For this reason, we cannot fix the value of the constant J_1^*, and the simplification $\nu_1^* = 0$, used in Delaunay's theory, cannot be used here, before the transformation is obtained. However, once the transformation is known, the adoption of $\nu_1^* = 0$ is possible. The order-half approximation

---

[1] This is a key difference between the classical theories and the Lie series theory for resonant system of Chap. 8. There, and in [33], the order in $\sqrt{\varepsilon}$ is replaced by the degree of homogeneity in the elements of the set $\mathcal{S} \equiv (J_1 - J_1^*, \sqrt{\varepsilon})$, thus allowing us to control the order of the terms and avoiding any mixing.

$$J_1 = J_1^* + \frac{\partial S_1}{\partial \theta_1} + \mathcal{O}(\varepsilon)$$

$$= J_1^* + \frac{1}{\nu_{11}^*\sqrt{\varepsilon}}\left(-\nu_1^* \pm \sqrt{\nu_1^{*2} - 2\varepsilon\nu_{11}^*(H_2 - H_2^*)}\right) + \mathcal{O}(\varepsilon)$$

shows that, in the case of libration, a choice of ν_1^* not equal to zero introduces an offset in the mean value of the oscillation of J_1, with respect to J_1^*.

A.3 Poincaré Singularity

As pointed out by Poincaré, the most serious difficulty in the study of the general Bohlin's problem by means of the perturbation equations (A.12) occurs in the case of librations. In that case, the homological equation is, generally, singular. For instance, the equation giving S_2 is

$$\left(\frac{\nu_1^*}{\sqrt{\varepsilon}} + \nu_{11}^*\frac{\partial S_1}{\partial \theta_1}\right)\frac{\partial S_2}{\partial \theta_1} + \Psi_3 = H_3^*, \tag{A.19}$$

where

$$\Psi_3 = G_{2,3} + \mathcal{E}_3' - G_3'^*$$
$$= \frac{\partial H_2}{\partial J_1^*}\frac{\partial S_1}{\partial \theta_1} + \frac{1}{6}\nu_{111}^*\left(\frac{\partial S_1}{\partial \theta_1}\right)^3 + \sum_{\varrho=2}^{N}\left(\frac{\partial H_2}{\partial J_\varrho^*}\frac{\partial S_1}{\partial \theta_\varrho} - \frac{\partial H_2^*}{\partial \theta_\varrho}\frac{\partial S_1}{\partial J_\varrho^*}\right). \tag{A.20}$$

Equation (A.19) is singular when

$$\frac{\nu_1^*}{\sqrt{\varepsilon}} + \nu_{11}^*\frac{\partial S_1}{\partial \theta_1} = 0$$

and this condition always happens at the border of the librations. In the particular case of the problem initially proposed by Bohlin (A.5) and in Garfinkel's Ideal Resonance Problem, the last term of (A.20) does not exist. Therefore, a suitable choice of H_3^* can be made so that the right-hand side also appears multiplied by $\left(\frac{\nu_1^*}{\sqrt{\varepsilon}} + \nu_{11}^*\frac{\partial S_1}{\partial \theta_1}\right)$ and the equation is no longer singular. However, in the general Bohlin's problem, the singularity cannot be eliminated just through an appropriate choice of H_3^* because some terms of the last summation depend on θ_1 and H_3^* cannot depend on θ_1.

Notwithstanding several attempts, these difficulties were not properly solved in the frame of classical theories, except for systems with just one degree of freedom, like Garfinkel's Ideal Resonance Problem, or reduced to one degree of freedom, such as the restricted problem proposed in Bohlin's paper.

A.4 An Extension of Delaunay Theory

A classical alternative to Bohlin theory is the Delaunay theory presented in Sect. 4.3 for one degree of freedom. It may be generalized to be used in the study of Bohlin's problem. To do this, we introduce the canonical transformation

$$(\theta_i, J_i) \Rightarrow (\alpha, \theta_\varrho^*, E, J_\varrho^*) \qquad (i = 1, 2, \cdots, N; \varrho = 2, \cdots, N)$$

defined by the Jacobian generating function

$$S = (\theta \mid J^*) + \sum_{k=1}^{n} \varepsilon^{k/2} S_k(\theta, E, J_\varrho^*), \tag{A.21}$$

where J_1^* is the solution of the equation giving the exact resonance:

$$\nu_1(J_1^*) = \left(\frac{dH_0}{dJ_1}\right)_{J_1=J_1^*} = 0. \tag{A.22}$$

The equations of the canonical transformation are

$$\alpha = \frac{\partial S}{\partial E}; \qquad \theta_\varrho^* = \frac{\partial S}{\partial J_\varrho^*}; \qquad J_i = \frac{\partial S}{\partial \theta_i} \tag{A.23}$$

and the transformed Hamiltonian is assumed to have a main part

$$\varepsilon E + H^*(\theta_\varrho^*, E, J_\varrho^*),$$

independent of α, and a remainder $\mathcal{R}_{n+1}$ divisible by $\varepsilon^{(n+1)/2}$. As in the von Zeipel–Brouwer theory (for $M = 1$), the main part of the transformed Hamiltonian defines a canonical system reduced to $N - 1$ degrees of freedom:

$$\dot{\theta}_\varrho^* = \frac{\partial H^*}{\partial J_\varrho^*}, \qquad \dot{J}_\varrho^* = -\frac{\partial H^*}{\partial \theta_\varrho^*}. \tag{A.24}$$

The solution is completed with the integral

$$E = \text{constant} \tag{A.25}$$

and the quadrature

$$\alpha = \int \frac{\partial}{\partial E}(H^* + \varepsilon E) \, dt. \tag{A.26}$$

Following the same steps as in Sect. A.2, we obtain $H_0(J_1^*) = H_0^*$, $0 = H_1^*$, the particular Bohlin equation

$$\frac{1}{2}\nu_{11}^* \left(\frac{\partial S_1}{\partial \theta_1}\right)^2 + H_2(\theta, J_\varrho^*) - H_2^*(\theta, E, J_\varrho^*) = E, \tag{A.27}$$

and the homological equation ($k \geq 2$)

$$\nu_{11}^* \frac{\partial S_1}{\partial \theta_1} \frac{\partial S_k}{\partial \theta_1} + \Psi_{k+1}^*(\theta, E, J_\varrho^*) = H_{k+1}^*(\theta, E, J_\varrho^*), \tag{A.28}$$

where Ψ_{k+1}^* represents known functions. As in the previous section, $G_2'^* = 0$ because $H_1^* = 0$. These equations are very similar to those of Bohlin theory except for the introduction of E and for the fact that J_1^* is no longer a variable but a constant. The constant J_1^* allows ν_1^* to be fixed (as $\nu_1^* = 0$).

In contrast with Bohlin's theory, a particular solution of the fundamental equation is no longer sufficient to determine the canonical transformation. As in Hamilton–Jacobi theory, we need an integral depending on the non-trivial constant E, that is, we need a complete integral of the fundamental equation (A.27). When a complete integral is known, the generating function

$$S_{(1)} = (\theta \mid J^*) + \sqrt{\varepsilon} S_1(\theta, E, J_\varrho^*)$$

defines a canonical transformation leading to a transformed Hamiltonian independent of α, except for terms factored by, at least, $\varepsilon^{3/2}$. The same *weak* averaging rule of Bohlin's theory is used.

The homological equation (A.28) defining the transformation at higher orders is linear and it is sufficient to obtain particular solutions of them. However, for librations, (A.28) is singular for $\partial S_1/\partial \theta_1 = 0$ (Poincaré's singularity). It is possible to use this method for one-degree-of-freedom problems [88], but, in the general case, it is not possible to eliminate this singularity through just an appropriate choice of H_{k+1}^*.

B

The Simple Pendulum

B.1 Equations of Motion

The simple pendulum is a one-dimensional mechanical system with a force field whose potential is

$$U(q) = -k \cos q \qquad (k > 0). \tag{B.1}$$

Its Hamiltonian is

$$H(q, p) = \frac{p^2}{2m} - mk \cos q \tag{B.2}$$

and the corresponding differential equations are

$$\frac{dq}{dt} = \frac{\partial H}{\partial p} = \frac{p}{m} \tag{B.3}$$

$$\frac{dp}{dt} = -\frac{\partial H}{\partial q} = -mk \sin q \tag{B.4}$$

or

$$\frac{d^2 q}{dt^2} = -k \sin q. \tag{B.5}$$

The last equation is the Newtonian equation of a point of mass m attracted by a center at $q = 0$ with a force $mk \sin q$. In a real pendulum, $k = g/\ell$, where g is the local acceleration of gravity and ℓ is the length of the pendulum.

The solution of (B.5) is classical. First of all, we have the energy integral

$$\frac{1}{2}\left(\frac{dq}{dt}\right)^2 - k \cos q = \frac{E}{m} \tag{B.6}$$

from which we obtain

$$\frac{dq}{dt} = \pm \sqrt{2\left(\frac{E}{m} + k \cos q\right)} \tag{B.7}$$

and the elliptic integral

$$t - t_0 = \int_{q_0}^{q} \frac{dq}{\pm\sqrt{2(\frac{E}{m} + k \cos q)}}. \tag{B.8}$$

The trajectories in the phase space are obtained just by plotting the curves $H = E$ (constant) for various values of E. They are shown in Fig. B.1.

We have the possibilities

- $\frac{E}{m} > k > 0$ (circulations)

  In this case, the square root in (B.7) is real for all values of q. The sign in front of the square root depends on the branch chosen for dq/dt and does not change for all t. Thus, the angle q is a monotonically increasing (or decreasing) time function. To each value of E, two solutions may be found, one in the upper half of the phase plane ($m\dot{q} > 0$) and another in the lower half of it ($m\dot{q} < 0$). These solutions of the pendulum motion are called *circulations*.

- $E = mk$ (separatrices)

  Again, the square root in (B.7) is real for all values of q but becomes zero at $q = \pm\pi$. At this point we also have $d^2q/dt^2 = 0$. The solutions corresponding to this energy level are the unstable equilibrium point U at $q = \pm\pi$ and the two *separatrices* starting in one unstable point and ending at its congruent, 2π away.

- $|E| < |mk|$ (librations)

  In this case, the square root in (B.7) is real only in the interval $|q| < \arccos(-E/mk)$ (since otherwise $\frac{E}{m} + k \cos q < 0$). The motion is a limited oscillation about $q = 0$. That the motion is an oscillation is an immediate

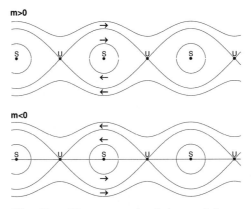

Fig. B.1. Phase portrait of the pendulum

consequence of having $\mathrm{d}^2q/\mathrm{d}t^2 \neq 0$ at the two interval boundaries $q = \pm \arccos(-E/mk)$. These solutions of the pendulum motion are called *librations*.

- $E = -mk$

  In this case, the square root in (B.7) is real only when $q = 0$. The corresponding solution is the stable equilibrium point S.

The square root is not real when $E/m < -k < 0$ and no motion corresponds to this case.

B.1.1 Circulation

In order to use standard elliptic integrals and elliptic functions in the integration of (B.7), we introduce

$$\cos q = 1 - 2\sin^2 \frac{q}{2}$$

and write (B.7) as

$$\frac{\mathrm{d}q}{\mathrm{d}t} = \pm \eta \sqrt{1 - \kappa^2 \sin^2 \frac{q}{2}}, \quad (B.9)$$

where

$$\kappa = \sqrt{\frac{2mk}{E+mk}} \quad (0 < \kappa < 1) \quad (B.10)$$

and

$$\eta = \sqrt{\frac{2}{m}(E+mk)} = \frac{2\sqrt{k}}{\kappa}. \quad (B.11)$$

The positive sign corresponds to the prograde circulations ($\mathrm{d}q/\mathrm{d}t > 0$) and the negative sign to retrograde circulations ($\mathrm{d}q/\mathrm{d}t < 0$). According to (B.3) p has the same sign as m in the prograde motion branch and the opposite sign in the retrograde motion branch (see Fig. B.1). The integration of (B.9) gives

$$\eta t = \pm 2\mathcal{F}\left(\frac{q}{2}, \kappa\right), \quad (B.12)$$

where $\mathcal{F}(\frac{q}{2}, \kappa)$ is the elliptic integral[1] of first kind with modulus κ. We assume $t = 0$ at $q = 0$. The period of the circulation is the time to go from $q = 0$ to $q = 2\pi$, that is,

$$T_C = \frac{4}{\eta}\mathbb{K}(\kappa), \quad (B.13)$$

---

[1] Calculations with elliptic integrals are easy when good tables are available. In order of increasing completeness, we cite [25], [1], [17]. The slight change of the usual notation for the elliptic integrals introduced here ($\mathcal{F}$ and $\mathcal{E}$ instead of F and E) is necessary to avoid confusion with other functions in the book. $\mathbb{K}$ and $\mathbb{E}$ are the corresponding complete elliptic integrals.

where $\mathbb{K}(\kappa) = \mathcal{F}(\frac{\pi}{2}, \kappa)$ is the complete elliptic integral of the first kind.

The inversion of (B.12) gives

$$q = \pm 2\,\mathrm{am}\frac{\eta t}{2}, \tag{B.14}$$

where *am* stands for the Jacobian *amplitude*. It is worth recalling that we also have

$$\cos\frac{q}{2} = \mathrm{cn}\frac{\eta t}{2} \qquad \sin\frac{q}{2} = \pm\mathrm{sn}\frac{\eta t}{2},$$

where *cn* and *sn* stand for the Jacobian *cosine amplitude* and *sine amplitude* elliptic functions, respectively. The modulus of these elliptic functions is κ.

For practical purposes, it is important to have the expression for $\sin q$. This may be easily obtained from the classical formulas for trigonometric functions and their equivalent for elliptic functions:

$$\sin q = 2\sin\frac{q}{2}\cos\frac{q}{2} = \pm 2\,\mathrm{sn}\frac{\eta t}{2}\,\mathrm{cn}\frac{\eta t}{2} = \mp\frac{4}{\eta\kappa^2}\frac{\mathrm{d}}{\mathrm{d}t}\mathrm{dn}\frac{\eta t}{2} \tag{B.15}$$

where *dn* is the Jacobian *delta amplitude* elliptic function.

The Fourier expansions of the solution are[2]

$$q = \pm w \pm \sum_{\ell=1}^{\infty}\frac{2}{\ell}\,\mathrm{sech}\,\ell\chi\,\sin\ell w, \tag{B.16}$$

$$\sin q = \pm\frac{2\pi^2}{\kappa^2\mathbb{K}^2}\sum_{\ell=1}^{\infty}\ell\,\mathrm{sech}\,\ell\chi\,\sin\ell w, \tag{B.17}$$

where

$$w = \frac{\pi\eta t}{2\mathbb{K}} = \frac{2\pi t}{T_C} \tag{B.18}$$

and

$$\chi(\kappa) = \frac{\pi\mathbb{K}(\sqrt{1-\kappa^2})}{\mathbb{K}(\kappa)}. \tag{B.19}$$

We recall that the quantity $e^{-\chi}$ is the so-called Jacobi's *nome* (designated by q in tables of elliptic functions).

B.1.2 Libration

In this case, $\kappa > 1$ and the left-hand side of (B.9) is real only for $|\sin\frac{q}{2}| < \frac{1}{\kappa}$. This difficulty is circumvented by means of the reciprocal modulus transformation. We introduce

---

[2] These formulas are found in [17]. There is a printing flaw there in equation 908.03. The right denominator is $1 + q^{(2m+1)}$ (see [98], p. 511). For a different approach see [82].

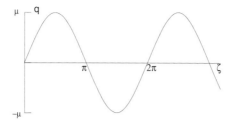

Fig. B.2. The function $q(\zeta)$. ($\mu = 2 \arcsin \kappa^{-1}$)

$$\sin \zeta = \kappa \sin \frac{q}{2} \tag{B.20}$$

or, after differentiation,

$$\cos \zeta \, d\zeta = \frac{\kappa}{2} \cos \frac{q}{2} \, dq$$

and

$$\frac{d\zeta}{dt} = \frac{\kappa \eta}{2} \sqrt{1 - \frac{1}{\kappa^2} \sin^2 \zeta}, \tag{B.21}$$

where the double sign disappears because the branches of $\arcsin(\alpha \sin \frac{q}{2})$ are chosen such that $\zeta(t)$ is monotonic. The integration of this equation gives

$$\frac{\kappa \eta t}{2} = \mathcal{F}\left(\zeta, \frac{1}{\kappa}\right), \tag{B.22}$$

where we assumed $\zeta = 0$ and $\dot{\zeta} > 0$ (i.e. $q = 0$ and $\dot{q} > 0$) at $t = 0$. The period of libration is equal to four times the time to go from $q = 0$ to the boundary of the libration at $q = 2 \arcsin \frac{1}{\kappa}$ (or $\zeta = \frac{\pi}{2}$) (see Fig. B.2). Thus

$$T_L = 4 \sqrt{\frac{1}{k}} \, \mathbb{K}\left(\frac{1}{\kappa}\right). \tag{B.23}$$

The inversion of (B.22) gives

$$\zeta = \text{am} \frac{\kappa \eta t}{2}. \tag{B.24}$$

We also have

$$\sin \zeta = \text{sn} \frac{\kappa \eta t}{2}$$

and, therefore,

$$\sin \frac{q}{2} = \frac{1}{\kappa} \, \text{sn} \frac{\kappa \eta t}{2}. \tag{B.25}$$

In the above calculations (and in the forthcoming ones), one must bear in mind that the modulus of the elliptic functions is now $1/\kappa$ (instead of κ). Using the fundamental relations of elliptic functions, we obtain

$$\cos\frac{q}{2} = \sqrt{1 - \frac{1}{\kappa^2}\operatorname{sn}^2\frac{\kappa\eta t}{2}} = \operatorname{dn}\frac{\kappa\eta t}{2} \qquad (B.26)$$

$$\sin q = \frac{2}{\kappa}\operatorname{sn}\frac{\kappa\eta t}{2}\operatorname{dn}\frac{\kappa\eta t}{2} = -\frac{4}{\kappa^2\eta}\frac{d}{dt}\operatorname{cn}\frac{\kappa\eta t}{2}. \qquad (B.27)$$

The Fourier expansions of the solution are

$$\zeta = w + \sum_{\ell=1}^{\infty}\frac{1}{\ell}\operatorname{sech}\ell\chi\,\sin 2\ell w, \qquad (B.28)$$

$$\sin q = \frac{\pi^2}{\mathbb{K}^2}\sum_{\ell=1}^{\infty}(2\ell-1)\operatorname{sech}(2\ell-1)\frac{\chi}{2}\sin(2\ell-1)w, \qquad (B.29)$$

where

$$w = \frac{\kappa\pi\eta t}{4\mathbb{K}} = \frac{2\pi t}{T_L} \qquad (B.30)$$

and

$$\chi = \chi\left(\frac{1}{\kappa}\right) = \frac{\pi\mathbb{K}\left(\sqrt{1-\frac{1}{\kappa^2}}\right)}{\mathbb{K}(\frac{1}{\kappa})}. \qquad (B.31)$$

Since we adopted $\dot{q} > 0$ at $t = 0$, the above equations become insensitive to the individual sign of m. One should bear in mind, however, that differences will appear when the variation of p is considered (see B.4). At $t = 0$, p may have the same sign as m. Then, the motion in the phase plane is clockwise when $m > 0$ and counterclockwise when $m < 0$ (see Fig. B.1).

B.1.3 The Separatrix

In this case, $\kappa = 1$ and the integrals are no longer elliptic. Equation (B.7) becomes

$$\frac{dq_{\text{sx}}}{dt} = \pm\sqrt{4k}\cos\frac{q_{\text{sx}}}{2}, \qquad (B.32)$$

where we use q_{sx} to denote the angle q over the separatrix. The solution of this equation is

$$t = \pm\sqrt{\frac{1}{k}}\ln\tan\left(\frac{q_{\text{sx}}}{4} + \frac{\pi}{4}\right) \qquad (-\pi < q_{\text{sx}} < \pi), \qquad (B.33)$$

or, after inversion,

$$q_{\text{sx}} = 4\arctan e^{\pm\sqrt{k}t} - \pi, \qquad (B.34)$$

where we have fixed the initial condition $q_{\text{sx}} = 0$ at $t = 0$. The positive sign in the above equations corresponds to a prograde motion along the separatrix (q_{sx} increases with t) and the negative sign corresponds to a retrograde motion along the separatrix (q_{sx} decreases as t increases). When $m > 0$, the prograde motion occurs on the upper separatrix and the retrograde motion on the lower separatrix. When $m < 0$, this situation is reversed.

From (B.3) and (B.32), we obtain

$$p_{sx} = \pm 2|m|\sqrt{k} \cos \frac{q_{sx}}{2}; \tag{B.35}$$

The width of the libration zone is given by the maximum separation of the two separatrices, that is, twice the value of $|p_{sx}|$ at $q_{sx} = 0$, or $4|m|\sqrt{k}$.

Equation (B.33) shows that the motion along the separatrices is asymptotic to the unstable equilibrium points $q_{sx} = \pm \pi$, since

$$\lim_{q_{sx} \to \pm \pi} |t| = \infty.$$

Exercise B.1.1. Show that the quantities w defined by (B.18) and (B.30) are the same angle variables defined in Sects. B.2.1 and B.2.2.

Exercise B.1.2. Explain, kinematically, why the argument of the periodic terms of q in (B.16) is w, while that of ζ in (B.28) is $2w$.

B.2 Angle–Action Variables of the Pendulum

Let us use the basic equations of Sects. 2.1 and 2.2 to construct the angle–action variables of the simple pendulum:

$$S = \int p(q, E) \, dq = m \int \left(\frac{dq}{dt}\right) dq, \tag{B.36}$$

$$J \stackrel{\text{def}}{=} \frac{1}{2\pi} \oint p(q, E) \, dq \tag{B.37}$$

and

$$w \stackrel{\text{def}}{=} \frac{\partial S(q, E(J))}{\partial J} = \frac{\partial S}{\partial E} \frac{dE}{dJ}. \tag{B.38}$$

We have to consider separately the cases of circulation ($E/m > k > 0$) and libration ($|E| < |mk|$).

B.2.1 Circulation

In this case, the integration of (B.36) gives

$$S = \pm m\eta \int \sqrt{1 - \kappa^2 \sin^2 \frac{q}{2}} \, dq = \pm 2m\eta \, \mathcal{E}\left(\frac{q}{2}, \kappa\right), \tag{B.39}$$

where we have introduced κ and η as defined in Sect. B.1.1. $\mathcal{E}(\frac{q}{2},\kappa)$ is the elliptic integral of the second kind with modulus κ. The $+$ sign corresponds to the branch of the circulation where $\dot{q} > 0$, and the $-$ sign to that where $\dot{q} < 0$. Differentiating S with respect to E, it follows that

$$\frac{\partial S}{\partial E} = \pm \frac{2}{\eta} \mathcal{F}\left(\frac{q}{2}, \kappa\right). \tag{B.40}$$

Since the angle q is circulating, the function $S(q)$ is 2π-periodic, and $\oint$ is just an integral from $q = 0$ to $q = \pm 2\pi$. Then

$$J = \frac{1}{2\pi} S(\pm 2\pi) = \frac{2m\eta}{\pi} \mathbb{E}(\kappa), \tag{B.41}$$

where $\mathbb{E}(\kappa) = \mathcal{E}(\frac{\pi}{2}, \kappa)$ is the complete elliptic integral of the second kind with modulus κ (the double signs of S and of its argument cancel themselves):

$$\frac{dJ}{dE} = \frac{2}{\eta\pi} \mathbb{K}(\kappa). \tag{B.42}$$

Hence, the angle variable is

$$w = \pm \frac{\pi \mathcal{F}(\frac{q}{2}, \kappa)}{\mathbb{K}(\kappa)} = \frac{\pi \mathcal{F}(\pm \frac{q}{2}, \kappa)}{\mathbb{K}(\kappa)}, \tag{B.43}$$

or $w = \dot{w}(t - t_0)$, where

$$\dot{w} = \frac{\pi \eta}{2\mathbb{K}} = \frac{\pi \sqrt{k}}{\kappa \mathbb{K}} > 0. \tag{B.44}$$

B.2.2 Libration

We have to introduce, first, the reciprocal modulus transformation, and replace q by the variable ζ defined by (B.20). It follows that

$$S = 4m\sqrt{k} \left[\mathcal{E}\left(\zeta, \frac{1}{\kappa}\right) + \beta \mathcal{F}\left(\zeta, \frac{1}{\kappa}\right) \right] \tag{B.45}$$

and

$$J = \frac{8m}{\pi} \sqrt{k} \left[\mathbb{E}\left(\frac{1}{\kappa}\right) + \beta \mathbb{K}\left(\frac{1}{\kappa}\right) \right], \tag{B.46}$$

where

$$\beta = \frac{1 - \kappa^2}{\kappa^2} = \frac{E - km}{2km}. \tag{B.47}$$

The calculation of the derivatives of $S(q, E)$ is, now, more cumbersome than in the previous case because the above functions depend on E also through the variable ζ. However, all calculations are elementary and the results are

$$\frac{\partial S}{\partial E} = \frac{1}{\sqrt{k}} \mathcal{F}\left(\varsigma, \frac{1}{\kappa}\right) \tag{B.48}$$

and

$$\frac{dJ}{dE} = \frac{2}{\pi\sqrt{k}} \mathbb{K}\left(\frac{1}{\kappa}\right). \tag{B.49}$$

As a consequence,

$$w = \frac{\pi \mathcal{F}(\varsigma, \kappa^{-1})}{2\mathbb{K}(\kappa^{-1})}, \tag{B.50}$$

or $w = \dot{w}(t - t_0)$, where

$$\dot{w} = \frac{\pi\sqrt{k}}{2\mathbb{K}(\kappa^{-1})} > 0. \tag{B.51}$$

B.3 Small Oscillations of the Pendulum

In the case of small oscillations, the results of Sect. B.1.2 can be written in a more explicit way. In this case, $1/\kappa$ is a small quantity ($\kappa \to \infty$) and we can use power series to express the hyperbolic functions appearing in the coefficients of the Fourier series. The key series is the one giving Jacobi's *nome* with modulus $1/\kappa$:

$$e^{-\chi(\kappa^{-1})} = \alpha^2(1 + 2\alpha^2 + 15\alpha^4 + 150\alpha^6 + 1707\alpha^8 + \cdots)^4,$$

where

$$\alpha = \frac{1}{4\kappa}; \tag{B.52}$$

or, computing the fourth power indicated in the right parenthesis,

$$e^{-\chi(\kappa^{-1})} = \alpha^2(1 + 8\alpha^2 + 84\alpha^4 + 992\alpha^6 + 12514\alpha^8 + \cdots). \tag{B.53}$$

Using the series for $\mathbb{K}$:

$$\mathbb{K}\left(\frac{1}{\kappa}\right) = \frac{\pi}{2}(1 + 4\alpha^2 + 36\alpha^4 + 400\alpha^6 + 4900\alpha^8 + \cdots),$$

and the elementary equation

$$\text{sech}\, b\chi = \frac{2}{e^{b\chi} + e^{-b\chi}} = 2e^{-b\chi}(1 + e^{-2b\chi})^{-1},$$

we obtain

$$\begin{aligned}\sin q = \ &8\alpha(1 - 5\alpha^2 - 25\alpha^4 - 219\alpha^6)\sin w + 24\alpha^3(1 + 4\alpha^2 + 30\alpha^4)\sin 3w \\ &+ 40\alpha^5(1 + 12\alpha^2)\sin 5w + 56\alpha^7 \sin 7w + \mathcal{O}(\alpha^9).\end{aligned} \tag{B.54}$$

It is also useful to have

280   B The Simple Pendulum

$$\dot{w} = \sqrt{k}(1 - 4\alpha^2 - 20\alpha^4 - 176\alpha^6 + \cdots) \tag{B.55}$$

and

$$p = m\frac{dq}{dt} = 8m\sqrt{k}\left[\alpha\,(1-\alpha^2-9\alpha^4-99\alpha^6)\cos w \right.$$
$$+\alpha^3(1+8\alpha^2+82\alpha^4)\cos 3w + \alpha^5(1+16\alpha^2)\cos 5w$$
$$\left. + \alpha^7\cos 7w\right] + \mathcal{O}(\alpha^9). \tag{B.56}$$

Exercise B.3.1. Write the small-amplitude oscillations of the simple pendulum as

$$q \simeq \sin q \simeq \Theta \sin w, \tag{B.57}$$

where Θ is the oscillation's half-amplitude and w is the angle variable of the simple pendulum. Since, by definition, $\{q,p\} = 1$, show that the action conjugate to w, in this approximation, is

$$J = \frac{m\dot{w}\Theta^2}{2}. \tag{B.58}$$

Hint: The small oscillations are isochronous.

B.3.1 Angle–Action Variables

The equations of Sect. B.2 may be approximated with elementary functions in the case of small-amplitude librations. To obtain J, besides the approximate formulas already introduced in Sect. B.3, we need:

$$\mathbb{E}\left(\frac{1}{\kappa}\right) = \frac{\pi}{2}(1 - 4\alpha^2 - 12\alpha^4 - 80\alpha^6 - 700\alpha^8 + \cdots). \tag{B.59}$$

Hence

$$J = 32m\sqrt{k}\,\alpha^2(1 + 2\alpha^2 + 12\alpha^4 + 100\alpha^6 + \cdots) \tag{B.60}$$

and its inverse

$$\alpha^2 = \Upsilon(1 - 2\Upsilon - 4\Upsilon^2 - 20\Upsilon^3 + \cdots), \tag{B.61}$$

where

$$\Upsilon = \frac{J}{32m\sqrt{k}} = \frac{1}{4\pi}[\mathbb{E}(\kappa^{-1}) + \beta\mathbb{K}(\kappa^{-1})] > 0. \tag{B.62}$$

With angle–action variables, the oscillations are given by:

$$\sin q = 8\sqrt{\Upsilon}\left[\left(1 - 6\Upsilon - \frac{25}{2}\Upsilon^2 - 84\Upsilon^3\right)\sin w + 3\Upsilon\left(1 + \Upsilon + \frac{11}{2}\Upsilon^2\right)\sin 3w\right.$$
$$\left. + 5\Upsilon^2(1+7\Upsilon)\sin 5w + 7\Upsilon^3\sin 7w\right] + \mathcal{O}(\Upsilon^{9/2}). \tag{B.63}$$

The angle variable is uniform and has the variation rate

$$\dot{w} = \sqrt{k}\left(1 - 4\Upsilon - 12\Upsilon^2 - 80\Upsilon^3 + \cdots\right) \tag{B.64}$$

and the libration period is

$$T_L = \frac{2\pi}{\sqrt{k}}(1 + 4\Upsilon + 28\Upsilon^2 + 240\Upsilon^3 + \cdots). \tag{B.65}$$

Similarly, for the momentum p, we obtain

$$p = 8m\sqrt{k\Upsilon}\left[\left(1 - 2\Upsilon - \frac{17}{2}\Upsilon^2 - 62\Upsilon^3\right)\cos w + \Upsilon\left(1 + 5\Upsilon + \frac{75}{2}\Upsilon^2\right)\cos 3w\right.$$
$$\left. + \Upsilon^2(1 + 11\Upsilon)\cos 5w + \Upsilon^3 \cos 7w\right] + \mathcal{O}(\Upsilon^{9/2}). \tag{B.66}$$

Finally, we may express the energy (that is, the Hamiltonian) with the angle–action variables. It is

$$E = -km + 32km\alpha^2 = -km + 32km\Upsilon(1 - 2\Upsilon - 4\Upsilon^2 - 20\Upsilon^3 + \cdots). \tag{B.67}$$

B.4 Direct Construction of Angle–Action Variables

In the case of small oscillations, the alternative formulations with undetermined coefficients (see Sect. 2.2) may be used to construct the angle–action variables of the pendulum. The first and most lengthy step in that formulation is the construction of the series representing the periodic solutions.

Let us represent the solutions of the pendulum by the series

$$q = \sum_{i=1}^{n} a_i \gamma^i,$$

where γ is a free parameter of the order of the amplitude of the oscillations ($\gamma = 0$ corresponds to the stable equilibrium solution $q = 0$) and a_i are undetermined periodic functions in the angles w. It is important to keep in mind that $\dot{w}$ is not the same in all solutions and is itself also a function of the parameter γ with undetermined coefficients. Let it be written

$$\dot{w} = \omega_0 + \sum_{i=1}^{n} o_i \gamma^i.$$

These series are then substituted into the differential equation. The identification following the powers of γ gives the equations allowing the determination of a_i and o_i. In the case of the pendulum, the differential equation is $\ddot{q} = q''\dot{w}^2 = -k\sin q$, where the primes indicate differentiation with respect to w. The equations resulting from the identification in the powers of γ are:

- $i = 1$
$$a_1'' \omega_0^2 + k a_1 = 0. \tag{B.68}$$

The solution is
$$a_1 = \sin w \qquad \omega_0 = \sqrt{k}. \tag{B.69}$$

Only the particular solutions of these equations are required. The arbitrary integration constants are the parameter γ and the phase of w (which does not need to appear explicitly).

- $i = 2$
$$a_2'' \omega_0^2 + k a_2 = -2 a_1'' \omega_0 o_1. \tag{B.70}$$

When the results for $i = 1$ are substituted into the right-hand side of the above equation, it becomes $-2\omega_0 o_1 \sin w$. The resulting equation is then a non-homogeneous ordinary differential equation with constant coefficients with equal proper and forced frequencies. The existence in the forced part of terms with the same frequency as the proper frequency of the associated homogeneous equation leads to unwanted unbounded terms in the solution. The identificsation to zero of the terms proportional to $\sin w$ in the right-hand side gives a rule to determine the o_i. In this simple case, $o_1 = 0$. For the resulting homogeneous equation we have the particular solution $a_2 = 0$.

- $i = 3$
$$a_3'' \omega_0^2 + k a_3 = \frac{1}{6} k a_1^3 - a_1''(2\omega_0 o_2 + o_1^2) - 2 a_2'' \omega_0 o_1 \tag{B.71}$$

or, after substitution of the previous results,
$$a_3'' k + k a_3 = 2\sqrt{k}\, o_2 \sin w + \frac{1}{8} k \sin w - \frac{1}{24} k \sin 3w. \tag{B.72}$$

Again, the terms in $\sin w$ in the right-hand side must vanish to avoid unbounded terms in the solution. This gives
$$o_2 = -\frac{\sqrt{k}}{16}. \tag{B.73}$$

The particular solution of the equation formed by the remaining terms is now easily obtained:
$$a_3 = \frac{1}{192} \sin 3w. \tag{B.74}$$

□

The next orders are treated in exactly the same way and we do not need to give the details of the calculations here. Collecting the results up to $n = 7$, we obtain

$$\begin{aligned} q &= \gamma \sin w + \frac{\gamma^3}{192} \sin 3w + \frac{\gamma^5}{20480}(5 \sin 3w + \sin 5w) \\ &\quad + \frac{\gamma^7}{1835008}(21 \sin 3w + 7 \sin 5w + \sin 7w) + \mathcal{O}(\gamma^9) \end{aligned}$$

and
$$\dot w = \sqrt{k} - \frac{1}{16}\sqrt{k}\,\gamma^2 + \frac{1}{1024}\sqrt{k}\gamma^4 - \frac{1}{65536}\sqrt{k}\gamma^6 + \mathcal{O}(\gamma^8). \tag{B.75}$$

In the sequence, we have, by mere application of the given expressions,

$$p = m\dot q = m\sqrt{k}\,\Big(\gamma\cos w + \frac{\gamma^3}{64}(-4\cos w + \cos 3w)$$
$$+ \frac{\gamma^5}{4096}(4\cos w - \cos 3w + \cos 5w)$$
$$+ \frac{\gamma^7}{262144}(-4\cos w + \cos 3w + \cos 5w + \cos 7w)\Big) + \mathcal{O}(\gamma^9),$$

$$J = \frac{1}{2\pi}\int_0^{2\pi} p q'\, dw = \frac{1}{2} m\sqrt{k}\,\gamma^2\left(1 - \frac{1}{16}\gamma^2 + \frac{5}{4096}\gamma^4 - \frac{1}{131072}\gamma^6\right) + \mathcal{O}(\gamma^8) \tag{B.76}$$

and

$$H = -mk + mk\left(\frac{1}{2}\gamma^2 - \frac{3}{64}\gamma^4 + \frac{17}{8192}\gamma^6 - \frac{13}{262144}\gamma^8\right) + \mathcal{O}(\gamma^{10}). \tag{B.77}$$

To obtain the Hamiltonian as a function of the actions, we need to solve (B.76) with respect to γ. We obtain

$$\gamma^2 = \frac{2J}{m\sqrt{k}} + \left(\frac{J}{2m\sqrt{k}}\right)^2 + \frac{27}{64}\left(\frac{J}{2m\sqrt{k}}\right)^3 + \frac{111}{512}\left(\frac{J}{2m\sqrt{k}}\right)^4 + \mathcal{O}(J^5), \tag{B.78}$$

hence

$$H = -mk + \sqrt{k}\,J - \frac{1}{16m}J^2 - \frac{1}{256\sqrt{k}m^2}J^3 - \frac{5}{8192km^3}J^4 + \mathcal{O}(J^5). \tag{B.79}$$

Exercise B.4.1. Show that the solutions obtained above are equivalent to those given in Sects. B.3 and B.3.1. *Hint.* Show first that

$$\gamma = 8\alpha + 24\alpha^3 + 184\alpha^5 + 1832\alpha^7 + \mathcal{O}(\alpha^9).$$

B.5 The Neighborhood of the Pendulum Separatrix

In studies of diffusion or capture into resonance, when separatrix crossings play a major role, we need the expressions of the action–angle variables in the neighborhood of the separatrix. These estimations are easy to obtain using the asymptotic expansions of the elliptic integrals for $\kappa \to 1$. These expansions are found in the literature [25], [17] as functions of $\kappa'^2 = 1 - \kappa^2$. In this application, it is convenient to have them written as functions of the normalized energy[3]:

---
[3] In diffusion studies, this quantity is often represented by w instead of h.

$$h = \frac{E}{mk} - 1. \tag{B.80}$$

We remember that at the separatrix, $E = mk$. Thus, $h \to 0$ when the trajectory approaches the separatrix. (Just for completeness, note that $h = -2$ at the stable equilibrium point O.) Some useful auxiliary expressions are $\eta = \sqrt{2k(h+2)}$, $\kappa = \sqrt{2/(h+2)}$ and

$$1 - \kappa^2 = \frac{h}{h+2} = \frac{h}{2} - \frac{h^2}{4} + \mathcal{O}(h^3). \tag{B.81}$$

In solutions outside the pendulum separatrix (that is, circulations), we use

$$\ln \frac{4}{\sqrt{1-\kappa^2}} = \frac{1}{2} \ln \frac{32}{h} + \frac{h}{4} + \frac{h^2}{16} + \mathcal{O}(h^3)$$

to obtain

$$\mathbb{K}(\kappa) = \frac{1}{2} \ln \frac{32}{h} + \frac{h}{16} \left(\ln \frac{32}{h} + 2 \right) + \mathcal{O}(h^2 \ln h)$$

$$\mathbb{E}(\kappa) = 1 + \frac{h}{8} \left(\ln \frac{32}{h} - 1 \right) + \mathcal{O}(h^2 \ln h)$$

$$\mathcal{F}(\phi, \kappa) = \ln \frac{1+\sin\phi}{\cos\phi} - \frac{h}{4} \left(\frac{\sin\phi}{\cos^2\phi} - \ln \frac{1+\sin\phi}{\cos\phi} \right) + \mathcal{O}(h^2)$$

$$\mathcal{E}(\phi, \kappa) = \sin\phi - \frac{h}{4} \left(\sin\phi - \ln \frac{1+\sin\phi}{\cos\phi} \right) + \mathcal{O}(h^2).$$

For solutions in the domain inside the separatrix (that is, librations), the arguments of the elliptic integrals are $1/\kappa$ instead of κ (in this case, $h < 0$ and $\kappa > 1$). We then use the exact expressions

$$1 - \left(\frac{1}{\kappa}\right)^2 = -\frac{h}{2} \tag{B.82}$$

and

$$\ln \frac{4}{\sqrt{1-\kappa^{-2}}} = \frac{1}{2} \ln \frac{32}{|h|}$$

to obtain

$$\mathbb{K}(\kappa^{-1}) = \frac{1}{2} \ln \frac{32}{|h|} - \frac{h}{16} \left(\ln \frac{32}{|h|} - 2 \right) + \mathcal{O}(h^2 \ln |h|)$$

$$\mathbb{E}(\kappa^{-1}) = 1 - \frac{h}{8} \left(\ln \frac{32}{|h|} - 1 \right) + \mathcal{O}(h^2 \ln |h|)$$

$$\mathcal{F}(\zeta, \kappa^{-1}) = \ln \frac{1+\sin\zeta}{\cos\zeta} + \frac{h}{4} \left(\frac{\sin\zeta}{\cos^2\zeta} - \ln \frac{1+\sin\zeta}{\cos\zeta} \right) + \mathcal{O}(h^2)$$

$$\mathcal{E}(\zeta, \kappa^{-1}) = \sin\zeta + \frac{h}{4}\left(\sin\zeta - \ln\frac{1+\sin\zeta}{\cos\zeta}\right) + \mathcal{O}(h^2).$$

The calculation of the action corresponding to one solution in the neighborhood of the separatrix is, now, easy to do. With the equations corresponding to circulation, we get

$$J = \frac{4m\sqrt{k}}{\pi}\left[1 + \frac{h}{8}\left(1 + \ln\frac{32}{|h|}\right)\right] + \mathcal{O}(h^2 \ln h). \quad \text{(B.83)}$$

With the equations corresponding to libration, we obtain twice the above value. This is due to the fact that the angle–action variables are not continuous at the separatrix between librations and circulations. It is enough to recall that the action is the area under the trajectory in the phase space ($\frac{1}{2\pi}\oint p\,dq$) to understand the doubling of the factor when passing from circulation to libration

B.5.1 Motion near the Separatrix

Let us consider the motion along a circulating or librating trajectory near the pendulum separatrix. We know that the time to go from one end of the separatrix to the other is infinite. We will calculate, in this section, the time spent in going from the vicinity of A to that of B along one of the two trajectories adjacent to the separatrix, shown in Fig. B.3. The travel times are long, but finite. In fact, the periods of circulation and libration tend to infinity, as the trajectory tends to the separatrix, with the same speed as $|\ln x|$ tends to ∞ when $x \to 0$. This means that it is enough to be a tiny distance away from the separatrix to have finite periods which are indeed large, but whose order of magnitude is not very different from that of the period of motions close to the stable equilibrium point O.

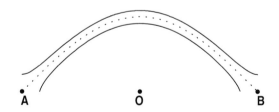

Fig. B.3. Trajectories adjacent to the separatrix

Let us calculate, separately, the times spent to go from one extremity to another along the two adjacent trajectories shown in Fig. B.3. Let us start with the outside one. The motion on this trajectory is a circulation. If we replace the complete elliptic integral $\mathbb{K}(\kappa)$ by its asymptotic value for $\kappa \to 1^-$,

$$\mathbb{K}(\kappa) \approx \ln \frac{4}{\sqrt{1-\kappa^2}}, \tag{B.84}$$

a simple calculation shows that

$$t_{\approx AB} = T_C \approx \sqrt{\frac{1}{k}} \ln \frac{32}{h}.$$

In the adjacent trajectory inside the separatrix, we similarly have

$$t_{\approx AB} = \frac{1}{2} T_L \approx \sqrt{\frac{1}{k}} \ln \frac{32}{|h|}. \tag{B.85}$$

The only differences with respect to the previous case are that h is, now, negative, T_L is proportional to $\mathbb{K}(\frac{1}{\kappa})$ instead of $\mathbb{K}(\kappa)$, and $t_{\approx AB}$ is equal to only half of T_L because the trajectory considered is only half of a complete libration.

B.6 The Separatrix or Whisker Map

One important characteristic of the motion near the separatrix is that even a small perturbation is enough to allow the motion to change from libration to circulation (or vice versa). The statistical study of these separatrix crossings is a necessary step to understanding the transport of trajectories through the phase space. To see this, it is convenient to obtain a map able to describe the evolution of the trajectories.

Let us consider a parametrically perturbed pendulum whose Hamiltonian is

$$H = \frac{p^2}{2m} - mk(1 - \varepsilon \cos \theta) \cos q, \tag{B.86}$$

where $\theta = \omega t + \theta°$, and let us calculate the variation of the energy when the system evolves on one trajectory adjacent to the separatrix. We have

$$\frac{dE}{dt} = \frac{\partial H}{\partial t} = -\varepsilon m k \omega \cos q \sin \theta.$$

The variation of E during the travel over the whole arc may be approximated by the corresponding variation along the separatrix:

$$\Delta E_{\text{sx}} = \int_{-\infty}^{\infty} -\varepsilon m k \omega \cos q_{\text{sx}} \sin \theta \, dt,$$

where q_{sx} is the function defined by (B.34). We may decompose $\sin \theta$ and discard from the integrand the term $-\varepsilon A \omega \cos q_{\text{sx}} \sin \omega t \cos \theta°$, since it is odd with respect to the central point of the integration interval. Then

$$\Delta E_{\text{sx}} = -\frac{1}{2} \varepsilon m k \omega \sin \theta° \int_{-\infty}^{\infty} [\cos(q_{\text{sx}} - \omega t) + \cos(q_{\text{sx}} + \omega t)] \, dt. \tag{B.87}$$

The integral on the right-hand side is improper (no limit exist). We may write it using the Melnikov integrals (see [21])

$$\mathcal{A}_\nu(\lambda) \stackrel{\text{def}}{=} \int_{-\infty}^{\infty} \cos\left(\frac{\nu}{2} q_{\text{sx}}^+ - \lambda \tau\right) d\tau, \tag{B.88}$$

where $q_{\text{sx}}^+ = 4\arctan(e^\tau) - \pi$ and $\tau = \sqrt{k}\, t$. We also have

$$\mathcal{A}_\nu(-\lambda) = (-1)^\nu \mathcal{A}_\nu(\lambda) e^{-\pi\lambda}. \tag{B.89}$$

Hence, since $\tau = \sqrt{k}\, t$,

$$\Delta E_{\text{sx}} = -\frac{1}{2}\varepsilon m k \lambda \left[\mathcal{A}_2(\lambda) + \mathcal{A}_2(-\lambda)\right] \sin\theta°, \tag{B.90}$$

where

$$\lambda = \frac{\omega}{\sqrt{k}}.$$

When $\lambda \gg 1$,

$$|\mathcal{A}_2(-\lambda)| \ll |\mathcal{A}_2(\lambda)|$$

and

$$\Delta E_{\text{sx}} \approx -\frac{1}{2}\varepsilon m k \lambda\, \mathcal{A}_2(\lambda) \sin\theta°$$

or, after introducing the normalized energy h defined by (B.80),

$$\Delta h_{\text{sx}} \approx -\frac{1}{2}\varepsilon\lambda\, \mathcal{A}_2(\lambda) \sin\theta°.$$

The separatrix map is a model giving, step by step, the variations of h and $\theta°$ when the system evolves from the vicinity of A to that of B near the separatrix (or from B to A, in the contrary direction, close to equivalent trajectories existing in the lower half of the phase portrait). The variation of the phase θ is given by $t_{\approx AB}$ as calculated in the previous section:

$$\Delta\theta = \omega t_{\approx AB} = \frac{\omega}{\sqrt{k}} \ln \frac{32}{|h|} = \lambda \ln \frac{32}{|h|}.$$

Therefore in the next step of the mapping, the initial phase $\theta°$ may be changed to $\theta° + \Delta\theta$; the corresponding variation of the energy is approximated by Δh_{sx}. Since we have assumed $\lambda \gg 1$, $\mathcal{A}_2(\lambda)$ is small, the variation of h during the motion on the given trajectory is small and h may be approximated by its value in any point of the interval, e.g. the final value. We thus obtain the *separatrix* or *whisker map*

$$\begin{aligned} h_{n+1} &= h_n - W \sin\theta°_n \\ \theta°_{n+1} &= \theta°_n + \lambda \ln \frac{32}{|h_{n+1}|}, \end{aligned} \tag{B.91}$$

where
$$W = \frac{\varepsilon\lambda}{2}\mathcal{A}_2(\lambda) \approx 4\pi\varepsilon\lambda^2 e^{-\pi\lambda/2};$$
to calculate the Melnikov integral, we have used the asymptotic ($\lambda \gg 1$) estimate of the non-oscillatory terms [21]:

$$\mathcal{A}_\nu(\lambda) \approx \frac{4\pi(2\lambda)^{\nu-1}}{(\nu-1)!} e^{-\pi\lambda/2} \qquad (\nu > 0). \tag{B.92}$$

We should emphasize that we have chosen a perturbation even with respect to q to avoid having to discuss the change of the arguments of $\mathcal{A}_2$ in ΔE_{sx} when the motion is near the lower separatrix instead of the upper one shown in Fig. B.3. When the perturbation is even with respect to q, the Melnikov integrals appear in pairs as in (B.90) and the result is invariant with respect to the considered sign changes.

B.7 The Standard Map

A very important form of the separatrix map is obtained through the linearization about a suitably chosen value h_r. In the neighborhood of $h = h_r$, we have
$$\ln\frac{32}{|h_{n+1}|} \approx \ln\frac{32}{|h_r|} - \frac{h_{n+1} - h_r}{h_r}.$$
We then introduce, in the separatrix map, the variable
$$I = -\lambda\frac{h - h_r}{h_r}$$
and the constant
$$K = \frac{\lambda W}{h_r} \approx \frac{4\pi\varepsilon\lambda^3}{h_r} e^{-\pi\lambda/2}.$$
The map becomes
$$I_{n+1} = I_n + K\sin\theta_n^\circ$$
$$\theta_{n+1}^\circ = \theta_n^\circ + \lambda\ln\frac{32}{|h_r|} + I_{n+1}.$$

The reference value h_r is chosen to be such that $\lambda\ln\frac{32}{|h_r|}$ is an integer multiple of 2π. Then, since $\theta^\circ \in \mathbf{S}^1$, this term has no influence on the result and may be neglected.

The resulting map is the *standard map*
$$\begin{aligned}I_{n+1} &= I_n + K\sin\theta_n^\circ \\ \theta_{n+1}^\circ &= \theta_n^\circ + I_{n+1}.\end{aligned} \tag{B.93}$$

C

Andoyer Hamiltonian with $k = 1$

C.1 Andoyer Hamiltonians

Andoyer Hamiltonians are one-degree-of-freedom Hamiltonians of the form

$$H_k = aJ + bJ^2 + \varepsilon\tau(2J)^{k/2}\cos k\sigma \qquad (k = 1, 2, 3, 4), \qquad (C.1)$$

where a, b, τ are constants and ε is a constant small parameter ($\varepsilon \ll |b|$). $\sigma \in \mathbf{S}$ and $J \in \mathbf{R}_0^+$ are two canonically conjugate variables. Without loss of generality, we may assume $b > 0$ and $\varepsilon\tau > 0$. (Otherwise it is enough to change the variables t into $-t$ or σ into $\sigma + \pi/k$, respectively, to have those conditions satisfied.) The restriction $J \geq 0$ may be easily replaced by $J \leq 0$. Indeed the whole geometrical study of the surfaces $H_k = $ const is done with non-singular variables and (C.7) is valid for either $J \in \mathbf{R}_0^+$ or $J \in \mathbf{R}_0^-$, if the Andoyer Hamiltonian is written as

$$H_k = a|J| + bJ^2 + \varepsilon\tau|2J|^{k/2}\cos k\sigma \qquad (k = 1, 2, 3, 4) \qquad (C.2)$$

The only real restriction is that J is sign-definite.

Andoyer Hamiltonians arise naturally as the Hori kernels in the analysis of resonant systems in the neighborhood of the origin. The case $k = 1$ was introduced in modern literature by Sessin, in 1981 [86], [89] and the composed case $k = 1, 2$ by Gerasimov, in 1982 [38]. It is found since then, in the literature, under the name "second fundamental resonance model" [49]. (The "first" fundamental model is the pendulum, which arises as the Hori kernel in the study of systems with a resonance occurring at a finite J.) It is worth stressing the fact that σ, J are _not_ the angle–action variables of the Andoyer Hamiltonian, but those of the non-singular differential rotator Hamiltonian H_0 obtained from H_k when $\varepsilon = 0$.

The importance of these Hamiltonians was first recognized by Poincaré and Andoyer. Poincaré pointed out the existence of the two stable solutions when $k = 1$, but did not notice that they occur, one for $\sigma = 0$ and the other for $\sigma = \pi$. In fact, he did not pay much attention to the center close to the origin (the one that is misplaced in his figures) since it does not fulfill completely the requisites of what is called a resonance center [81]. The first complete analysis is due to Andoyer [2], who considered the composed Hamiltonian $k = 1, 2$,

$$aJ + bJ^2 + \varepsilon\tau\sqrt{|2J|}\cos\sigma + \varepsilon\tau' J\cos 2\sigma, \tag{C.3}$$

and correctly pointed out the possibilities of oscillations about $\sigma = 0$ and $\sigma = \pi$ when the term $\varepsilon\tau\sqrt{|2J|}\cos\sigma$ dominates over the last one. This Hamiltonian also appeared when averaging the conservative Duffing equation [44]. Differential equations identical to those resulting from the Andoyer Hamiltonian were also considered by Pars [79] in his study of forced oscillations of small amplitude. Extended models including J^3 were considered by Breiter [16].

C.2 Centers and Saddle Points

In this section, we determine the singular points of Andoyer Hamiltonians, a necessary step to study their phase portraits. Because of the singularity at $J = 0$, it is usual to introduce Poincaré's non-singular variables[1] (see Chap. 7):

$$x = \sqrt{|2J|}\cos\sigma \qquad y = \sqrt{|2J|}\sin\sigma. \tag{C.4}$$

However, to have a unified geometric analysis of Andoyer Hamiltonians, we prefer, here, to introduce a new angular variable

$$\theta = k\sigma \tag{C.5}$$

and the set of Cartesian-like variables defined by

$$X = \sqrt{|2J|}\cos k\sigma \qquad Y = \sqrt{|2J|}\sin k\sigma. \tag{C.6}$$

The main property of these transformations, in what concerns the study of the singular points of (C.1), is the k–folding due to (C.5). However, it is important to keep in mind that, when $k \neq 1$, the transformation to (X, Y) is a canonical transformation, but with a valence (multiplier) sk (where s is the sign of J). However, we do not intend to use

$$H_k = \frac{1}{2}a(X^2 + Y^2) + \frac{1}{4}b(X^2 + Y^2)^2 + \varepsilon\tau(X^2 + Y^2)^{(k-1)/2}X, \tag{C.7}$$

as a Hamiltonian function, but only as a constant of the motion (energy) that can be used to determine geometrical properties of the orbits, except at the origin.

---

[1] When canonical equations in non-singular variables are used, it is important to recall that the corresponding canonical transformations are $(\sigma, J) \to (x, y)$ when $J < 0$ and $(\sigma, J) \to (y, x)$ when $J > 0$.

C.2 Centers and Saddle Points

The locus of the singular points of the system of curves defined by $H_k = $ const is given by
$$\frac{\partial H_k}{\partial X} = \frac{\partial H_k}{\partial Y} = 0,$$
that is,
$$\frac{\partial H_k}{\partial X} = 2DX + \varepsilon\tau(X^2+Y^2)^{(k-1)/2} = 0 \tag{C.8}$$
$$\frac{\partial H_k}{\partial Y} = 2DY = 0, \tag{C.9}$$
where D is the partial derivative of H_k with respect to (X^2+Y^2):
$$D = \frac{1}{2}a + \frac{1}{2}b(X^2+Y^2) + \varepsilon\tau\frac{k-1}{2}(X^2+Y^2)^{(k-3)/2}X.$$

An immediate consequence of these equations is that $Y=0$ and the singular points lie necessarily on the X-axis. Indeed, the other possibility from (C.9) is $D=0$. However, when this condition is introduced into (C.8), the resulting equation may only be satisfied if $\varepsilon = 0$, that is, if the Hamiltonian is reduced to the undisturbed differential rotator H_0.

Introducing $Y=0$ into (C.8), it becomes
$$aX + bX^3 + k\varepsilon\tau |X|^{k-1} = 0, \tag{C.10}$$
where the introduction of $|X|$ instead of X comes from the fact that this quantity is the square root of X^2, which is taken as positive, whichever is the sign of X.

In order to know the nature of the singularity, we have to compute the Hessian of H_k at the singular point:
$$\Delta_k = [a + 3bX^2 + \varepsilon\tau k(k-1)X|X|^{k-3}][a + bX^2 + \varepsilon\tau(k-1)X|X|^{k-3}],$$
where, as before, the same precautions concerning the sign of some terms were taken. The singular point is a center or a saddle point according as Δ_k is positive or negative, respectively. For $\Delta_k = 0$ the singularity is a double point.

Taking into account that the roots must satisfy (C.10), the previous equation may be reduced to
$$\Delta_k = \left[2a + \varepsilon\tau k(4-k)X|X|^{k-3}\right]\varepsilon\tau X|X|^{k-3}. \tag{C.11}$$

We note that the origin $(X=0, Y=0)$ is always a singular point for $k \neq 1$. However, its nature cannot be obtained from the sign of Δ_k, since the change of variables defined by (C.6) is singular at the origin. In order to determine the nature of the singular point at the origin, we may note that for $k \leq 4$ and under the condition $\varepsilon\tau \ll b$, the curves $H_k = cte$ tend to regular closed curves enveloping the origin as $(X^2+Y^2) \to \infty$. Thus, if there are no other singular points of higher orders, the topological nature of the singular point at the origin is obtained from the simple rule *"Number of centers = Number of saddle points + 1"*.

C.2.1 The Case $k = 1$

In this case, we have

$$H_1 = \frac{1}{2}a(X^2 + Y^2) + \frac{1}{4}b(X^2 + Y^2)^2 + \varepsilon\tau X \tag{C.12}$$

and (C.10) is a cubic equation,

$$aX + bX^3 + \varepsilon\tau = 0, \tag{C.13}$$

whose solution may be easily found using classical formulas.

Equation (C.13) has one, two or three real solutions according to the value of a. We normalize the value of X using the factor

$$A_1 = \sqrt[3]{\frac{4\varepsilon\tau}{b}} > 0,$$

and introduce the critical value

$$a_1^* = -\frac{3}{2}\sqrt[3]{2b\varepsilon^2\tau^2} = -\frac{3bA_1^2}{4} < 0. \tag{C.14}$$

Then, (C.13) becomes

$$4\xi^3 - 3\alpha\xi + 1 = 0, \tag{C.15}$$

where

$$\xi = \frac{X}{A_1} \tag{C.16}$$

and

$$\alpha = \frac{a}{a_1^*}. \tag{C.17}$$

When $\alpha \geq 1$, that is, $a \leq a_1^*$, (C.15) admits three real roots[2]:

$$\xi = \sqrt{\alpha}\cos\left[\frac{1}{3}\arccos\left(-\alpha^{-3/2}\right) + \frac{2n\pi}{3}\right] \quad (n = 0, 1, 2). \tag{C.19}$$

Figure C.1 shows the sectors in which the arguments of ξ are defined for $n = 0, 1, 2$; the figure corresponds to the case $\pi/2 \leq \arccos\left(-\alpha^{-3/2}\right) \leq 3\pi/2$.

---

[2] In solving (C.15), we prefer to use analogies with some elementary transcendental functions instead of the classical algebraic formula. The formulas giving these analogies are

$$\begin{array}{lll} 4\cos^3\chi - 3\cos\chi = \cos 3\chi & \text{when} & \alpha \geq 1 \\ 4\cosh^3\chi - 3\cosh\chi = \cosh 3\chi & \text{when} & 0 < \alpha < 1 \\ 4\sinh^3\chi + 3\sinh\chi = \sinh 3\chi & \text{when} & \alpha < 0 \end{array} \tag{C.18}$$

and the analogy with (C.15) is obtained by making $\xi = \sqrt{\alpha}\cos\chi$, $\xi = -\sqrt{\alpha}\cosh\chi$ or $\xi = \sqrt{-\alpha}\sinh\chi$, respectively.

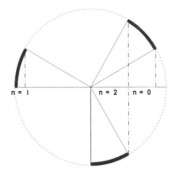

Fig. C.1. Argument of the roots on the unit circle

In the other possible case, $3\pi/2 \leq \arccos(-\alpha^{-3/2}) \leq 5\pi/2$, the figure has a counterclockwise shift of 30 degrees and the order of the solutions corresponding to $n = 2$ and $n = 0$ is interchanged.

When $\alpha = 1$, the two real roots on the right-hand side coalesce into $\xi = \frac{1}{2}$, and (C.15) has two roots: one single and one double.

For $\alpha < 1$, that is, $a > a_1^*$, (C.15) admits only one real solution, which is written

$$\xi = -\sqrt{\alpha}\cosh\left[\frac{1}{3}\operatorname{arccosh}(\alpha^{-3/2})\right], \qquad (C.20)$$

when $0 < \alpha < 1$, or

$$\xi = -\sqrt{-\alpha}\sinh\left[\frac{1}{3}\operatorname{arcsinh}(-\alpha)^{-3/2}\right] \qquad (C.21)$$

when $\alpha < 0$. If $\alpha = 0$, the solution is $\xi = -4^{-1/3}$.

The nature of these singular points is fixed by the sign of the Hessian

$$\Delta_1 = a_1^{*2}(\alpha - 4\xi^2)\left(\alpha - \frac{4}{3}\xi^2\right).$$

Some easy calculations allow one to see that, for $\alpha \geq 1$, the rightmost root is a saddle point and the two others are centers. When $\alpha = 1$, the two rightmost roots coalesce into a double point ($\Delta_1 = 0$). For $\alpha < 1$, the only remaining root is a center at the left of the origin.

C.3 Morphogenesis

The morphogenesis of the curves $H_1 = \text{const}$ may be studied using the parameter a (or α). This parameter defines the existence, or non-existence, of a resonance. For $\varepsilon = 0$, the Andoyer Hamiltonian degenerates into the non-singular differential rotator studied in Sect. 7.4, whose solutions are circular motions with frequency

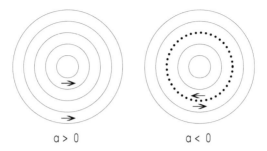

Fig. C.2. Phase portraits of the undisturbed differential rotator H_0. The arrows show the direction of the motion when $J > 0$

$$\nu_0 = \left.\frac{\partial H_k}{\partial J}\right|_{\varepsilon=0} = \left(\frac{a}{|J|} + 2b\right) J. \tag{C.22}$$

All motions have the same direction when $a \cdot b > 0$ (or $a > 0$, since we assumed $b > 0$). A zero frequency, that is, a resonance, located at

$$|J_r| = -\frac{a}{2b}, \tag{C.23}$$

occurs when $a \cdot b < 0$ (or $a < 0$). These *undisturbed* solutions are shown in Fig. C.2.

The orbits defined by the Andoyer Hamiltonian H_1 are shown in Fig. C.3 starting from values $\alpha < 1$ (or $a > a_1^*$), crossing the critical value $\alpha = 1$ (or $a = a_1^* < 0$), and going up to positive values of α.

The nomenclature of the different regimes of motion in this resonance is not unambiguously established. In the case $\alpha > 1$, we have to distinguish between oscillations about two centers.

(a.) The center appearing near the origin, at its right in Fig. C.3 (bottom right), which, using a purely topological point of view, may be seen as the new position of the center of the *unperturbed* differential rotator (H_0), just slightly shifted to the right of the origin because of the linear perturbation. The motions around this center are called *circulations* or, more exactly, *inner* circulations, to distinguish them from the large orbits enclosing the whole resonance region (the *outer* circulations).

(b.) The new center and the domain surrounding it, appearing at the left of the origin for $\alpha > 1$ ($a < a_1^*$) – see Fig. C.3 (bottom) – which comes from a qualitative change of the phase portrait of the regular differential rotator H_0 due to the action of a perturbation on the zero frequencies located at $|J_r|$. This region, with its center and saddle point, is the kind of modification expected in these circumstances (Poincaré–Birkhoff theorem). The motions inside it, around this new center are called *librations*.

Thus, from a topological point of view, *circulations* are all motions topologically equivalent to the motions of the undisturbed differential rotator while

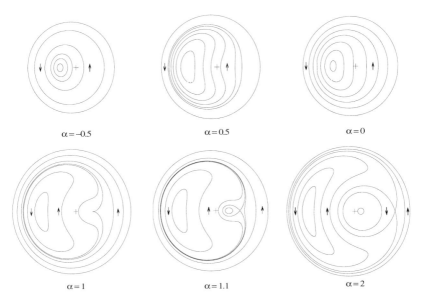

Fig. C.3. Phase portraits of the Andoyer Hamiltonian H_1 ($b > 0, \varepsilon\tau > 0$). The origin of the (X, Y) plane is marked with $+$. The arrows show the direction of the motions when $J > 0$

librations are those that happen in the region of the phase portrait created by the perturbation $\varepsilon\tau X$.

However, the names circulation and libration have kinematical origin and, according to their original definitions, *circulations* are motions on orbits going around the origin of the (X, Y) plane while *librations* are motions on closed orbits not including the origin. Since the center of the inner circulations does not coincide with the origin of the (X, Y) plane, the kinematical and topological definitions are not equivalent, thus giving rise to ambiguities.

For instance, in asteroid dynamics, the asteroids known as "apocentric librators" have a regime of motion in which the critical angle oscillates about π but, topologically, these *librations* are just inner circulations with amplitudes small enough to allow the orbits not to include the origin of the (X, Y) plane. In this book, when it is necessary to consider these regimes of motion, they will be referred to explicitly, to avoid ambiguities. For instance, the region whose size is discussed hereafter is that of the motions about the center at the left of the origin, that is, the topologically defined libration zone.

When $\alpha < 1$, we have only one center and, from the topological point of view, all motions are *circulations*. But, kinematically, we always have circulating and librating solutions.

The regimes of motion and bifurcations of H_1 are schematically represented in Fig. C.4, where we have introduced the energy unit

C Andoyer Hamiltonian with $k = 1$

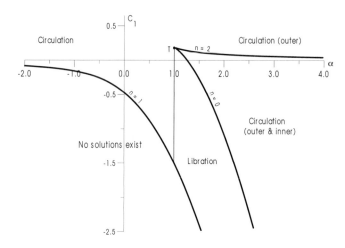

Fig. C.4. Morphogenesis of the solutions of H_1. The lines $n = 1$ and $n = 2$ correspond to the centers and the line $n = 0$ to the saddle point. T corresponds to the cusp

$$H_1^* = \frac{1}{4} b A_1^4 \tag{C.24}$$

and considered (C.12) restricted to the axis $Y = 0$ as

$$C_1 = \left.\frac{H_1}{H_1^*}\right|_{Y=0} = \xi^4 - \frac{3}{2}\alpha\xi^2 + \xi. \tag{C.25}$$

It is worth noting that the catastrophe set separating the families of curves with three singular points from those having only one singular point does not happen for $\alpha = 0$, but for $\alpha = 1$ ($a = a_1^*$). For $\alpha = 1$, the frequency of the undisturbed differential rotator H_0 vanishes at

$$|J_r| = \frac{3}{2}\left(\frac{\varepsilon\tau}{2b}\right)^{2/3} = \frac{3}{8}A_1^2.$$

For $0 < \alpha < 1$, the resonance (or zero frequency) still appears in the undisturbed differential rotator H_0 at $J = J_r$. Nevertheless, in this interval, the only qualitative effect visible in the phase portraits of the Andoyer Hamiltonian H_1 is the shift of the center to the left of the origin.

Exercise C.3.1. Show that, for $\alpha = 0$, the normalized energy C_1 corresponding to the center is $C_1 = -3 \times 4^{-4/3}$.

Exercise C.3.2. Show that, at the cusp, $C_1 = 0.1875$.

C.4 Width of the Libration Zone

The width of the libration zone is given by the intersections with the X-axis of the two separatrix branches emanating from the saddle point. Let ξ_1 be the

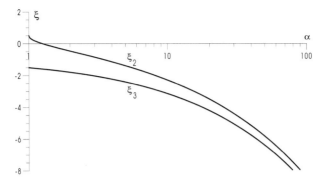

Fig. C.5. Width and location of the libration zone

value of ξ at the saddle point and ξ_2, ξ_3 its values at the other intersections. The d'Alembert form of the polynomial (C.25) is

$$(\xi - \xi_1)^2(\xi - \xi_2)(\xi - \xi_3) = 0;$$

hence

$$\xi^4 - (2\xi_1 + \xi_2 + \xi_3)\xi^3 + [\xi_1^2 + 2\xi_1(\xi_2 + \xi_3) + \xi_2\xi_3]\xi^2 \\ - [\xi_1^2(\xi_2 + \xi_3) + 2\xi_1\xi_2\xi_3]\xi + \xi_1^2\xi_2\xi_3 = 0. \quad (C.26)$$

Comparing the coefficients of ξ^3, ξ^2 and ξ^0 in (C.25) and (C.26), we obtain

$$2\xi_1 + \xi_2 + \xi_3 = 0,$$

$$\xi_1^2 + 2\xi_1(\xi_2 + \xi_3) + \xi_2\xi_3 = -\frac{3\alpha}{2}$$

and

$$\xi_1^2\xi_2\xi_3 = -C_1,$$

respectively. The two first results may be combined to give the width of the resonance

$$\Delta\xi = \xi_2 - \xi_3 = \sqrt{6\alpha - 8\xi_1^2}. \quad (C.27)$$

ξ_1 is the normalized abscissa of the saddle point and, for $\varepsilon\tau > 0$, the rightmost of the three roots of (C.26):

$$\xi_1 = \sqrt{\alpha}\cos\left[\frac{1}{3}\arccos\left(-\alpha^{-3/2}\right)\right] \quad (C.28)$$

(note that ξ_2 and ξ_3 are not roots of (C.15) and that $\xi_3 < \xi_2 < \xi_1$).

The functions $\xi_2(\alpha)$ and $\xi_3(\alpha)$ are shown in Fig. C.5 for $\alpha > 1$.

In terms of $X_j = A_1\xi_j$, we have

$$(X_2 - X_3)^2 = -\frac{8a}{b} - 8X_1^2. \quad (C.29)$$

If we prefer to measure the size of the libration zone using the actions J, we use the immediate relation

$$\Delta J = |J_3 - J_2| = \frac{1}{2}|X_2 + X_3|(X_3 - X_2) = X_1(X_2 - X_3).$$

When ξ_2 becomes positive, it is more appropriate to define the resonance width through $\Delta \xi = |\xi_3|$ or, in the actions:

$$\Delta J = |J_3| = \frac{1}{2}X_3^2.$$

Two limits are important:

(a.) $\alpha = 1$ $(a = a_1^*)$.
In this case, one center and the saddle point coalesce into a cusp and $\xi_1 = \xi_2$. The cusp abscissa is $\xi_1 = \frac{1}{2}$. Then $X_1 = A_1 \xi_1 = \frac{1}{2}A_1 = \sqrt[3]{\varepsilon\tau/2b}$ and

$$X_2 - X_3 = 2A_1 = 4\sqrt[3]{\frac{\varepsilon\tau}{2b}}; \qquad \Delta J = \frac{1}{2}X_3^2 = \frac{9}{8}A_1^2 = \frac{9}{4}\sqrt[3]{\frac{2\varepsilon^2\tau^2}{b^2}}.$$

The width of the libration zone in the (X, Y) plane is of order $\mathcal{O}(\varepsilon^{1/3})$ and, in the action J, of order $\mathcal{O}(\varepsilon^{2/3})$.

(b.) $\alpha \gg 1$ $(a \ll a_1^* < 0)$.
In this case, we may expand the trigonometric and inverse trigonometric functions of (C.28) to obtain

$$\xi_1 = \frac{\sqrt{3\alpha}}{2} - \frac{1}{6\alpha} + \mathcal{O}(\alpha^{-5/2}).$$

Then, $X_1 = \sqrt{-a/b}$ and, from (C.27), $\Delta \xi \sim 2 \times (3\alpha)^{-\frac{1}{4}}$, or

$$\Delta X \approx \sqrt{\frac{8\varepsilon\tau}{\sqrt{-ab}}}; \qquad \Delta J \approx \sqrt{\frac{8\varepsilon\tau}{b}}\sqrt{\frac{-a}{b}}.$$

The width of the libration zone is of order $\mathcal{O}(\sqrt{\varepsilon})$ in the (X, Y) plane as well as in the action J.

Exercise C.4.1. Show that, when $\alpha = \sqrt[3]{2}$, the inner branch of the separatrix passes through the origin.

C.5 Integration

Let us consider the first Andoyer Hamiltonian:

$$H_1 = a|J| + bJ^2 + \varepsilon\tau\sqrt{|2J|}\cos\sigma. \qquad (C.30)$$

To reduce the number of free constants, we introduce the same quantities used in the study of the morphogenesis of H_1, viz.

$$A_1 = \sqrt[3]{\frac{4\varepsilon\tau}{b}} > 0, \qquad H_1^* = \frac{1}{4}bA_1^4, \qquad a_1^* = -\frac{3}{2}\sqrt[3]{2b\varepsilon^2\tau^2} = -\frac{3bA_1^2}{4} < 0,$$

the adimensional quantities

$$\alpha = \frac{a}{a_1^*}, \qquad C_1 = \frac{H_1}{H_1^*},$$

and the scaled action

$$I = \frac{|J|}{A_1^2} > 0. \tag{C.31}$$

Equation (C.30) then becomes

$$C_1 = -3\alpha I + 4I^2 + \sqrt{2I}\cos\sigma. \tag{C.32}$$

$C_1(\sigma, I)$ is the Hamiltonian of a dynamical system whose equations are

$$\frac{d\sigma}{dt'} = -3\alpha + 8I + \frac{1}{\sqrt{2I}}\cos\sigma$$

$$\frac{dI}{dt'} = \sqrt{2I}\sin\sigma. \tag{C.33}$$

The definition of the new independent variable t' must take into account that:

(1) the Hamiltonian was scaled by the factor H_1^*;
(2) the valence (multiplier) of the canonical transformation $(\sigma, J) \to (\sigma, I)$ is $\lambda = sA_1^{-2}$ (where $s = \pm 1$ is the sign of J) (see Sect. 1.5.2).

Then

$$t' = \frac{1}{4}sbA_1^2 t = -\frac{1}{3}sa_1^* t. \tag{C.34}$$

Equation (C.33) may be combined with the energy integral $C_1 = \text{const}$ to give

$$\left(\frac{dI}{dt'}\right)^2 = 2I - (C_1 + 3\alpha I - 4I^2)^2 \equiv P(I) \tag{C.35}$$
$$= -16I^4 + 24\alpha I^3 - (9\alpha^2 - 8C_1)I^2 + (2 - 6\alpha C_1)I - C_1^2.$$

To proceed with the calculation, we need to know the behavior of the polynomial $P(I)$. First of all, we notice that $P(I) < 0$ for all $I < 0$ (since, in this case, $P(I)$ is the sum of two negative parts). Therefore, all roots are positive, except when $C_1 = 0$, in which case one root is zero.

Let us now consider two consecutive real roots of $P(I)$ delimiting one interval where $P(I) > 0$. The motion in this interval may be obtained from the integration of (C.35) and is given by

$$t' - t'_0 = \int_{I_0}^{I} \frac{dI}{\sqrt{P(I)}}, \qquad (C.36)$$

where I_0 is the smaller of the two real roots of $P(I)$. The inversion of this integral gives

$$I = I_0 + \frac{1}{4} P'(I_0) \left[\wp(t' - t'_0; g_2, g_3) - \frac{1}{24} P''(I_0) \right]^{-1}, \qquad (C.37)$$

where $\wp$ is Weierstrass' elliptic function and g_2, g_3 are its invariants ([98], Sect. 20.6)

$$g_2 = \frac{1}{3} \left(8C_1 + \frac{9}{2}\alpha^2 \right)^2 - 12\alpha$$

$$g_3 = \frac{1}{27} \left(8C_1 + \frac{9}{2}\alpha^2 \right)^3 - 2\alpha \left(8C_1 + \frac{9}{2}\alpha^2 \right) + 4.$$

The three constants appearing in this solution, C_1, I_0, t'_0, are not independent; indeed, the constants C_1 and I_0 are related by $P(I_0) = 0$. To avoid the solution of the quartic equation

$$C_1 = 4I_0^2 - 3\alpha I_0 + \sqrt{2I_0} \cos \sigma_0,$$

it is wise to keep I_0 as an arbitrary integration constant and use it to obtain C_1. We recall that I_0 is a point of minimum of $I(t)$ and, therefore, $\sin \sigma_0 = 0$ (see C.33); in addition, if $b > 0, \varepsilon\tau > 0$ and the solution is a true libration, $\sigma_0 = \pi$ (see Fig. C.3).

To make applications easier, Jacobian elliptic functions are introduced instead of the Weierstrass function. To do this, we need to consider the sign of

$$\Delta = g_2^3 - 27g_3^2, \qquad (C.38)$$

the discriminant of the cubic resolvent of the equation $P(I) = 0$:

$$4e^3 - g_2 e - g_3 = 0. \qquad (C.39)$$

Δ is equal to zero for the values of C_1 for which a double root occurs, that is, at the values of C_1 corresponding to centers and saddle points.

Exercise C.5.1. Show that the roots of the system formed by the equations $P(I) = 0$ and $P'(I) = 0$ are the singular points of the Hamiltonian H_1. *Hint:* Start with the definition of singular points.

Exercise C.5.2. Identify the domains where $\Delta > 0$ and $\Delta < 0$ in Fig. C.4.

C.5.1 The Case $\Delta > 0$

In this case, the polynomial $P(I)$ has four real positive roots and to each value of C_1 and α there corresponds one outer and one inner circulation[3]. In this case, $g_2 > 0$ while g_3 may be either positive or negative.

When $\Delta > 0$, the Weierstrassian function $\wp$ is related to Jacobian elliptic functions (see [1], Sect. 18.9) through

$$\wp(t' - t'_0) = e_3 + \frac{e_1 - e_3}{\operatorname{sn}^2 z}, \tag{C.40}$$

where e_1, e_2, e_3 are the roots of (C.39) in decreasing order ($e_1 > e_2 > e_3$),

$$z = \sqrt{e_1 - e_3}\,(t' - t'_0)$$

and sn is the Jacobian *sine amplitude* elliptic function of modulus

$$\kappa_1 = \sqrt{\frac{e_2 - e_3}{e_1 - e_3}}.$$

It then follows that

$$I - I_0 = \frac{A\,\operatorname{sn}^2 z}{1 - B\,\operatorname{sn}^2 z}, \tag{C.41}$$

where

$$A = \frac{P'(I_0)}{4(e_1 - e_3)}$$

and

$$B = \frac{P''(I_0) - 24 e_3}{24(e_1 - e_3)}. \tag{C.42}$$

It may also be convenient to introduce the Jacobian *amplitude*

$$\phi = \operatorname{am} z$$

and (C.41) becomes

$$I - I_0 = \frac{A \sin^2 \phi}{1 - B \sin^2 \phi}. \tag{C.43}$$

This formulation brings the additional problem of having to solve the algebraic equation (C.39). This may be easily done using the first of (C.18). Then

$$e_i = \sqrt{\frac{g_2}{3}} \cos\left[\chi\left(\operatorname{mod}\frac{2\pi}{3}\right)\right],$$

where

---

[3] One should remember that $P(I)$ is not an integral of the motion. The classification of the roots of $P(I)$ and, correspondingly, of the solution types, following the values of two parameters, is however possible. See [87].

$$\chi = \frac{1}{3} \arccos \sqrt{\frac{27g_3^2}{g_2^3}} \qquad \left(0 \leq \chi \leq \frac{\pi}{6}\right).$$

Then, in the order requested by their definitions,

$$\begin{aligned} e_1 &= \sqrt{\frac{g_2}{3}} \cos \chi \\ e_2 &= \sqrt{\frac{g_2}{3}} \cos\left(\chi + \frac{4\pi}{3}\right) \\ e_3 &= \sqrt{\frac{g_2}{3}} \cos\left(\chi + \frac{2\pi}{3}\right). \end{aligned} \qquad (C.44)$$

One may note that the well-known relations $e_2 \cdot g_3 < 0$ and $e_1 + e_2 + e_3 = 0$ are satisfied.

C.5.2 The Case $\Delta < 0$

In this case $P(I)$ has only two real roots. When $\alpha > 1$ (or $a < a_1^*$), the corresponding motions are outer circulations or librations; otherwise, they belong to the only regime of motion existing for $\alpha < 1$ (or $a > a_1^*$). All sign combinations of g_2 and g_3 are possible.

When $\Delta < 0$,
$$\wp(t' - t_0') = e_2 + \eta \frac{1 + \mathrm{cn}\, z}{1 - \mathrm{cn}\, z}, \qquad (C.45)$$

where e_2 is the only real root of (C.39),
$$z = 2\sqrt{\eta}\,(t' - t_0'),$$

$$\eta = \sqrt{3e_2^2 - \frac{g_2}{4}}$$

and cn is the Jacobian *cosine amplitude* elliptic function of modulus
$$\kappa_2 = \sqrt{\frac{1}{2} - \frac{3e_2}{4\eta}}.$$

It then follows that
$$I - I_0 = \frac{A'(1 - \mathrm{cn}\, z)}{2 - B'(1 - \mathrm{cn}\, z)}, \qquad (C.46)$$

where
$$A' = \frac{P'(I_0)}{4\eta}$$

and
$$B' = \frac{P''(I_0) + 24(\eta - e_2)}{24\eta}. \qquad (C.47)$$

If we introduce the Jacobian amplitude

$$\phi' = \frac{1}{2}\operatorname{am} z,$$

(C.46) becomes

$$I - I_0 = \frac{A' \sin^2 \phi'}{1 - B' \sin^2 \phi'}. \tag{C.48}$$

Transforming back to elliptic functions one may see that (C.46) can be written in the same form as (C.41):

$$I - I_0 = \frac{A' \operatorname{sn}^2 z'}{1 - B' \operatorname{sn}^2 z'},$$

where $z' = \operatorname{am}^{-1}\phi' = \operatorname{am}^{-1}(\frac{1}{2}\operatorname{am} z)$ is no longer a function linear with respect to t', at variance with z.

Once more, we have to solve the algebraic equation (C.39), which is done using the second and third of (C.18) according to $g_2 > 0$ and $g_2 < 0$, respectively. We then have, for $g_2 > 0$,

$$e_2 = \pm \sqrt{\frac{g_2}{3}} \cosh \chi, \tag{C.49}$$

where

$$\chi = \frac{1}{3} \cosh^{-1} \sqrt{\frac{27 g_3^2}{g_2^3}}$$

and, for $g_2 < 0$,

$$e_2 = \pm \sqrt{\frac{-g_2}{3}} \sinh \chi, \tag{C.50}$$

where

$$\chi = \frac{1}{3} \sinh^{-1} \sqrt{\frac{27 g_3^2}{-g_2^3}}.$$

In (C.49) and (C.50), the sign in front of the square root is to be chosen equal to the sign of g_3. One should keep in mind that these solutions are numerically unstable in the neighborhood of $g_2 = 0$, in which case their limits for $g_2 \to 0$ should be used.

C.5.3 The Separatrices

Let us consider, now, the motion along the separatrices. In this case, $\Delta = 0$, the integral given by (C.36) is only pseudo-elliptic and the Weierstrassian function of (C.37) degenerates into elementary transcendental functions. The motions along the separatrices are asymptotic to the unstable equilibrium. A numerical study shows that $g_3 < 0$. Then (see [1], Sect. 18.12),

$$\wp(t' - t'_0) = c + \frac{3c}{\sinh^2 z}, \tag{C.51}$$

where

$$c = \sqrt{\frac{g_2}{12}} = \sqrt[3]{\frac{-g_3}{8}} = -\frac{3g_3}{2g_2}$$

and

$$z = \sqrt{3c}(t' - t'_0).$$

Hence

$$I - I_0 = \frac{A'' \sinh^2 z}{1 - B'' \sinh^2 z}, \tag{C.52}$$

where

$$A'' = \frac{P'(I_0)}{12c}$$

and

$$B'' = \frac{P''(I_0)}{72c} - \frac{1}{3}.$$

In the specific case of the separatrix asymptotic to the cusp, occurring when $\alpha = 1$ ($a = a_1^*$), we have $g_2 = g_3 = 0$ and then

$$\wp(t' - t'_0) = \frac{1}{(t - t_0)^2}. \tag{C.53}$$

Therefore, the motion asymptotic to the cusp is given by

$$I - I_0 = \frac{6P'(I_0)(t' - t'_0)^2}{24 - P''(I_0)(t' - t'_0)^2}. \tag{C.54}$$

C.5.4 The Angle σ

The canonical equations issued from the Hamiltonian C_1 were reduced to only one differential equation in I, which was integrated to give $I(t')$. To complete the solution of the canonical system it is also necessary to obtain $\sigma(t)$. This can be obtained from

$$\sin \sigma = \frac{1}{\sqrt{2I}} \frac{dI}{dt'} = \frac{d}{dt'} \sqrt{2I}. \tag{C.55}$$

One should keep in mind that, in this equation, $\sqrt{2I}$ means the positive branch of the square root. Thus, $I(t')$ increases monotonically when $\sin \sigma > 0$ and decreases when $\sin \sigma < 0$. This is also the behavior of $J(t)$ as one may observe, in the case $J > 0$, in Fig. C.3. (When $J < 0$, $J(t)$ increasing means $|J(t)|$ decreasing and the arrows in Fig. C.3 are inverted.)

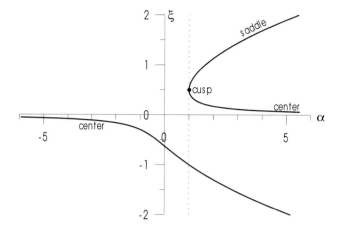

Fig. C.6. Location of the equilibrium points. Case $k = 1$

C.6 Equilibrium Points

Let us study the stable equilibrium points of H_1. The location of these points on the horizontal axis of the (X, Y) plane, in terms of the scaled variables ξ, is shown in Fig. C.6.

At the equilibrium points we have, simultaneously, $P(I_0) = 0$ and $P'(I_0) = 0$ (see Exercise C.5.1). Solving these two equations with respect to the parameters α and C_1, we obtain

$$C_1 = -4I_0^2 \pm \frac{1}{2}\sqrt{2I_0}$$

$$\alpha = \frac{1}{3}\left(8I_0 \pm \frac{1}{\sqrt{2I_0}}\right), \tag{C.56}$$

where the double sign is taken as positive when $\sigma = 0$ and negative when $\sigma = \pi$.

The corresponding values of g_2 and g_3 are

$$g_2 = \frac{1}{12}\left(\frac{1}{2I_0} \mp 8\sqrt{2I_0}\right)^2$$

$$g_3 = \frac{1}{216}\left(\frac{1}{2I_0} \mp 8\sqrt{2I_0}\right)^3.$$

We note that the double sign is, in these equations, the opposite of that of the preceding ones. One may also note that g_2 and g_3 are such that $\Delta = g_2^3 - 27g_3^2 = 0$.

At the equilibrium points, we also obtain

$$P''(I_0) = -\frac{1}{I_0} \pm 16\sqrt{2I_0} = -12\sqrt{\frac{g_2}{3}}.$$

It is worthwhile noting that the final result is such that $P''(I_0) < 0$ at both centers, a result in agreement with the fact that these points correspond to maxima of $P(I)$

C.6.1 The Inner Circulations Center

This center is located on the positive semi-axis ($\xi > 0$) and is the $\Delta \to 0$ limit of the case $\Delta > 0$. The discriminant of the cubic resolvent has, in this case, one single and one double real root. Using (C.44), we obtain

$$e_1 = \sqrt{\frac{g_2}{3}} \tag{C.57}$$

and

$$e_2 = e_3 = -\frac{1}{2}\sqrt{\frac{g_2}{3}}$$

(since $\chi = 0$ when $\Delta = 0$).

B is given by (C.42). Therefore, from the equations given above, we obtain $P''(I_0) - 24e_3 = 0$ and, then, $B = 0$.

C.6.2 The Libration Center

This center is located on the negative semi-axis ($\xi < 0$) and is the $\Delta \to 0$ limit of the case $\Delta < 0$. The roots e_2 and e_3 of the discriminant of the cubic resolvent are complex and the only remaining real root is

$$e_2 = \sqrt{\frac{g_2}{3}}.$$

We also have

$$\eta = \sqrt{3e_2^2 - \frac{g_2}{4}} = \frac{3}{2}e_2. \tag{C.58}$$

B' is given by (C.47) and, from the equations given above, we obtain $P''(I_0) + 24(\eta - e_2) = 0$ and, then, $B' = 0$.

C.7 Proper Periods

With the values of the several parameters determined in the previous section, we may calculate the proper frequencies at the two centers, or, equivalently, the proper periods. The actions I are π-periodic functions with respect to the arguments ϕ and ϕ'. However, ϕ and ϕ' are different functions of t' and the periods of the oscillations about the two equilibrium points need to be calculated separately.

C.7.1 Inner Circulations

In this case, $\Delta > 0$ and $\phi = \operatorname{am} z$. The variation of z in one period is $2\mathbb{K}$ (since $\operatorname{am} 0 = 0$ and $\operatorname{am} 2\mathbb{K} = \pi$). Then

$$T' = \frac{2\mathbb{K}(\kappa_1)}{\sqrt{e_1 - e_3}},$$

where we have indicated the period by T' since the time variable of the given solution is t'. To have the period in the actual timescale of the given Hamiltonian, we first note that, from (C.34),

$$T = \frac{3T'}{|a_1^*|} = \frac{4T'}{bA_1^2}. \tag{C.59}$$

Hence

$$T = \frac{8\mathbb{K}(\kappa_1)}{bA_1^2 \sqrt{e_1 - e_3}}. \tag{C.60}$$

In the limit, when the oscillation amplitude goes to zero, the e_j have the limits given by (C.57) and $\kappa_1 \to 0$. To find the period as a function of the initial conditions, we recall that, in the limit, $P''(I_0) = 24 e_3$. Hence

$$T \to \frac{8\pi\sqrt{2}}{bA_1^2 \sqrt{-P''(I_0)}}. \tag{C.61}$$

We note that ϕ is a monotonically increasing or decreasing function according to $J > 0$ or $J < 0$, respectively.

C.7.2 Librations

In this case, $\Delta < 0$ and $\phi' = \frac{1}{2} \operatorname{am} z$. The variation of z in one period is $4\mathbb{K}$ (since $\operatorname{am} 0 = 0$ and $\operatorname{am} 4\mathbb{K} = 2\pi$). Then

$$T' = \frac{4\mathbb{K}(\kappa_2)}{2\sqrt{\eta}}$$

and

$$T = \frac{8\mathbb{K}(\kappa_2)}{bA_1^2 \sqrt{\eta}}. \tag{C.62}$$

In the limit, when the libration amplitude goes to zero, η has the limit given by (C.58), $\kappa_2 \to 0$ and

$$T \to \frac{4\pi\sqrt{2}}{bA_1^2 \sqrt{3e_1}}. \tag{C.63}$$

T tends to a limit given by the same equation as the period of small-amplitude librations (C.61). However, the limits of the two cases are not equal since I_0 (and $P''(I_0)$) are not the same at both equilibrium points

C.8 The Angle Variable w

The angle variable is, by definition,

$$w = \frac{2\pi(t-t_0)}{T} = \frac{2\pi|t'-t'_0|}{T'}. \tag{C.64}$$

The origin of the angle w is taken at the lowest point of the solution in the (σ, I) plane. The modulus in (C.64) is used because $t'(t)$ is a monotonically decreasing function when $J < 0$. (In the standard case $b > 0, \tau > 0$.)

Exercise C.8.1. Show that

$$\phi = \pm \frac{w}{2} + \mathcal{O}(\kappa_1^2) \tag{C.65}$$

and the same for ϕ'. Discuss the double sign on the right-hand side.

C.9 Small-Amplitude Librations

We may use the expansions of the elliptic functions to obtain harmonic approximations of the solutions in the case of small oscillations about one center. However, the alternative formulation presented in Sect. 2.2, to obtain angle–action variabes, is more convenient. We present, in this section, the direct construction of the angle–action variables of the Andoyer Hamiltonian in the case $k = 1$, for small-amplitude librations, using Fourier series to represent its periodic solutions. It is worthwhile noting that this technique can be easily extended to the Andoyer Hamiltonians with $k > 1$ and even to more complex Hamiltonians including several trigonometric terms[4].

The first and lengthy step is the construction of the periodic solutions themselves. We represent the solutions of the Andoyer Hamiltonian by the series

$$\sigma = \sum_{i=0}^{n} s_i \gamma^i \qquad J = \sum_{i=0}^{n} a_i \gamma^i, \tag{C.66}$$

where s_i and a_i are undetermined periodic functions in the angle w and γ is a free parameter of the order of the amplitude of the oscillations. ($\gamma = 0$ corresponds to one of the stable equilibria discussed in Sect. C.6.) For simplicity, we assume $s_0 = 0$, that is, the oscillations are around one center in the right-hand side of the X-axis. There is no loss of generality in this choice since a center on the negative semi-axis is moved to the right-hand semi-axis when we perform the transformation $\sigma \to \sigma' = \sigma + \pi$. It is important to keep in mind that $\dot{w}$ is not the same in all solutions and is itself also a function of the parameter γ. We assume:

---

[4] For the angle–action variables of the composed Andoyer Hamiltonian (C.3) with the two harmonics $k = 1, 2$, see [74].

$$\dot{w} = \omega_0 + \sum_{i=1}^{n} o_i \gamma^i. \tag{C.67}$$

The given series must satisfy the differential equations (assuming $J > 0$):

$$\frac{d\sigma}{dt} = a + 2bJ + \frac{\varepsilon\tau}{\sqrt{2J}} \cos\sigma$$

$$\frac{dJ}{dt} = \varepsilon\tau\sqrt{2J} \sin\sigma.$$

However, when the series representing the solutions are substituted into the differential equations written in this form, the results are cumbersome and the task of identification of the undetermined coefficients is very complex. A great deal of simplicity results from the introduction of the auxiliary variable $h = \sqrt{2J}$ and by multiplying the first equation by h to avoid denominators. The above equations then become:

$$h\frac{d\sigma}{dt} = ah + bh^3 + \varepsilon\tau \cos\sigma$$

$$\frac{dh}{dt} = \varepsilon\tau \sin\sigma. \tag{C.68}$$

In what follows, we substitute the given series for σ and the series

$$h = \sum_{i=0}^{n} h_i \gamma^i$$

in the above equations. The identification in powers of γ gives the equations allowing for the determination of h_i, s_i and o_i. The equations resulting from the identification are:

(a.) $i = 0$

$$ah_0 + bh_0^3 + \varepsilon\tau = 0, \tag{C.69}$$

$$h_0' = 0,$$

where the prime indicates differentiation with respect to w. The second equation shows that h_0 is a constant and (C.69) allows this constant to be determined. Equation (C.69) is the same cubic equation discussed in Sect. C.2.1. The only difference is that, there, the unknown is the non-singular variable $X \in \mathbf{R}$ while, here, it is $h = \sqrt{2J}$. To be consistent with the choice $s_0 = 0$, we assume that $h > 0$ and play with the sign of τ to recover all solutions. Equation (C.69) may have up to three real roots according to the value of α (see C.17). When $\alpha > 1$, we have three real roots as indicated in Fig. C.1 by $n = 0, 1, 2$. $n = 1$ corresponds to the libration center; to have it on the positive semi-axis, we choose σ so that $\tau < 0$. This is the case considered in this section. $n = 0$ corresponds to the center of the inner circulations and we should choose $\tau > 0$ to study this case. $n = 2$

corresponds to a saddle point and is of no interest here. When $\alpha < 1$, the only remaining root is a continuation of the libration center that will be on the positive side for $\tau < 0$. In the latest cases, the solutions are in the neighborhood of the origin and we need to use non-singular variables. This is the subject of Exercises C.9.2 and C.9.3.

(b.) $i = 1$

$$\omega_0 h_0 s_1' - \Xi h_1 = 0, \tag{C.70}$$
$$\omega_0 h_1' - \varepsilon \tau s_1 = 0,$$

where

$$\Xi = a + 3bh_0^2.$$

These two equations are equivalent to the second-order equation

$$\omega_0 h_0 s_1'' - \frac{\varepsilon \tau \Xi}{\omega_0} s_1 = 0. \tag{C.71}$$

Since we have assumed that w is the angle variable, the proper frequency of the solution may be equal to 1. Therefore,

$$\omega_0 = \sqrt{\frac{-\varepsilon \tau \Xi}{h_0}} \tag{C.72}$$

and the system has the particular solution

$$s_1 = \sin w$$

and, from (C.70),

$$h_1 = \frac{\omega_0 h_0}{\Xi} \cos w = -\frac{\varepsilon \tau}{\omega_0} \cos w.$$

The role of integration constants is played by the parameter γ and the initial phase of w.

The above equations introduced the condition $\tau \Xi < 0$, which is the condition for which h_0 is a stable equilibrium point. (See Sect. C.9.1. By hypothesis, $\varepsilon > 0$, $\tau < 0$.) The solutions are librations.

(c.) $i = 2$

$$\omega_0 h_0 s_2' - \Xi h_2 = -o_1 h_0 \cos w + \frac{a \varepsilon \tau}{\Xi} \cos^2 w - \frac{1}{2} \varepsilon \tau \sin^2 w,$$

$$\omega_0 h_2' - \varepsilon \tau s_2 = -\frac{\varepsilon o_1}{\omega_0} \sin w.$$

These equations are transformed into the second-order equation

$$\omega_0 h_0 (s_2'' + s_2) = 2 o_1 h_0 \sin w - \frac{3 \varepsilon \tau \omega_0^2}{2 \Xi^2} \sin 2w. \tag{C.73}$$

The existence, in the forced part, of terms with the same frequency as the proper frequency of the associated homogeneous equation leads to non-acceptable unbounded solutions. The elimination of the terms proportional to $\sin w$ in the right-hand side of the equation is necessary and gives the rule to determine the o_i. In this simple case, we get $o_1 = 0$. Integration gives

$$s_2 = -\frac{(\Xi - 2bh_0^2)^2}{2\omega_0 \Xi} \sin 2w = -\frac{\omega_0^3}{2\Xi^3} \sin 2w,$$

$$h_2 = -\varepsilon\tau \frac{\Xi - 6bh_0^2}{4\Xi^2} + \varepsilon\tau \frac{\Xi - 2bh_0^2}{4\Xi^2} \cos 2w.$$

We note the existence of a non-periodic term in h_2 showing that the large-amplitude oscillations do not have the same symmetry center as the small-amplitude ones.

(d.) $i = 3$

Proceeding in exactly the same way, we obtain

$$o_2 = -\frac{\varepsilon\tau b h_0}{2\omega_0 \Xi^2}(2\Xi^2 - 12\Xi bh_0^2 + 15 b^2 h_0^4),$$

$$s_3 = \frac{8\Xi^3 - 36\Xi^2 bh_0^2 + 54\Xi b^2 h_0^4 - 27 b^3 h_0^6}{24\Xi^3} \sin 3w$$

and

$$h_3 = \varepsilon\tau \frac{\Xi^3 + 8\Xi^2 bh_0^2 - 48\Xi b^2 h_0^4 + 60 b^3 h_0^6}{8\omega_0 \Xi^3} \cos w$$

$$-\varepsilon\tau \frac{\Xi^3 - 4\Xi^2 bh_0^2 + 6\Xi b^2 h_0^4 - 3 b^3 h_0^6}{8\omega_0 \Xi^3} \cos 3w.$$

The next orders are analogous. However, the results become cumbersome and we omit them here. It deserves to be noted that we have to go to order $k+1$ to determine o_k. From the equations for $i = 4$ and $i = 5$, we obtain $o_3 = 0$ and

$$o_4 = \frac{\varepsilon\tau b^2 h_0^3}{16\omega_0 \Xi^5}(48\Xi^4 - 648\Xi^3 bh_0^2 + 2816\Xi^2 b^2 h_0^4 - 5052\Xi b^3 h_0^6 + 3255 b^4 h_0^8).$$

It is worthwhile mentioning that the five parameters appearing in the above expressions $(b, \tau, \omega_0, h_0, \Xi)$ are not independent, and these expression can be written in many other equivalent ways.

The quality of the periodic solution thus constructed may be assessed from Fig. C.7, which shows exact and approximated solutions of the Hamiltonian shown in Fig. C.3 in the case $\alpha = 2$, $|\varepsilon\tau| = -0.3$. The agreement is very good up to libration amplitudes as high as 45 degrees, but deteriorates for larger amplitudes.

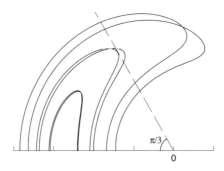

Fig. C.7. Solutions of the Hamiltonian shown in Fig. C.3 in the case $\alpha = 2$, $|\varepsilon\tau| = 0.3$ (dash-dot lines) and neighboring solutions of the same Hamiltonian constructed with the series, up to γ^5 (solid lines). The dashed line corresponds to a libration amplitude of 60 degrees

Exercise C.9.1. Show that, when $\tau\Xi < 0$, the positive roots of the polynomial $\mathcal{P} = bh^3 + ah + \varepsilon\tau$ correspond to stable equilibrium points. *Hint:* Ξ is the derivative of $\mathcal{P}$ at $h = h_0$.

C.9.1 The Action Λ

Using the solutions up to γ^5, the definition of the action variable leads to

$$\Lambda = \frac{\omega_0 h_0^2}{2\Xi}\gamma^2 - \frac{bh_0^4}{4\omega_0\Xi^3}(6\Xi^3 - 33\Xi^2 bh_0^2 + 61\Xi b^2 h_0^4 - 38b^3 h_0^6)\gamma^4. \qquad \text{(C.74)}$$

In the construction of the new Hamiltonian, we need to solve this equation with respect to γ. At this order, we obtain,

$$\gamma^2 = \frac{2\Xi}{\omega_0 h_0^2}\Lambda + \frac{2b}{h_0^2\Xi\omega_0^2}(6\Xi^2 - 21\Xi bh_0^2 + 19\,b^2h_0^4)\Lambda^2 + \mathcal{O}(\Lambda^3). \qquad \text{(C.75)}$$

C.9.2 The New Hamiltonian

In terms of the action Λ, the Hamiltonian may be obtained directly by substituting h and σ into the given Andoyer Hamiltonian (recalling that $J = h^2/2$). The result is:

$$H = -\frac{1}{2}h_0\varepsilon\tau\gamma^2 + \frac{\varepsilon\tau\, bh_0^3}{8\Xi^3}(10\Xi^2 - 30\Xi bh_0^2 + 23\,b^2h_0^4)\gamma^4 + \mathcal{O}(\gamma^6), \qquad \text{(C.76)}$$

where the γ-independent term $-\frac{1}{4}h_0^2(a+\Xi)$ was discarded. Using (C.75), the above equation becomes

$$H = \omega_0\Lambda + \frac{b}{2\Xi^2}(2\Xi^2 - 12\Xi bh_0^2 + 15b^2h_0^4)\Lambda^2 + \mathcal{O}(\Lambda^3). \qquad \text{(C.77)}$$

As expected, periodic terms are absent from H. Since the six parameters used $(a, b, \tau, \omega_0, h_0, \Xi)$ are not independent, we have to use the definitions of some of them (e.g. ω_0, h_0, Ξ) to check the cancellation of the periodic terms.

A better approximation for H, using the solutions at the same order of approximation as above, may be obtained from the integration of the equation $\omega = \dot{w} = \frac{\partial H}{\partial \Lambda}$. From the solution up to $i = 5$, we obtain $\omega = \omega_0 + o_2\gamma^2 + o_4\gamma^4 + \mathcal{O}(\gamma^6)$, whose integration with respect to Λ gives the Hamiltonian up to $\mathcal{O}(\Lambda^3)$ (one order more than the Hamiltonian obtained from the direct substitution).

The explicit calculation gives, for the additional term of the Hamiltonian,

$$-\frac{b^3 h_0^2}{4\omega_0 \Xi^4} (64\Xi^3 - 432\Xi^2 b h_0^2 + 960\Xi b^2 h_0^4 - 705\, b^3 h_0^6) \Lambda^3.$$

Exercise C.9.2. Obtain the solutions of the equations of the motion in the non-singular canonical variables $x = h\cos\sigma$, $y = h\sin\sigma$.

Exercise C.9.3. Adapt the solutions in non-singular canonical variables to the case $\tau > 0$ (small-amplitude *inner circulation*). *Hint*: See Exercise C.9.1.

Exercise C.9.4. Show that, in the case of inner circulations, the actions Λ are negative.

D

Andoyer Hamiltonians with $k \geq 2$

D.1 Introduction

Andoyer Hamiltonians with $k = 1$ are important integrable systems that are used as Hori kernels in the construction of the solutions of problems involving first-order resonances[1]. In the study of higher-order resonances, we need the Andoyer Hamiltonians with $k > 1$ [60].

The study of the centers and saddle points of these Hamiltonians can be done using the general equations derived in Sect. C.2.

D.2 The Case $k = 2$

The energy of the second Andoyer Hamiltonian is

$$H_2 = \frac{1}{2} a(X^2 + Y^2) + \frac{1}{4} b(X^2 + Y^2)^2 + \varepsilon\tau \sqrt{X^2 + Y^2}\, X \tag{D.1}$$

and (C.10) becomes

$$aX + bX^3 + 2\varepsilon\tau\,|X| = 0, \tag{D.2}$$

whose roots may be easily found.

We normalize the value of X using the factor

$$A_2 = \sqrt{\frac{4\varepsilon\tau}{b}} \tag{D.3}$$

and introduce the critical value

$$a_2^* = -2\varepsilon\tau = -\frac{bA_2^2}{2}. \tag{D.4}$$

---

[1] In the study of planetary motions, the order of a resonance $(\bar{h}\,|\omega) \simeq 0$ is defined as $k = |\sum_{i=1}^{N} \bar{h}_i|$. (The ω_i are the high frequencies, the so-called mean motions.)

Equation (D.2) then becomes

$$-\alpha\xi + 2\xi^3 + |\xi| = 0, \tag{D.5}$$

where

$$\xi = \frac{X}{A_2} \tag{D.6}$$

and

$$\alpha = \frac{a}{a_2^*}. \tag{D.7}$$

Equation (D.5) has one root at the origin and the two others are given by

$$\xi = s\sqrt{\frac{\alpha - s}{2}} \qquad (s = \pm 1),$$

whose condition of existence is $\alpha - s > 0$, that is $\alpha > -1$ for the root in the negative semi-axis and $\alpha > 1$ for the root in the positive semi-axis.

The nature of these singular points is fixed by the sign of the Hessian

$$\Delta_2 = -s a_2^{*2}(\alpha - s).$$

Since the condition of existence of the roots is $(\alpha - s) > 0$, these singular points are one center and one saddle point. The center is the root in the negative semi-axis (in which case $s = -1$ and $\Delta_2 > 0$). The nature of the singularity lying at the origin of the (X, Y) plane is given by the topological rule given in Sect. C.2: when $|\alpha| \leq 1$, it is a cusp, and, when $|\alpha| > 1$, it is a center.

D.2.1 Morphogenesis

The morphogenesis of the curves $H_2 = \text{const}$, when α varies, is shown in Figs. D.1 and D.2. These curves are shown in two different planes: (a) the plane of the coordinates (X, Y) defined by (C.6); (b) the plane of the Poincaré canonical variables (x, y) defined by (C.4).

In the (x, y) plane, the cusp at the origin of the (X, Y) plane (when $|\alpha| \leq 1$) gives rise to a saddle point while the center situated away from the origin moves to the vertical axis and is duplicated. (One should keep in mind that this picture corresponds to $\varepsilon\tau > 0$; when $\varepsilon\tau < 0$, the center lies on the positive X semi-axis and, as a consequence, the centers resulting from the duplication in the (x, y) plane remain on the horizontal axis.)

There are, in this case, two catastrophe sets, respectively at $\alpha = -1$ and $\alpha = +1$. One of the catastrophe sets appears for $\alpha = 1$, when the singularity at the origin changes from a center ($\alpha > 1$) to a saddle point ($\alpha \leq 1$). In the Andoyer Hamiltonian $k = 1$, only orbit deformations are seen for $\alpha \leq 1$. In this Hamiltonian, however, we have the formation of the figure eight shaped separatrix and, inside it, the family of librations continue to exist up to $\alpha = -1$. At this point, a second catastrophe set appears and,

D.2 The Case $k = 2$ 317

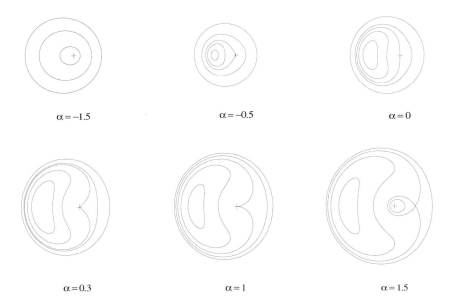

Fig. D.1. Regimes of motion of H_2 represented in the (X, Y) plane. The origin of the (X, Y) plane is marked with $+$

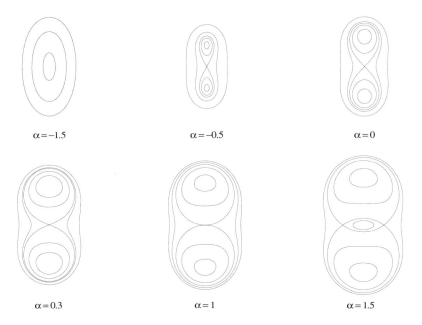

Fig. D.2. Phase portraits of the Andoyer Hamiltonian H_2

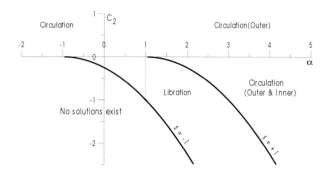

Fig. D.3. Morphogenesis of the solutions of H_2

for $\alpha < -1$, the origin becomes a center again and the flow of trajectories is topologically equivalent to that of the undisturbed differential rotator. We may also note that the effects of the resonance appear in the phase portrait even in the interval $-1 < \alpha < 0$, when no change in the direction of the motions is seen in the phase portrait of the undisturbed differential rotator H_0. (The resonance is "virtual". See Sect. 9.1.)

The morphogenesis is schematically represented in Fig. D.3 where we introduced the energy unit

$$H_2^* = \frac{1}{4}bA_2^4 \tag{D.8}$$

and considered (D.1) restricted to the axis $Y = 0$, as

$$C_2 = \left.\frac{H_2}{H_2^*}\right|_{Y=0} = \xi^4 - \alpha\xi^2 + \xi\,|\xi|. \tag{D.9}$$

When $k \geq 2$ the ambiguities of the nomenclature discussed in Sect. C.3 no longer exist. In these cases, the origin of the (x,y) plane is always a singular point (center or saddle point), thus, topological and kinematical points of view lead to equal classifications.

Exercise D.2.1. Show that, in the (X,Y) plane, when $|\alpha| < 1$, the angle between the tangents to each branch of the cusp and the X-axis is

$$\Phi_{\text{cusp}} = \lim_{J \to 0} \Phi = \arccos(\alpha)$$

and interpret the cases $\alpha = -1$, $\alpha = 0$ and $\alpha = +1$.

D.2.2 Width of the Libration Zone

The width of the libration zone is given by the intersection of the separatrix and the X-axis. In the calculations of the boundaries of ΔX, we have to consider two different subcases following $|\alpha| < 1$ or $\alpha > 1$. These boundaries, as functions of α, are shown in Fig. D.4

D.2 The Case $k = 2$

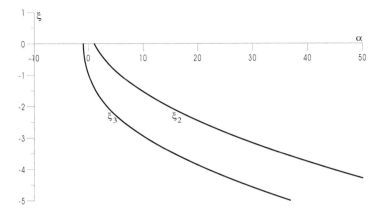

Fig. D.4. Width and location of the libration zone. Case $k = 2$

(a.) $|\alpha| < 1$ ($|a| < 2\varepsilon\tau$)

In this case, since the separatrix ends at the origin, we have $H_{2\text{sep}} = 0$. The other intersection with the X-axis is given by the other root of $H_2(X, 0) = 0$ or $C_2(\xi) = 0$. Hence

$$\Delta\xi = \sqrt{1 + \alpha}; \qquad \Delta X = \sqrt{\frac{4\varepsilon\tau - 2a}{b}}; \qquad \Delta J = \frac{2\varepsilon\tau - a}{b}.$$

The width of the resonance zone is zero when $a = -a_2^*$ ($\alpha = -1$) and increases to reach

$$\Delta\xi = \sqrt{2}; \qquad \Delta X = \sqrt{\frac{8\varepsilon\tau}{b}}; \qquad \Delta J = \frac{4\varepsilon\tau}{b}$$

at $a = a_2^*$ ($\alpha = +1$). It is of order $\mathcal{O}(\sqrt{\varepsilon})$ in the (X, Y) plane and of order $\mathcal{O}(\varepsilon)$ in the action J.

(b.) $\alpha > 1$ ($a < -2\varepsilon\tau$)

This case is very similar to the case $k = 1$ considered in Sect. C.4, but it cannot be studied by just using some elementary theorems on polynomials, as was done there, because $H_2(X, 0)$ is not a polynomial, but a matching of two bi-quadratic polynomials, one for X positive ($s = +1$) and the other for X negative ($s = -1$). However, since only even powers of X appear in $H_2(X, 0)$, the direct calculation is simple. The intersections of the separatrix with the X-axis are given by $H_2(X, 0) = H_{2\text{sep}}$. The value of $H_{2\text{sep}}$ is the value of H_2 at the saddle point, which is known, since the value of X at the saddle point is the root of (D.2) corresponding to $s = +1$. In normalized variables

$$\xi_{\text{sad}} = \sqrt{\frac{\alpha - 1}{2}}$$

and
$$C_{2\text{sep}} = C_2(\xi_{\text{sad}}) = -\left(\frac{\alpha-1}{2}\right)^2.$$

The boundaries of the libration zone are the two negative roots of the equation
$$\xi^4 - (\alpha+1)\xi^2 - C_{2\text{sep}} = 0,$$
that is,
$$\xi_{2,3} = -\frac{\sqrt{\alpha}\pm 1}{\sqrt{2}}.$$

Therefore[2]

$$\Delta\xi = \sqrt{2}; \qquad \Delta X = \sqrt{\frac{8\varepsilon\tau}{b}}; \qquad \Delta J = \frac{\sqrt{-8a\varepsilon\tau}}{b}.$$

It is noteworthy that, in this case, ΔX is independent of α and is of order $\mathcal{O}(\sqrt{\varepsilon})$. In the action J, the width of the libration zone is of order $\mathcal{O}(\varepsilon)$, at the limit value $a = a_2^* = -2\varepsilon\tau$, and becomes of order $\mathcal{O}(\sqrt{\varepsilon})$ for a finite. The functions $\xi_2(\alpha)$ and $\xi_3(\alpha)$ are shown in Fig. D.4

D.3 The Case $k = 3$

The energy of the third Andoyer Hamiltonian is
$$H_3 = \frac{1}{2}a(X^2+Y^2) + \frac{1}{4}b(X^2+Y^2)^2 + \varepsilon\tau(X^2+Y^2)X \tag{D.10}$$

and the equation giving the singular points is
$$aX + bX^3 + 3\varepsilon\tau X^2 = 0. \tag{D.11}$$

We normalize X using the factor
$$A_3 = \frac{3\varepsilon\tau}{2b} \tag{D.12}$$

and introduce the critical value
$$a_3^* = -\frac{9\varepsilon^2\tau^2}{4b} = -bA_3^2. \tag{D.13}$$

Equation (D.11) then becomes
$$-\alpha\xi + \xi^3 + 2\xi^2 = 0, \tag{D.14}$$

where

---
[2] $\Delta J = \frac{1}{2}|X_2^2 - X_3^2| = \frac{1}{2}\Delta X |X_2 + X_3|.$

$$\xi = \frac{X}{A_3} \tag{D.15}$$

and

$$\alpha = \frac{a}{a_3^*}. \tag{D.16}$$

When $\alpha > -1$, (D.14) has one solution at the origin and the others at the points

$$\xi = -1 \pm \sqrt{1+\alpha}. \tag{D.17}$$

When $\alpha = -1$, the two roots away from the origin coalesce into a cusp at $\xi = -1$ and, for $\alpha < -1$ these two roots are no longer real and the origin is the only remaining real root.

The Hessian of H_3 is

$$\Delta_3 = -\frac{4}{3} a_3^{*2} \xi(\alpha - \xi)$$

or, taking (D.14) and (D.17) into account,

$$\Delta_3 = \pm \frac{4}{3} a_3^{*2} \xi^2 \sqrt{1+\alpha}.$$

Therefore, the singular points situated away from the origin of the (X, Y) plane are, always, one center and one saddle point. The origin is a center. The only exception occurs when $\alpha = 0$, in which case one of the roots defined by (D.17) (the saddle point) coalesces with the center located at the origin giving rise to a cusp. This cusp is not apparent in the corresponding phase portrait, in Fig. D.1, because the two cusp branches have vertical tangent at the cusp. After the triplication of the polar angle, the origin becomes, in the (x, y) plane, a second-order saddle point with three stable and three unstable branches concurring there.

D.3.1 Morphogenesis

The curves $H_3 = \text{const}$ are shown in Figs. D.5 and D.6 for several values of the parameter α. They are shown in two different planes: the (X, Y) plane and the (x, y) plane. There are, again, two catastrophe sets: one at $\alpha = 0$, when the saddle point, whose abscissa is

$$\xi_{\text{sad}} = -1 + \sqrt{1+\alpha},$$

crosses the origin, passing from one side of the X-axis to another; the other at $\alpha = -1$, when the number of singular points in the (X, Y) plane changes from 1 to 3.

The morphogenesis of these orbits is schematically represented in Fig. D.7 where we introduced the energy unit

$$H_3^* = \frac{1}{4} b A_3^4$$

D Andoyer Hamiltonians with $k \geq 2$

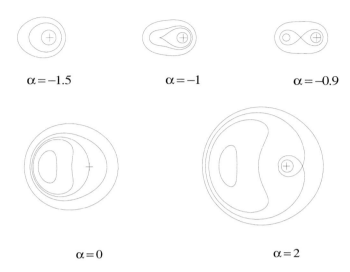

Fig. D.5. Regimes of motion of H_3 represented in the (X, Y) plane. The origin of the (X, Y) plane is marked with $+$

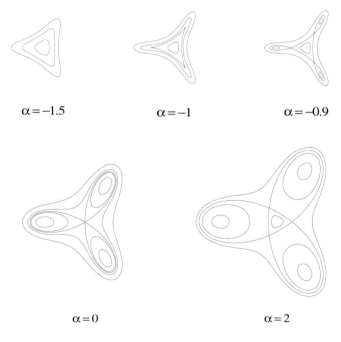

Fig. D.6. Phase portraits of the Andoyer Hamiltonian H_3

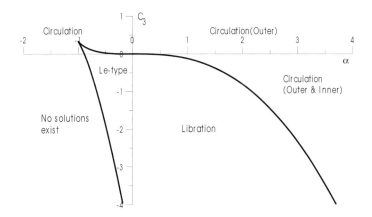

Fig. D.7. Morphogenesis of the solutions of H_3

and considered (D.10), restricted to the axis $Y = 0$, as

$$C_3 = \left.\frac{H_3}{H_3^*}\right|_{Y=0} = \xi^4 + \frac{8}{3}\xi^3 - 2\alpha\xi^2. \tag{D.18}$$

Exercise D.3.1. Show that, in the (X, Y) plane, when the saddle point is not too close to the origin, the angle Φ between the tangents to the branches of the saddle point and the X-axis is

$$\Phi = \arctan\left(\lim_{X \to X_{\text{sad}}} \frac{Y}{X - X_{\text{sad}}}\right) = \pm \arctan\sqrt{3(\xi_{\text{sad}} + 1)} \tag{D.19}$$

and interpret the limit $\alpha \to -1$. *Hint:* The expansion of H_3 about the saddle point is

$$H_3 = H_3(X_{\text{sad}}, 0) + \left(bX_{\text{sad}}^2 + \frac{3}{2}\varepsilon\tau X_{\text{sad}}\right)(X - X_{\text{sad}})^2 - \frac{1}{2}\varepsilon\tau X_{\text{sad}} Y^2 + \cdots$$

Exercise D.3.2. Show that (D.19) is not valid in the limit $\alpha = 0$.

Exercise D.3.3. Show that, in the case $\alpha = 0$, when the saddle point coalesces with the center situated at the origin, the angle Φ between the X-axis and the tangents to the singular point thus formed is $90°$. *Hint:* When $\alpha = 0$, the curve $H_3 = 0$ degenerates into a circle with center at $\left(-\frac{2\varepsilon\tau}{b}, 0\right)$ and radius $2\varepsilon\tau/b$.

D.3.2 Width of the Libration Zone

The intersection of the separatrix with the X-axis are given by the roots of $C_3 = C_{3\text{sad}}$. Since ξ_{sad} is a double root of $C_3 - C_{3\text{sad}} = 0$, we may divide the left-hand side of this equation by $(\xi - \xi_{\text{sad}})^2$. The result is

$$\xi^2 + \left(\xi_{\text{sad}} + \frac{4}{3}\right)(2\xi + \xi_{\text{sad}}) = 0, \qquad (D.20)$$

where (D.14) was used to simplify ξ_{sad} in the result. The other intersections of the separatrix with the axis $Y = 0$ are given by the solutions of (D.20):

$$\xi_{2,3} = -\xi_{\text{sad}} - \frac{4}{3}\left(1 \pm \sqrt{1 + \frac{3\xi_{\text{sad}}}{4}}\right).$$

The width of the resonance zone is calculated differently in the cases $-1 < \alpha < 0$ and $\alpha > 0$.

(a.) $-1 < \alpha < 0$ $(0 < a < |a_3^*|)$

In this case the libration zone is the lobe not including the origin. These particular librations are indicated with the label *Le–type*[3] in Fig. D.7. Adopting $\xi_2 > \xi_3$, it follows that

$$\Delta\xi = \xi_{\text{sad}} - \xi_3 = 2\xi_{\text{sad}} + \frac{4}{3}\left(1 + \sqrt{1 + \frac{3\xi_{\text{sad}}}{4}}\right),$$

$$\Delta X = 2X_{\text{sad}} + \frac{2\varepsilon\tau}{b}\left(1 + \sqrt{1 + \frac{bX_{\text{sad}}}{2\varepsilon\tau}}\right)$$

and

$$\Delta J = \frac{3\varepsilon\tau X_{\text{sad}}}{b} + \frac{4\varepsilon^2\tau^2}{b^2}\left(1 + \frac{bX_{\text{sad}}}{2\varepsilon\tau}\right)^{\frac{3}{2}}.$$

The width of the libration zone is zero when $\alpha = -1$ ($\xi_3 = \xi_2 = -1$) and it grows up to $4\varepsilon\tau/b$ as $\alpha \to 0$, in the (X, Y) plane, or $4\varepsilon^2\tau^2/b^2$, in the action J. The boundaries ξ_{sad} and ξ_3 delimiting the libration lobe are given by the leftmost arc in Fig. D.8. The vertex A represents the point where this lobe vanishes and merges with the origin.

(b.) $\alpha > 0$ $(a < 0)$

In this case the libration zone is delimited by the two branches of the separatrix. Then

$$\Delta\xi = \xi_2 - \xi_3 = \frac{8}{3}\sqrt{1 + \frac{3\xi_{\text{sad}}}{4}};$$

$$\Delta X = \frac{4\varepsilon\tau}{b}\sqrt{1 + \frac{bX_{\text{sad}}}{2\varepsilon\tau}}$$

and

$$\Delta J = \frac{8\varepsilon^2\tau^2}{b^2}\left(1 + \frac{bX_{\text{sad}}}{2\varepsilon\tau}\right)^{\frac{3}{2}} \qquad (D.21)$$

---

[3] Mnemonic for the lemniscata-like appearance of the separatrix.

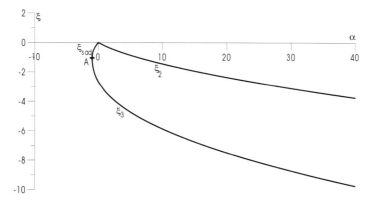

Fig. D.8. Width and location of the libration zone. Case $k = 3$

The width of the libration zone is of order $\mathcal{O}(\varepsilon)$ in the (X, Y) plane and $\mathcal{O}(\varepsilon^2)$ in the action J. The functions $\xi_2(\alpha)$ and $\xi_3(\alpha)$ are shown in Fig. D.8.

(c.) $\alpha \gg 1$ $(a \ll a_3^*)$

Now, the distance of the saddle point to the origin may be approximated by $X_{\text{sad}} \approx \sqrt{-a/b}$, that is, X_{sad} is finite and $|bX_{\text{sad}}/2\varepsilon\tau| \gg 1$. Hence

$$\Delta X \approx \sqrt{\frac{8\varepsilon\tau X_{\text{sad}}}{b}},$$

which is of order $\mathcal{O}(\sqrt{\varepsilon})$. In the same way

$$\Delta J \approx \sqrt{\frac{8\varepsilon\tau X_{\text{sad}}^3}{b}},$$

which is also of order $\mathcal{O}(\sqrt{\varepsilon})$.

D.4 The Case $k = 4$

The energy of the fourth Andoyer Hamiltonian is

$$H_4 = \frac{1}{2}a(X^2 + Y^2) + \frac{1}{4}b(X^2 + Y^2)^2 + \varepsilon\tau(X^2 + Y^2)\sqrt{X^2 + Y^2}X \quad \text{(D.22)}$$

and (C.10) becomes

$$-\alpha X + X^3 + \varepsilon_1 X^2 |X| = 0, \quad \text{(D.23)}$$

where we normalized the parameters by introducing

Fig. D.9. Regimes of motion of H_4 represented in the (X,Y) plane. The origin of the (X,Y) plane is marked with $+$

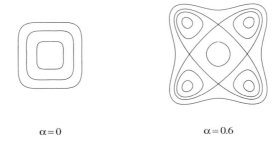

Fig. D.10. Phase portraits of the Andoyer Hamiltonian H_4

$$\alpha = -\frac{a}{b} \quad \text{and} \quad \varepsilon_1 = \frac{4\varepsilon\tau}{b}.$$

(X is not normalized since in this case we would have $A_4 = 1$.) The only critical value, in this case, is $\alpha = 0$. When $\alpha = 0$, the equation has just a triple singular point at the origin. When $\alpha > 0$ ($a < 0$), the equation has one root at the origin and the others are given by

$$X_i = s\sqrt{\frac{\alpha}{1+s\varepsilon_1}} \quad (s = \pm 1)$$

(note that $0 < \varepsilon_1 < 1$ if $b > 4\varepsilon\tau > 0$). When $\alpha < 0$ ($a > 0$), there is only one singular point, which is a center at the origin.

The Hessian of H_4 is

$$\Delta_4 = 2a\varepsilon\tau X\,|X| = -\frac{1}{2}ab^2\varepsilon_1 X\,|X|. \tag{D.24}$$

Therefore, the singular points situated away from the origin (when $\alpha > 0$) are one center and one saddle point. (The center is the singular point lying on the negative semi-axis.)

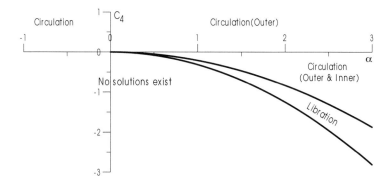

Fig. D.11. Morphogenesis of the solutions of H_4

D.4.1 Morphogenesis

The morphogenesis of the curves $H_4 = $ const, is simpler than those shown in the previous sections. The regimes of motion and bifurcations are shown in Figs. D.9 and D.10, in the (X, Y) and (x, y) planes, respectively. The only catastrophe set corresponds to $\alpha = 0$. In this set, the origin is a triple root of (D.23), where the saddle points and the two centers coalesce into a center at the origin. The two rightmost portraits in Fig. D.9 show how the whole structure inside the separatrix is squeezed into a vanishing neighborhood of the origin when $\alpha \to 0$ ($\alpha > 0$) with almost no changes. The morphogenesis is schematically represented in Fig. D.11 where we have introduced the energy unit

$$H_4^* = \frac{b}{4}.$$

On the axis $Y = 0$, we have

$$C_4 = \left.\frac{H_4}{H_4^*}\right|_{Y=0} = X^4 - 2\alpha X^2 + \varepsilon_1 |X| X^3. \tag{D.25}$$

The curves separating the regimes of motion in the half-plane $\alpha > 0$ are arcs of the parabolas $C_4 = -\alpha^2/(1 \pm \varepsilon_1)$. The lower parabola corresponds to a center.

D.4.2 Width of the Libration Zone

The intersections of the separatrix with the X-axis are given by the roots of $C_4 = C_{4\text{sep}}$. As in the case $k = 2$, we have to remember that C_4 is not a polynomial, but a matching of two bi-quadratic polynomials, one for X positive ($s = +1$) and the other for X negative ($s = -1$).

According to (D.24), the saddle point is the singularity for which $s = +1$ and the values of C_4 at the two other intersections of the separatrix are given by

$$C_{4\text{sep}} = C_4(X_{\text{sad}}) = -\frac{\alpha^2}{1+\varepsilon_1}.$$

The boundaries of the resonance zone are the two negative roots of the bi-quadratic equation $C_4(X) = C_{4\text{sep}}$ (see Fig. D.9), that is,

$$(1-\varepsilon_1)X^4 - 2\alpha X^2 + \frac{\alpha^2}{1+\varepsilon_1} = 0. \tag{D.26}$$

They are

$$X_i = -\sqrt{\frac{\alpha}{1-\varepsilon_1}\left(1 \pm \sqrt{\frac{2\varepsilon_1}{1+\varepsilon_1}}\right)}.$$

Then

$$\Delta X = \sqrt{\frac{\alpha}{1-\varepsilon_1}}\left(\sqrt{1+\sqrt{\frac{2\varepsilon_1}{1+\varepsilon_1}}} - \sqrt{1-\sqrt{\frac{2\varepsilon_1}{1+\varepsilon_1}}}\right)$$

and

$$\Delta J = \frac{2\alpha}{1-\varepsilon_1}\sqrt{\frac{2\varepsilon_1}{1+\varepsilon_1}}.$$

If $\varepsilon\tau \ll b$, then $\varepsilon_1 \ll 1$ and we may write the approximate equations

$$\Delta X = \sqrt{2\alpha\varepsilon_1} \qquad \Delta J = 2\alpha\sqrt{2\varepsilon_1}. \tag{D.27}$$

The limits in this case are immediate. The width of the resonance is 0 when $\alpha = 0$ and is of order $\mathcal{O}(\sqrt{\varepsilon})$ when α is a finite quantity.

Exercise D.4.1. Modify (D.26) to make it valid in the positive semi-axis $X > 0$ and show that the only roots of $C_4 = C_{4\text{sep}}$ are, indeed, the above calculated two roots and the saddle point.

D.5 Comparative Analysis

Let us summarize the topological evolution of the solutions. In all cases, when $\alpha \gg 0$ ($a \ll 0$), $H_k(X,Y)$ has three singular points, all of them in the X-axis, in the order center–center–saddle while, when $\alpha \ll 0$ ($a \gg 0$), only one singular point – one center – remains. The transitions follow different routes. For $k = 1$, one center and the saddle point coalesce into a cusp when α decreases and reaches $\alpha = 1$ ($a = a_1^*$), disappearing for $\alpha < 1$ ($a > a_1^*$). For $k = 2$ the transition is similar. However, what occurred in only one point for $k = 1$, now occurs in the whole interval $|\alpha| \leq 1$ ($|a| \leq |a_2^*|$), where the singular points are one center and one cusp. For $k = 3$, the transition is more complex: when α decreases and reaches $\alpha = 0$ ($a = 0$), a double point forms, but the two singular points forming it are not destroyed, subsisting for $\alpha < 0$ ($a > 0$), with only the change in their relative positions. In the transition zone $-1 < \alpha < 0$ ($a_3^* < a < 0$), the three singular points in the X-axis are in the order center–saddle–center. Only when the two singular points coalesce for the second time, do they disappear. For $k = 4$, there is no cusp formation; in the transition at $\alpha = 0$, all three singular points coalesce at the origin.

The other point deserving a comparison is the width of the libration zone. First of all, we mention that for finite a ($a < 0$), this width is of order $\mathcal{O}(\sqrt{\varepsilon})$ in the action J, as well as in the Poincaré canonical variables defined by (C.4) for all k. This result is related to the introduction of the square root of the small parameter discussed in Sect. 4.2. A similar study of the singularities of the solution of the Hamilton–Jacobi equation in this case has not been done. For practical purposes, the order of magnitude of the width of the libration zones near the catastrophes gives the necessary indication. They are $\mathcal{O}(\sqrt[3]{\varepsilon})$, $\mathcal{O}(\sqrt[2]{\varepsilon})$ and $\mathcal{O}(\varepsilon)$, for $k = 1, 2$ and 3, respectively. As a consequence, the series solution of problems involving such resonances near the origin are calculated using the powers of the cube root of the parameter, the square root of the parameter or the parameter itself, according to the order of magnitude of the width of the libration zone. If the square root of the small parameter used far from the origin were adopted also for resonances near the origin, the rigorous identification of the orders in the separation of perturbation equations would be impaired. For instance, in the case of a first-order resonance, the adoption of $\sqrt[2]{\varepsilon}$ as the small parameter leads to terms of higher fractional orders (*e.g.* 5/4), while the adoption of $\sqrt[3]{\varepsilon}$, as indicated by the width of the resonance, does not introduce spurious fractional orders and the perturbation equations may be established without ambiguities, as seen in Chap. 9.

For $k = 4$, the libration zone width starts from zero and is of order $\mathcal{O}(a\sqrt{\varepsilon})$. A perturbation theory for this case should consider this particular width.

Andoyer Hamiltonians with $k > 4$ were not considered. Indeed, in real problems, $aJ + bJ^2$ is the leading part of the Taylor expansion of a function $H_0(J)$ regular at the origin and higher-order terms ought also to be included in $H_0(J)$ when $k \geq 5$.

D.5.1 Virtual Resonances

Virtual resonances are present in the cases $k = 2$ and $k = 3$. In these cases, topological changes are seen near the origin in the interval $-1 < \alpha < 0$. If the last term of the Andoyer Hamiltonians is seen as a perturbation of a differential rotator, in the two quoted cases, this perturbation creates libration zones for some negative values of α, notwithstanding the fact that no zero frequencies appear in the phase portrait of the undisturbed differential rotator in that cases (see Sect. 9.1). Indeed, the right-hand side of (C.23), giving the critical value $|J_r|$, becomes negative for $\alpha < 0$ and a circle where $\nu_0 = 0$ in the undisturbed $\varepsilon = 0$ case does not exist in the real (x, y) plane. However, it exists for "virtual" values of J, that is, for values of J corresponding to complex values of x, y and the disturbance created by the term $\varepsilon\tau X$ is wide enough to be seen in the portrait of the system in the real (x, y) plane.

References

1. M. Abramowitz, I. A. Stegun (eds.), *Handbook of Mathematical Functions* (NBS, Washington, 1964); *Pocketbook of Mathematical Functions* (Verlag Harri Deutsch, Thun, 1984)
2. H. Andoyer, "Contribution à la théorie des petites planètes dont le moyen mouvement est sensiblement double de celui de Jupiter". Bulletin Astronomique **20**, 321 (1903)
3. V. I. Arnold, "Small denominators and problems of stability of motion in Classical and Celestial Mechanics". In: *Hamiltonian Dynamical Systems*, ed. by R. S. MacKay and J. D. Meiss (Adam Hilger, Bristol, 1987), pp. 260–366 (transl. from Russ. Math. Surveys **18**, 85, 1963)
4. V. I. Arnold, *Mathematical Methods of Classical Mechanics*, 1st edn (Springer-Verlag, New York, 1978)
5. V. I. Arnold, V. V. Kozlov, A. I. Neishtadt, "Mathematical aspects of classical and celestial mechanics". In: *Dynamical Systems III*, ed. by V. I. Arnold (Springer-Verlag, Berlin, 1988)
6. D. Benest, Cl. Froeschlé (eds.), *Singularities in Gravitational Systems* (Springer, Berlin, 2002)
7. M. V. Berry, "Regular and irregular motions". In: *Topics in Nonlinear Dynamics. A tribute to Sir Edward Bullard*, ed. by S. Jorna (American Institute of Physics, New York, 1978), pp 16–120
8. K. Bohlin, "Über eine neue Annäherungsmetode in der Störungsteorie". Bihang till K. Svenska Vet.-Akad. **14** Afd. I, No. 5 (1888); almost completely reproduced in Astron. Nachrichten, **121**, 17 (1889)
9. H. Bohr, *Almost Periodic Functions* (Chelsea, New York, 1951)
10. N. Bohr, "On the quantum theory of line-spectra". D. Kgl. Danske Vidensk. Selsk. Skrifter, Naturvidensk. og Mathem. Afd. 8. Række, IV.1,1-3 (1918).
11. M. Boll and C. Salomon, *Introduction à la Théorie des Quanta* (O. Doin, Paris, 1928)
12. M. Born, *Vorlesungen über Atommechanik* (Springer, Berlin, 1925); *The Mechanics of the Atom* (F. Ungar, New York, 1927)
13. M. Born, *Problems of Atomic Dynamics* (MIT Press, Cambridge, 1926) (reprinted in 1970), Lectures 12–13
14. D. Brouwer, "Solution of the problem of artificial satellite theory without drag". Astron. J. **64**, 378 (1959); partly reproduced in [15], Sect. 17.12

15. D. Brouwer, G. Clemence, *Methods of Celestial Mechanics* (Academic Press, New York, 1961)
16. S. Breiter, "Extended fundamental model of resonance". Celest. Mech. Dynam. Astron. **85**, 209 (2003)
17. P. F. Byrd, M. D. Friedman, *Handbook of Elliptic Integrals for Engineers and Scientists* (Springer-Verlag, New York, 1971), Sect. 902
18. C. Carathéodory, *Variationsrechnung und Partielle Differentialgleichingen Erste Ordnung* (Teubner, Berlin, 1935); *Calculus of Variations and Partial Differential Equations of First Order*, Vol. I (Holden-Day, San Francisco, 1965)
19. A. Celletti, L. Chierchia, "KAM tori for N-body problems: a brief history". Celest. Mech. Dynam. Astron. **95**, 117 (2006)
20. C. V. L. Charlier, *Die Mechanik des Himmels*, vols 1–2 (W. De Gruyter, Leipzig, 1902–1906)
21. B. V. Chirikov, "A universal instability of many-dimensional oscillator systems". Phys. Reports, **52**, 263 (1979)
22. C. Delaunay, "Mémoire sur la théorie de la Lune". Mém. de l'Acad. des Sciences **28** (1860) and **29** (1867)
23. A. Deprit, "Canonical transformation depending on a small parameter". Celest. Mech. **1**, 12 (1969)
24. R. Dvorak, F. Freistetter, "Orbit dynamics, stability and chaos in planetary systems". In: *Chaos and Stability in Planetary Systems*, ed. by R. Dvorak, F. Freistetter and R. Kurths (Springer, Berlin Heidelberg, 2005) Sect. 5.3
25. H. B. Dwight, *Tables of Integrals and other Mathematical Data*, 4th. edn (Macmillan, New York, 1961)
26. A. Einstein, "Zum Quantensatz von Sommerfeld und Epstein". Verhandl. d. Deutsche Phys. Gesellsch. **19**, 82 (1917)
27. A. Elipe, V. Lanchares, T. López-Moratalla, A. Riaguas, "Nonlinear stability in resonant cases: A geometrical approach". J. Nonlinear Sci., **11**, 211 (2001)
28. P. S. Epstein, "Zur Quantentheorie". Ann. d. Phys. **51**, 168 (1916)
29. T. Feagin, R. G. Gotlieb, "Generalization of Lagrange's implicit function theorem to N dimensions". Celest. Mech. **3**, 2 (1971)
30. S. Ferraz-Mello, W. Sessin, "A note on resonance in regular variables and averaging". Celest. Mech. **34**, 453 (1984)
31. S. Ferraz-Mello, "Resonance in regular variables. I: Morphogenetic analysis of the orbits in the case of a first-order resonance". Celest. Mech. **35**, 209 (1985)
32. S. Ferraz-Mello, "On a class of integrable Hamiltonians". Ciência e Cultura, **40**, 598 (1988)
33. S. Ferraz-Mello, "On Hamiltonian averaging theories and resonance". Celest. Mech. Dynam. Astron. **66**, 39 (1997)
34. S. Ferraz-Mello, "Do average Hamiltonians exist?". Celest. Mech. Dynam. Astron. **73**, 243 (1999)
35. J. Ford, G. H. Lunsford, "Stochastic behavior of resonant nearly linear oscillator systems in the limit of zero nonlinear coupling". Phys. Rev. **A1**, 59 (1970)
36. B. Garfinkel, A. Jupp, C. Williams, "A recursive von Zeipel algorithm for the ideal resonance problem". Astron. J. **76**, 157 (1971)
37. B. Garfinkel, "On resonance in Celestial Mechanics (A survey)". Celest. Mech. **28**, 275 (1982)
38. I. A. Gerasimov, "A qualitative investigation of the motion of asteroids of the Hecuba type". Soviet Astron. **26**, 715 (1982)

39. A. Giorgilli, "Classical constructive methods in KAM theory". Planet. Space Sci. **46**, 1441 (1998)
40. A. Giorgilli, U. Locatelli, "A classical self-contained proof of Kolmogorov's theorem on invariant tori". In: *Hamiltonian Systems with Three or more Degrees of Freedom*, ed. by C. Simó (Kluwer, Dordrecht, 1999) pp.72–89
41. I. S. Gradstein, I. M. Ryshik, *Tables of Series, Products and Integrals*, vols 1–2 (Verlag Harri Deutsch, Thun Frankfurt/Main 1981)
42. A. Gramain, *Topologie des Surfaces* (Presses Univ. France, Paris, 1971); *Topology of Surfaces*, (BCS Associates, Moscow Idaho, 1984)
43. W. Gröbner, *Die Lie-Reihen und Ihre Anwendungen* (Deutsch Verlag d. Wiss., Berlin, 1960)
44. J. Guckenheimer, P. Holmes, *Nonlinear Oscillations, Dynamical Systems and Bifurcations of Vector Fields* (Springer-Verlag, New York, 1983)
45. Y. Hagihara, *Celestial Mechanics* (MIT Press, Cambridge, 1970)
46. M. Hénon, C. Heiles, "The applicability of the third integral of motion: Some numerical examples". Astron. J. **69**, 73 (1964)
47. J. Henrard, "Virtual singularities in artificial satellite theory". Celest. Mech. **10**, 437 (1974)
48. J. Henrard, "Orbital evolution of the Galilean satellites". In *The Motion of Planets and Natural and Artificial Satellites* Proc. CNPq–NSF Workshop (Embu, 1981), ed. by S. Ferraz-Mello and P. Nacozy (IAG-USP, São Paulo, 1983) pp. 233–244
49. J. Henrard, A. Lemaitre, "A second fundamental model for resonance". Celest. Mech. **30**, 218 (1983)
50. J. Henrard, A. Lemaitre, "A perturbation method for problems with two critical arguments". Celest. Mech. **39**, 213 (1986)
51. G. Herglotz, *Vorlesungen über die Theorie der Berührungstransformationen* (Göttingen, summer 1932, unpubl.) Cf. R. B. Guenther, C. M. Guenther, J. A. Gottsch, *The Herglotz Lectures on Contact Transformations and Hamiltonian Systems* (Julius Schauder Center, NCU, Toruń, 1996)
52. G.-I. Hori, "The motion of an artificial satellite in the vicinity of the critical inclination". Astron. J. **65**, 291 (1960)
53. G.-I. Hori, "Theory of general perturbations with unspecified canonical variables". Publ. Astronom. Soc. Japan, **18**, 287 (1966)
54. G.-I. Hori, "Theory of general perturbations". In: *Recent Advances in Dynamical Astronomy*, ed. by D. B. Tapley and V. Szebehely (Reidel, Dordrecht, 1973), pp 231–249
55. A. Jupp, "A solution of the ideal resonance problem for the case of libration". Astron. J. **74**, 33 (1968)
56. A. Jupp, "A comparison of the Bohlin–von Zeipel and Bohlin–Lie series methods in resonant systems". Celest. Mech. Dynam. Astron. **26**, 413 (1982)
57. A. N. Kolmogorov, "Preservation of conditionally periodic movements with small change in the Hamilton function". In: *Stochastic Behavior in Classical and Quantum Hamiltonian Systems*, ed. by G. Casati and J. Ford (Springer-Verlag, Berlin 1979) pp 51–56 (transl. from Dokl. Akad. Nauk SSSR, **98**, 527, 1954)
58. J. Kovalevsky, *Introduction à la Mécanique Céleste* (Armand Colin, Paris, 1963); *Introduction to Celestial Mechanics* (D. Reidel, Dordrecht, 1967)
59. C. Lanczos, *The Variational Principles of Mechanics*, 4th edn (University of Toronto Press, Toronto, 1970)

60. A. Lemaitre, "High-order resonances in the restricted three-body problem". Celest. Mech. **32**, 109 (1984)
61. A. Lemaitre, "Proper elements: What are they?". Celest. Mech. Dynam. Astron. **56**, 103 (1993)
62. A. Lemaitre, S. D'Hoedt, N. Rambaux, "The 3:2 spin-orbit resonant motion of Mercury". Celest. Mech. Dynam. Astron. **95**, 213 (2006)
63. A. J. Lichtenberg, M. A. Lieberman, *Regular and Stochastic Motion* (Springer-Verlag, New York, 1983)
64. A. Lindstedt, "Ueber die allgemeine Form der Integrale des Dreikörperproblems". Astron. Nachrichten, **105**, 97 (1882)
65. G. W. Mackey, *Mathematical Foundations of Quantum Mechanics* (W. A. Benjamin, New York, 1963)
66. C. Marchal, *The Three-Body Problem* (Elsevier, Amsterdam, 1990), Chap. 9
67. A. P. Markeev, "On the stability of the triangular libration points in the circular bounded three body problem". PMM – Appl. Math. Mech. **33**, 105 (1966)
68. A. P. Markeev, "Stability of the triangular Lagrangian solutions of the restricted three-body problem in the three-dimensional circular case". Soviet Astron. **15**, 682 (1972)
69. W. Mersman, "A new algorithm for the Lie transforms". Celest. Mech. **3**, 81 (1969)
70. P. J. Message, "Planetary perturbation theory from Lie series, including resonance and critical arguments". In: *Long-Term Dynamical Behaviour of Natural and Artificial N-Body Systems*, ed. by. A. E. Roy (Kluwer, Dordrecht, 1987) pp. 47–72
71. K. R. Meyer, G. R. Hall, *Introduction to Hamiltonian Dynamical Systems and the N-body Problem* (Springer-Verlag, New York, 1991)
72. K. R. Meyer, D. S. Schmidt, "The stability of the Lagrange triangular point and a theorem of Arnold". J. Diff. Equat. **62**, 222 (1986)
73. A. Milani, A. M. Nobili, M. Carpino, "Secular variations of the semi-major axes: Theory and experiment". Astron. Astrophys. 172, 265 (1987)
74. O. Miloni, S. Ferraz-Mello, "Analytical Proper Elements for the Hilda asteroids. II". Celest. Mech. Dynam. Astron. (to be published)
75. B. Morando, "Orbites de résonance des satellites de 24 h.". Bulletin Astronomique **24**, 47 (1962)
76. A. Morbidelli, "On the successive elimination of perturbation harmonics". Celest. Mech. Dynam. Astron. **55**, 101 (1993)
77. A. Morbidelli, *Modern Celestial Mechanics. Aspects of Solar System Dynamics* (Taylor and Francis, London, 2002), Chap. 11
78. A. H. Nayfeh, D. T. Mook, *Nonlinear Oscillations* (John Wiley, New York, 1979)
79. L. A. Pars, *A Treatise on Analytical Dynamics* (Heinemann, London, 1965)
80. H. Poincaré, *Les Méthodes Nouvelles de la Mécanique Céleste* (Gauthier-Villars, Paris, 1893), Vol. II
81. H. Poincaré, "Sur les planètes du type d'Hécube". Bulletin Astronomique **19**, 289 (1902)
82. J. L. Sagnier, "Méthodes de Perturbation en Mécanique Céleste". Notes Sc. Tech. Bureau des Longitudes, S032 (1991)
83. J. A. Sanders, F. Verhulst, *Averaging Methods in Nonlinear Dynamical Systems* (Springer-Verlag, New York, 1985)

84. K. Schwarzschild, "Zur Quantenhypothese". Sitz. Ber. Kgl. Preuss. Akad. d. Wiss., S. 548 (1916)
85. J. L. Sérsic, "Aplicaciones de un certo tipo de transformaciones canónicas a la Mecánica Celeste". Série Astronómica Observ. Astron. La Plata, No. 35 (1969) (reprint from *Dr. Thesis*, Univ. La Plata, 1956)
86. W. Sessin, "Estudo de um sistema de dois planetas com períodos comensuráveis na razão 2:1". *Dr. Thesis* (Univ. São Paulo, São Paulo, 1981)
87. W. Sessin, S. Ferraz-Mello, "Motion of two planets with periods commensurable in the ratio 2:1. Solutions of the Hori auxiliary system". Celest. Mech. **32**, 307 (1984)
88. W. Sessin, "Application of the extended Delaunay method to the ideal resonance problem". Celest. Mech. **40**, 293 (1987)
89. W. Sessin, M. Tsuchida, "2:1 commensurability in Uranus-Neptune system". In *The Motion of Planets and Natural and Artificial Satellites* Proc. CNPq–NSF Workshop (Embu, 1981), ed. by S. Ferraz-Mello and P. Nacozy (IAG-USP, São Paulo, 1983) pp. 263-272.
90. H. Shniad, "The equivalence of von Zeipel mappings and Lie transforms". Celest. Mech. **2**, 114 (1970)
91. C. L. Siegel, *Vorlesungen über Himmelsmechanik* (Springer-Verlag, Berlin, 1956); C. L. Siegel, J. Moser, *Lectures on Celestial Mechanics* (Springer-Verlag, Berlin, 1971)
92. A. Sommerfeld, "Zur Theorie der Balmerschen Serie". Sitzungsber. der Kgl. Bayer. Akad. d. Wiss. S.425 (1915)
93. A. Sommerfeld, *Atombau und Spektrallinien* (Vieweg, Braunschweig, 1922); *Atomic Structure and Spectral Lines* (E. P. Dutton, New York, 1922)
94. A. Sommerfeld, *Mechanics* (Academic Press, New York, 1952), Chap. 5
95. V. Szebehely, *Theory of Orbits* (Academic Press, New York, 1967), Chap. 3
96. H. von Zeipel, "Recherches sur le mouvement des petites planètes". Arkiv f. Mat. Astr. o. Fys. **11** No. 1 and No. 7 (1916)
97. E. T. Whittaker, *A Treatise on the Analytical Dynamics of Particles and Rigid Bodies* (Dover, New York, 1944)
98. E. T. Whittaker, G. N. Watson, *A Course of Modern Analysis* (Cambridge Univ. Press, Cambridge, 1952)
99. A. Wintner, *The Analytical Foundations of Celestial Mechanics* (Princeton Univ. Press, Princeton 1941), p. 46
100. J. Wisdom, "Canonical solution of the two critical arguments problem". Celest. Mech. **38**, 175 (1986)

Index

action–angle, *see* angle–action
algebraic manipulators
　branch choice, 202
almost periodic functions
　mean value theorem, 154
Andoyer, 290
Andoyer Hamiltonians, 289
　centers and saddle points, 290
　comparative analysis, 328
　libration width, 296, 318, 323, 327
　morphogenesis, 293, 316, 321, 327
　with $k = 1$, 192, 218, 226, 292–313
　　equilibrium points, 304
　　integration, 298–304
　　proper periods, 306
　　small-amplitude librations, 308
　with $k = 2$, 210, 261, 315–320
　with $k = 3$, 210, 320–325
　with $k = 4$, 325–328
angle–action variables, 30–60, 179
　in Lie series theory, 140
　of a quadratic Hamiltonian, 57
　of Andoyer Hamiltonian, 308
　of the ideal resonance problem, 115
　of the pendulum, 277, 280
apparent forces, 17
Arnold, 42, 69, 93, 241, 249
artificial satellite motion, 86, 265
asteroid motion, 77, 176, 191
asymptotic motions, 114, 182
averaging principle, 69
averaging rule, 68, 75, 106, 142
　weak, 266, 270

Birkhoff normalization, 236
　single resonance, 240
　theorem, 237
Bohlin, 63, 182, 263
　equation, 266, 269
　Hamiltonian, 264
　problem, 139, 181, 223, 263
　theory, 103, 263
Bohr, H., 154
Bohr, N., 42
Born, 32, 42
Brouwer, 70, 265
Burgers, 42

canonical condition, 12, 133
canonical transformations, 6–13
　conservative, 7
　infinitesimal, 127
　multiplier, 8
　valence, 8
Cartesian-like variables, 290
Cauchy existence theorem, 132
Cauchy–Darboux theory of characteristics, 154
central motions, 43, 56
centrifugal force, 17
characteristic curves, 152, 154
Charlier, 18, 42, 98
　theory, 37
circulation, 21, 29, 110, 115, 272, 273, 277
　inner, 294, 301, 302
　outer, 294, 301
conditionally periodic, *see* multiperiodic

338 Index

conservative systems, 4
convergence, 63
coordinate–momentum order, 30
Coriolis force, 17
critical
  angles, 75
  terms, 76, 174
cubic equation, 292
cusp, 296

d'Alembert property, 164, 236
degeneracy
  accidental, 50
  complete, 51
  essential, 50, 101, 235
  in the sense of Kolmogorov, 88
  in the sense of Schwarzschild, 50
  isoenergetic, 94
  proper, 93
degenerate
  angles, 76
  systems, 70, 93, 94
degree of homogeneity
  as order of magnitude, 168, 184, 185, 211, 267
Delaunay
  equation, 106
  lunar theory, 99
  operation, 99
  problem, 61
  theory, 103
    extension, 269
  variables, 1, 42, 52
    singularities, 161
  variation equations, 1
Delaunay–Morbidelli operation, 120
Delaunay–Poincaré equation, 106
Deprit, 137, 143
differential rotator, 166, 209, 293, 329
Diophantine condition, 89
divergence of the series, 88, 156
Duffing equation, 290

$\sqrt{\varepsilon}$, $\sqrt[3]{\varepsilon}$, see small parameter
eccentric anomaly, 48
eccentricity, 48
  forced, 83
Einstein, 35, 42
  theory, 35

elimination of harmonics, 118
energy
  in extended phase space, 14
  normalized, 284
  total mechanical, 4
Epstein, 42
equinoctial elements, 163
equivalence of Lie and Jacobian mappings, 129
Euler collinear solutions, 243
extended phase space, 13, 94

Ford–Lunsford Hamiltonian, 255
formal solution, 63, 232
frequency commensurability
  exact, 234
  multiple, 253
  third and fourth order, 242
frequency relocation, 63, 89

Garfinkel, 103, 107, 204, 267
Garfinkel–Jupp–Williams Integrals, 109
generalized coordinates and momenta, 2
generalized potential energy, 15
Gerasimov, 289
Giorgilli, 98
gyroscopic
  forces, 15
  systems, 15, 60, 242

Hénon–Heiles Hamiltonian, 250
Hamilton, 18
  equations, 4, 10
  principle, 3
Hamilton–Jacobi equation, 18–20
  complete solution, 19
Hamilton–Jacobi mapping, 20
Hamiltonian, 3
  average, 158
  quadratic, 57
  quasiharmonic, 231
  reduction, 4, 93, 154, 174
  regular, 165
harmonic oscillator, 23, 32
Helmholtz invariant, 5, 35
Henrard, 41, 147, 165
Hill equation, 255
homological equation, 67, 73, 107, 141, 152, 171, 190, 212, 216, 236, 263
  indetermination, 68

Hori, 139
  auxiliary equations, *see* Hori kernel
  formal first integral, 157
  general theory, 151
  kernel, 153, 155, 171, 183, 216, 289
    separated, 198
  theory, *see* Lie series perturbation
    theory

ideal resonance problem, 107–118,
  204–207, 267
inclination, 50
initial conditions diagram, 225
involution, 24
  Jacobi's lemma, 26
  Lie's theorem, 27
  Liouville's theorem, 26
  Mayer's lemma, 24

Jacobi, 18, 26
  identity, 130
  partial differential equation, 18
Jacobian canonical transformation
  inversion, 94
Jacobian generating function, 7
Jupp, 207

KAM
  theory, 42, 63
  tori, 89
Kepler motion, 47
Kolmogorov, 63
  theorem, 88–94
Kramers, 42

$\mathcal{L}_4$, 244
  stability, 249
Lagrange
  brackets, 9
  equilateral solutions, 244
  formula, 95
  implicit function theorem, 96
  point transformation, 86, 174
  variables, 162
Lagrangian, 3
law of structure, 226
least action principle, 4
Lemaitre, 41
libration, 22, 29, 112, 116, 272, 274,
  278, 302

Le-type, 324
  ambiguity in definition, 295, 318
  apocentric, 295
  center, 226
  domain, 192
  limits, 226
  lobes, 182, 210
  width, 277
Lie, 27
Lie derivative, 130
  homogeneity, 131
Lie generating function, 129
Lie mapping, 20, 129
  duality, 130
  inversion, 134
Lie series, 131
  canonical condition, 133
  commutation theorem, 134
  Deprit's recursion formula, 137, 143
  expansion, 136
    about the origin, 167
    in resonance neighborhood, 185, 187
  perturbation theory, 139
    comparison to classical, 144–151
    in angle–action variables, 140
    in non-singular variables, 169
    in unspecified variables, 151
Lie transformation, 127
Lindstedt, 97
  series, 98
  theory, 263
Liouville, 26, 54
long-period terms, 73, 76, 236
longitudes, 163
lunar theories, 87

$m < 0$, 23
Mathieu equation, 255, 261
Maupertuis principle, 4
Mayer's lemma, 24, 39
mean anomaly, 49
mean distance, 48
Mechanik des Himmels, 18
Melnikov integrals, 287
Message, 217
Méthodes Nouvelles, 103, 263
minimum principle, 267
Monge, 152
Morbidelli, 118

multiperiodic systems, 35–57

non-degenerate, 52
  actions, 62
  degrees of freedom, 70
  quasiperiodic solutions, 88
non-resonance condition, 69, 74, 88, 173, 181, 223, 233
non-singular variables, 162, 213, 290
nonlinear oscillators, 231
normal form, 237
normalization, 236

old Quantum Theory, 18, 31, 42
order of a resonance, *see* resonance order

pendulum, *see* simple pendulum
  parametrically perturbed, 286
periodic motions, 29
perturbation equations
  Bohlin, 266
  for resonant systems, 189, 215
  Hori, 141
  in non-singular variables, 170
  Poincaré, 67
  von Zeipel–Brouwer, 72
perturbation theory
  equivalence of Lie series to classical, 144
phase integral, 31
Planck, 42
planetary theories, 87
Poincaré, 62, 70, 103, 228, 290
  singularity, 182, 268, 270
  theorem, 88
  theory, 63, 263
    equivalence to Lie series, 144
  variables, 82, 162
Poincaré–Birkhoff theorem, 182, 209, 294
Poisson brackets, 11
  splitting, 186, 214
Poisson terms, 98
post-harmonic solution, 221
post-pendulum solution, 197, 200
post-post-pendulum solution, 201, 206
proper degeneracy, 93
proper elements, 83, 97, 159

proper frequencies, 89
pseudo time, 155

reciprocity relations, 12
repetition numbers, 38
resonance, 51, 74, 101, 234
  1:2:3, 255
  abnormal, 103
  higher-order, 315
  in a neighborhood of the origin, 209
  internal
    2:1, 246
    3:1, 247
  higher orders, 249
  non-central, 210
  order, 211, 237, 315
  overlap, 125
  parametric, 255
    nonlinear, 260
  second fundamental model, 289
  secondary, 183, 224, 226
  secular, 183, 197, 223, 226
  single, 213
  virtual, 209, 318, 329
resonant oscillator, 234
restricted three-body problem, 242
  elliptic, 77, 229
Riemann surface, 42, 48
rotating frames, 17
Routh critical mass ratio, 244
Routhian reduction, 4

Schwarzschild, 42, 50
  transformation, 51
scissors averaging, 69
secular terms, 73, 76, 236
secular theory, 81, 159
semi-major axis, *see* mean distance
separable systems, 37, 42, 53
  Liouville, 55
  Stäckel, 56
separatrix, 303
separatrix crossing, 283, 286
separatrix map, 286
Sérsic, 139
Sessin, 227, 289
  integral, 218, 227
  transformation, 217, 227
    asteroidal case, 229

short-period terms, 73, 76, 236
simple pendulum, 107, 192, 271–288
   separatrix, 272, 276
   separatrix neighborhood, 283
   small oscillations, 279
single-resonance problem, 181
singularities of the actions, 161
small-amplitude librations, 117, 280, 308
small divisors, 69, 74, 155
small parameter, 216
   cube root, 211, 216, 298
   square root, 101, 216, 263, 298, 319, 325
Sommerfeld, 31, 42, 48
Stäckel, 54
standard map, 288
super-convergent algorithm, 92
symplectic unit matrix, 10

three-phonon interaction, 254

Toda lattice, 252
topological constraint, 156
true anomaly, 49
twist mapping, 209

undetermined periodic components, 33, 281, 308
uniformized angles, 37
unspecified canonical variables, 151

virtual resonance, 209, 329
von Zeipel, 70
   averaging rule, 73, 174, 190, 223, 235, 265
von Zeipel–Brouwer theory, 70–87, 256
   equivalence to Lie series, 144
   iterative use, 86, 93

Weierstrass
   implicit functions theory, 103
whisker map, 286

Astrophysics and Space Science Library

Volume 340: *Plasma Astrophysics, Part I: Fundamentals and Practice*, by Boris V. Somov. Hardbound ISBN 0-387-34916-9, September 2006

Volume 339: *Cosmic Ray Interactions, Propagation, and Acceleration in Space Plasmas*, by Lev Dorman. Hardbound ISBN 1-4020-5100-X, August 2006

Volume 338: *Solar Journey: The Significance of Our Galactic Environment for the Heliosphere and the Earth*, edited by Priscilla C. Frisch. Hardbound ISBN 1-4020-4397-0, September 2006

Volume 337: *Astrophysical Disks*, edited by A.M. Fridman, M.Y. Marov, I.G. Kovalenko. Hardbound ISBN 1-4020-4347-3, June 2005

Volume 336: *Scientific Detectors for Astronomy 2005*, edited by J.E. Beletic, J.W. Beletic, P. Amico. Hardbound ISBN 1-4020-4329-5, December 2005

Volume 335: *Organizations and Strategies in Astronomy 6*, edited by A. Heck. Hardbound ISBN 1-4020-4055-5, November 2005

Volume 334: *The New Astronomy: Opening the Electromagnetic Window and Expanding our View of Planet Earth*, edited by W. Orchiston. Hardbound ISBN 1-4020-3723-6, October 2005

Volume 333: *Planet Mercury*, by P. Clark and S. McKenna-Lawlor. Hardbound ISBN 0-387-26358-6, November 2005

Volume 332: *White Dwarfs: Cosmological and Galactic Probes*, edited by E.M. Sion, S. Vennes, H.L. Shipman. Hardbound ISBN 1-4020-3693-0, September 2005

Volume 331: *Ultraviolet Radiation in the Solar System*, by M. Vázquez and A. Hanslmeier. Hardbound ISBN 1-4020-3726-0, November 2005

Volume 330: *The Multinational History of Strasbourg Astronomical Observatory*, edited by A. Heck. Hardbound ISBN 1-4020-3643-4, June 2005

Volume 329: *Starbursts – From 30 Doradus to Lyman Break Galaxies*, edited by R. de Grijs, R.M. González Delgado. Hardbound ISBN 1-4020-3538-1, May 2005

Volume 328: *Comets*, by J.A. Fernández. Hardbound ISBN 1-4020-3490-3, July 2005

Volume 327: *The Initial Mass Function 50 Years Later*, edited by E. Corbelli, F. Palla, H. Zinnecker. Hardbound ISBN 1-4020-3406-7, June 2005

Volume 325: *Kristian Birkeland – The First Space Scientist*, by A. Egeland, W.J. Burke. Hardbound ISBN 1-4020-3293-5, April 2005

Volume 324: *Cores to Clusters – Star Formation with next Generation Telescopes*, edited by M.S. Nanda Kumar, M. Tafalla, P. Caselli. Hardbound ISBN 0-387-26322-5, October 2005

Volume 323: *Recollections of Tucson Operations*, by M.A. Gordon. Hardbound ISBN 1-4020-3235-8, December 2004

Volume 322: *Light Pollution Handbook*, by K. Narisada, D. Schreuder Hardbound ISBN 1-4020-2665-X, November 2004

Volume 321: *Nonequilibrium Phenomena in Plasmas*, edited by A.S. Shrama, P.K. Kaw. Hardbound ISBN 1-4020-3108-4, December 2004

Volume 320: *Solar Magnetic Phenomena*, edited by A. Hanslmeier, A. Veronig, M. Messerotti. Hardbound ISBN 1-4020-2961-6, December 2004

Volume 319: *Penetrating Bars through Masks of Cosmic Dust*, edited by D.L. Block, I. Puerari, K.C. Freeman, R. Groess, E.K. Block. Hardbound ISBN 1-4020-2861-X, December 2004

Volume 318: *Transfer of Polarized light in Planetary Atmospheres*, by J.W. Hovenier, J.W. Domke, C. van der Mee. Hardbound ISBN 1-4020-2855-5. Softcover ISBN 1-4020-2889-X, November 2004

Volume 317: *The Sun and the Heliosphere as an Integrated System*, edited by G. Poletto, S.T. Suess. Hardbound ISBN 1-4020-2830-X, November 2004

Volume 316: *Civic Astronomy – Albany's Dudley Observatory, 1852–2002*, by G. Wise Hardbound ISBN 1-4020-2677-3, October 2004

Volume 315: *How does the Galaxy Work – A Galactic Tertulia with Don Cox and Ron Reynolds*, edited by E.J. Alfaro, E. Pérez, J. Franco Hardbound ISBN 1-4020-2619-6, September 2004

Volume 314: *Solar and Space Weather Radiophysics – Current Status and Future Developments*, edited by D.E. Gary and C.U. Keller Hardbound ISBN 1-4020-2813-X, August 2004

Volume 313: *Adventures in Order and Chaos,* by G. Contopoulos. Hardbound ISBN 1-4020-3039-8, January 2005

Volume 312: *High-Velocity Clouds,* edited by H. van Woerden, U. Schwarz, B. Wakker Hardbound ISBN 1-4020-2813-X, September 2004

Volume 311: *The New ROSETTA Targets – Observations, Simulations and Instrument Performances,* edited by L. Colangeli, E. Mazzotta Epifani, P. Palumbo Hardbound ISBN 1-4020-2572-6, September 2004

Volume 310: *Organizations and Strategies in Astronomy 5,* edited by A. Heck Hardbound ISBN 1-4020-2570-X, September 2004

Volume 309: *Soft X-ray Emission from Clusters of Galaxies and Related Phenomena,* edited by R. Lieu and J. Mittaz Hardbound ISBN 1-4020-2563-7, September 2004

Volume 308: *Supermassive Black Holes in the Distant Universe,* edited by A.J. Barger Hardbound ISBN 1-4020-2470-3, August 2004

Volume 307: *Polarization in Spectral Lines,* by E. Landi Degl'Innocenti and M. Landolfi Hardbound ISBN 1-4020-2414-2, August 2004

Volume 306: *Polytropes – Applications in Astrophysics and Related Fields,* by G.P. Horedt Hardbound ISBN 1-4020-2350-2, September 2004

Volume 305: *Astrobiology: Future Perspectives,* edited by P. Ehrenfreund, W.M. Irvine, T. Owen, L. Becker, J. Blank, J.R. Brucato, L. Colangeli, S. Derenne, A. Dutrey, D. Despois, A. Lazcano, F. Robert Hardbound ISBN 1-4020-2304-9, July 2004
Paperback ISBN 1-4020-2587-4, July 2004

Volume 304: *Cosmic Gammy-ray Sources,* edited by K.S. Cheng and G.E. Romero Hardbound ISBN 1-4020-2255-7, September 2004

Volume 303: *Cosmic rays in the Earth's Atmosphere and Underground,* by L.I. Dorman Hardbound ISBN 1-4020-2071-6, August 2004

Volume 302: *Stellar Collapse,* edited by Chris L. Fryer Hardbound, ISBN 1-4020-1992-0, April 2004

Volume 301: *Multiwavelength Cosmology,* edited by Manolis Plionis Hardbound, ISBN 1-4020-1971-8, March 2004

Volume 300: *Scientific Detectors for Astronomy,* edited by Paola Amico, James W. Beletic, Jenna E. Beletic Hardbound, ISBN 1-4020-1788-X, February 2004

Volume 299: *Open Issues in Local Star Fomation,* edited by Jacques Lépine, Jane Gregorio-Hetem Hardbound, ISBN 1-4020-1755-3, December 2003

Volume 298: *Stellar Astrophysics – A Tribute to Helmut A. Abt,* edited by K.S. Cheng, Kam Ching Leung, T.P. Li Hardbound, ISBN 1-4020-1683-2, November 2003

Volume 297: *Radiation Hazard in Space,* by Leonty I. Miroshnichenko Hardbound, ISBN 1-4020-1538-0, September 2003

Volume 296: *Organizations and Strategies in Astronomy, volume 4,* edited by André Heck Hardbound, ISBN 1-4020-1526-7, October 2003

Volume 295: *Intergrable Problems of Celestial Mechanics in Spaces of Constant Curvature,* by T.G. Vozmischeva Hardbound, ISBN 1-4020-1521-6, October 2003

Volume 294: *An Introduction to Plasma Astrophysics and Magnetohydrodynamics,* by Marcel Goossens Hardbound, ISBN 1-4020-1429-5, August 2003
Paperback, ISBN 1-4020-1433-3, August 2003

For further information about this book series we refer you to the following web site:
www.springer.com/astronomy

Printed in The United States of America

DATE DUE

SCI QB 361 .F47 2007

Ferraz-Mello, Sylvio.

Canonical perturbation theories